# IMPROVED
# GRASSLAND
# MANAGEMENT

And God said, Let the earth bring forth grass, the herb yielding seed and the fruit tree yielding fruit after his kind, whose seed is in itself, upon the earth; and it was so.

And the earth brought forth grass, and herbs yielding seed after his kind, and the tree yielding fruit, whose seed was in itself, after his kind; and God saw that it was good.

*Genesis* 1: 11–12

## *GRASS*

I have written of dawn, of the moon, and the trees;
Of people, and flowers, and the song of the bees.
But over these things my mind would pass,
And come to rest among the grass.

Grass so humble, that all things tread
Its tender blades. Grass—the bread,
The staff of life; a constant need
Of man and beast—a power indeed.

Grass so vagrant—does anything stray
With such gallant courage? The hardest way
Is coaxed and beguiled by the wayward grace
Of the constant friend of every space.

God in his wisdom gave many friends
To grace our way, as along it wends
But the grandeur of many, my mind would pass
And come to rest among the grass.

Mabel Duggan, *The Grasses and Pastures of South Africa* (1955)

Reprinted by permission of Central News Agency Limited

# IMPROVED GRASSLAND MANAGEMENT

*John Frame*

FARMING PRESS

First published 1992
Reprinted 1994
Reprinted with amendments 2000 and as a paperback

ISBN 0 85236 543 8

A catalogue record for this book is available
from the British Library

**Published by Farming Press**
**Miller Freeman UK Ltd, Miller Freeman House,**
**Sovereign Way, Tonbridge TN9 1RW, United Kingdom**

Distributed in North America by Diamond Farm Enterprises
Box 537, Alexandria Bay, NY 13607, USA

Cover design by Carol Jarvis
Typeset by Galleon Photosetting, Ipswich
Printed and bound in Great Britain by Butler and Tanner Ltd, Frome, Somerset

# Contents

Preface    ix
Acknowledgements    xi
Foreword    xiii

*Chapter 1*    TYPES OF BRITISH GRASSLAND: OVERVIEW    1
**Sward productivity ● Survey information ● Rough grazing**

*Chapter 2*    CHARACTERISTICS OF GRASSES AND LEGUMES    11
**Ryegrasses ● Other grass species ● Forage legumes ● Secondary grasses**

*Chapter 3*    HERBAGE VARIETY EVALUATION    24
**Classification of maturity groups ● National List varietal evaluation ● Recommended Lists**

*Chapter 4*    SEED MIXTURES    31
**Choosing seed mixtures ● Specimen mixtures**

*Chapter 5*    HERBAGE PLANT BREEDING    38
**Selection and genetic manipulation ● Development of herbage plant breeding ● Present breeding objectives ● Forage legume breeding**

*Chapter 6*    SEED PRODUCTION AND CERTIFICATION    46
**Seed grades and standards ● Seed production and usage ● Seed growing rules and regulations ● Management of seed crops ● Seed harvesting**

*Chapter 7*     SWARD ESTABLISHMENT AND RENOVATION     57
**The seedbed ● Fertiliser application
● Post-sowing management ● Weeds ● Pests
● Sward renovation**

*Chapter 8*     CONTROLLING WEEDS     68
**Docks ● Thistles ● Ragwort ● Common
chickweed ● Nettles ● Annual meadow grass
● Rushes ● Soft brome ● Bracken ● Gorse
and broom**

*Chapter 9*     EFFECT OF SOIL FACTORS ON GRASS     79
**Soil formation ● Soil components ● Soil
texture and structure ● Soil maps ● Soil
trampling and sward poaching ● Sod pulling
● Wheel-induced effects on soil**

*Chapter 10*     WEATHER AND WATER CONTROL: IRRIGATION,
DRAINAGE AND SUBSOILING     91
**Temperature ● Rainfall ● Irrigation
● Drainage ● Subsoiling**

*Chapter 11*     SOIL FERTILITY AND GRASS PRODUCTION: NITROGEN     101
**Soil nitrogen status ● Symbiotic N-fixation by
legumes ● Nitrogen in animal excreta ● Types
of nitrogen fertiliser ● Reducing N losses
● Response of grass swards to N ● Response
of grass/white clover swards to N ● Seasonal
distribution of nitrogen ● Nitrogen and
intensity of defoliation ● Prediction of N
requirement ● Nitrogen and cutting systems**

*Chapter 12*     SOIL FERTILITY AND GRASS PRODUCTION: LIME AND
MINERAL NUTRIENTS     119
**Lime ● Soil sampling and analysis
● Phosphorus ● Potassium ● Sulphur
● Magnesium ● Sodium ● Trace elements
● Blended solid fertilisers ● Liquid fertilisers**

CONTENTS

*Chapter 13*   USING ORGANIC MANURES                          136
Animal slurry ● Effective slurry application
● Farmyard manure ● Sewage sludge
● Silage effluent


*Chapter 14*   FEEDING VALUE OF GRASS                         146
Chemical composition ● Digestibility
● Metabolisable energy


*Chapter 15*   SWARD GROWTH AND DEVELOPMENT                   161
Grass tillering process ● Grass leaf
development ● White clover development
● Photosynthesis and tissue turnover
● Seasonal herbage production


*Chapter 16*   THE GRAZING PROCESS                            175
Selective grazing ● Excretal return
● Trampling


*Chapter 17*   METHODS AND SYSTEMS OF GRAZING                 187
Continuous stocking ● Rotational grazing
● Creep grazing ● 'Clean' grazing ● Mixed
grazing ● Zero grazing ● Storage feeding
● Choosing a grazing system ● Monitoring
sward state


*Chapter 18*   SEASONAL OBJECTIVES AND MANAGEMENT             199
Early bite ● Midsummer grazing ● Topping
grassland ● Late bite ● Winter management


*Chapter 19*   SILAGE MAKING                                  209
Silage in systems ● Fermentation
● Respiration losses ● Wilting ● Undesirable
clostridia ● Additive use ● Sealing ● Silage
effluent ● Aerobic deterioration ● Big bale
silage ● Silage feeding quality ● Health and
safety ● Silage competitions

*Chapter 20*  HAY MAKING                                                      233
             **Stage of growth at cutting ● Drying processes
             ● Influence of weather ● Baling ● Barn
             drying systems ● Storage losses ● Key points**

*Chapter 21*  GRASSLAND RECORDING                                             243
             **Animal unit methods ● Utilised metabolisable
             energy ● Interpretation**

*Chapter 22*  MANAGEMENT OF FORAGE LEGUMES                                     251
             **White clover ● Red clover ● Lucerne
             ● Legume use in the future**

*Chapter 23*  GRASSLAND NATURE CONSERVATION                                    272
             **Loss of species rich grassland ● Nature
             conservation schemes ● Gene survival
             ● Wildflower mixture productivity ● Forage
             herbs ● Future nature conservation**

*Chapter 24*  HILL LAND IMPROVEMENT                                           281
             **Pasture resources ● The two-pasture system
             ● Soil improvement ● Sward improvement
             ● Soil/sward improvement ● Forage
             conservation ● Future hill land use**

*Appendices*

             1  **Common and Latin names of selected
                plants, pests and diseases**                                 295
             2  **Classification of soil analysis results**                   301
             3  **Key for identification of vegetative grasses**              302
             4  **Species performing well in seed mixtures
                used in different soil types**                                306
             5  **Selected metric conversion factors**                       308
             6  **Useful addresses**                                          310
             7  **References cited and further reading**                      316

*Index*                                                                       338

Colour photographs appear between pages
82 and 83, 114 and 115, 178 and 179,
210 and 211, 274 and 275, 306 and 307

# Preface

In all its diverse forms, grass is the renewable foundation feed of the world's great ruminant production industry. It is literally green gold and we neglect its sustainability and the efficiency of its use at our peril. Yet grass is more than a feedstuff—it is Mother Earth's green carpet and plays an important role in the quality of life of mankind, whether for natural landscape beauty or for recreation and sport.

In British and many other temperate agricultural regions, a scenario exists of restrictions on output of animal products, demands for improvements in animal welfare, product quality influenced by the consumer market, the necessity for environmentally friendly production methods and stringent containment of production costs. In the face of a host of political and economic pressures, grassland farmers must make swift and flexible yet sensible responses and adaptations to changes, some of which will be unforeseen.

In 1933, The Right Honorable J. H. Thomas, then Secretary of State for the Dominions, wrote: 'In agriculture, as in every human activity, we seem to be passing into a new world. There never was a time when tremendous changes were more certain, when events were harder to forecast or when action was more difficult to plan.' These prophetic words are particularly relevant for today's rapidly changing world, a world 'growing ever smaller' in which competition for agricultural markets is intensifying. Precision of inputs, predictability of outputs and profitability are triple key objectives for the grassland farmer.

In Britain and many European countries, grassland is settling out into (1) specialist intensively managed grassland with relatively high fertiliser nitrogen inputs and associated stocking rates, e.g. dairy herds, but with great care to minimise adverse pollution effects on the environment; (2) less intensively stocked grassland, based on grass/forage legume swards receiving little or no fertiliser nitrogen or on long term/permanent grass receiving moderate inputs of fertiliser nitrogen, e.g. lowland sheep and beef interprises; (3) extensively utilised grassland, e.g. hill sheep and suckler cow enterprises, but with large tracts used for rural diversification, e.g. afforestation and non-traditional ruminant enterprises; and (4) grassland, much of it in the hills and uplands, diverted to non-agricultural uses, e.g. national parks, wildernesses, leisure and recreation areas.

Part of the second and third categories of grassland will be used for organic (biological) farming and for schemes with voluntary extensification measures, e.g. environmentally sensitive areas. Various forms of grassland, including vegetation euphemistically called 'grassland', will clothe arable set-aside land.

ix

The above background has been kept in mind when writing this book. A reader-friendly, advisory style is used to subdivide grassland management into its interrelated and interaction components before dealing with utilisation by grazing and conservation. Management in relation to the new surge of interest in extensification, nature conservation, forage legume use and environmental sustainability is also covered.

The approach taken is the application to practice of basic scientific principles, which are outlined in a series of insight chapters and sections using mainly British and European grassland as examples. Nevertheless the principles can be applied to grassland elsewhere in the temperate world. Grassland science is one of the youngest sciences but great strides have been made in recent decades. This is not to gainsay that there is a certain art in grassland management too, as many farmers will testify, but farmers who turn science into practice are taking advantage of what is a major 'resource' and acquiring a competitive edge.

The book is intended for grassland farmers, extension officers, students of agriculture and land use, plant scientists and scientists of many disciplines impinging on biological sciences—in fact for everyone with an interest in the most important crop in the world. The author will appreciate being notified of any errors or omissions and will welcome comment.

JOHN FRAME
Ayr
*September 1992*

# Acknowledgements

In the preparation of this book I have drawn heavily on my experiences as a grassland researcher and specialist adviser during my career with the Scottish Agricultural College, Auchincruive, Ayr, a career which also included consultancies and study visits to various parts of the world. I am grateful for the encouragement and wisdom given by past and present colleagues, farmers and ranchers over the years. I pay particular tribute to my early mentor and grassland enthusiast, Idris V. Hunt.

Thanks are given to Gordon Tiley who read through the manuscript with a constructively critical eye and to Graham Swift who did likewise on a major part of it. Comment and advice on specific chapters were also generously given by Scott Laidlaw, John Weddell, Gillian Whytock, Henry Murdoch and David Kidd. Nevertheless any shortcomings or omissions in the text are my own and comments or suggestions for improvement will be welcome.

I acknowledge the valuable advisory publications of SAC, ADAS and NIAB as sources of information and I strongly recommend them to readers. I also thank the College Library staff for their unstinting efforts in responding to my various requests for published information. I am grateful to the many sources, individuals and organisations for permission to reproduce copyright material and to colleagues who permitted me the use of their photographs. Every effort has been made to seek out the sources and acknowledge them in the course of the text and in the reference list; any oversight is unintentional. My family, to whom dedication is made, was a tower of support during the writing of the book, and my wife, Nancy, typed the manuscript.

I am delighted that Professor J. David Leaver, a former colleague and now Professor of Agriculture at Wye College, University of London kindly agreed to write the Foreword.

My thanks and appreciation are also given to the staff of Farming Press Books, particularly Roger Smith for his help and encouragement and Julanne Arnold for her meticulous attention to detail in producing the finished book.

Finally I acknowledge the inspiration and wonder of grass, earth's green gold.

*To Nancy, Shona, Sheelagh and Fiona*

# Foreword

Grassland has always played an important role in agricultural development by extending man's food supply to areas where crops cannot be grown for direct human consumption. It also has many other roles of benefit to man. In lowland areas, crop rotations benefit from the incorporation of a grass ley in the enhancement of soil fertility and structure and in breaking the cycle of pests and diseases associated with arable crops. Also the amenity value of grassland for recreation and as a contributor to the beauty of the landscape cannot be overemphasised.

Nevertheless, the balance is changing. Livestock production and consumption tends to increase as per capita income increases up to a certain point and then plateaus. In many developed countries with relatively static populations this has already occurred and the consumption of products from ruminant livestock is declining. The intensification of grassland which has been so ably supported by research into the understanding of the dynamic interrelationships of soil, plant, animal, climate and man is therefore unlikely to continue in such countries. Financial support for farmers will increasingly be channelled to those practising less intensive systems. Lower stocking rates and lower fertiliser inputs are a likely outcome together with a greater reliance on legumes.

Profit margins in agriculture will be squeezed for the foreseeable future in many countries as product prices increase at lower rates than inflation. The management of grassland under these very different constraints will thus be a major challenge. It is against this background that this book has been written.

John Frame has used his substantial research, development and extension background to cover the relevant areas of grassland, ranging from intensive to extensive systems. A science based approach is used to describe grassland production and its utilisation as grazing or as conserved forage. This is combined with sound management advice to produce a factually based and very readable text.

Grassland provides a low cost, high nutritional value feed for ruminants together with a high landscape and amenity value. It forms a complex ecosystem and its management requires an understanding of its many interacting components. This book provides a valuable contribution to this understanding.

J. D. LEAVER
Wye College, University of London

*October 1992*

*Chapter 1*

# Types of British Grassland: Overview

The study of grasses would be of great consequence to a northerly, and grazing kingdom. The botanist that could improve the sward of the district where he lived would be a useful member of society; to raise a thick turf on a naked soil would be worth volumes of systematic knowledge; and he would be the best commonwealth's man that could occasion the growth of *two blades of grass* where *one* alone was seen before'.

Gilbert White, *The Natural History of Selborne* (1788)

The area of agricultural land in the United Kingdom is 18.5 million ha, representing 77 per cent of the total land area, and grassland makes up nearly 13 million ha or 70 per cent of the agricultural land area. Grassland is the primary source of feed for ruminant livestock. As a proportion of their diet, grass provides 60 to 65 per cent for dairy cows, 80 to 85 per cent for beef cattle and 90 to 95 per cent for sheep. Total livestock numbers comprise 12 million cattle and 44 million sheep (see Table 1.1).

**Table 1.1  Livestock numbers in the United Kingdom, June 1991**

| *Cattle* | *'000 head* |
|---|---|
| Dairy cows | 2,847 |
| Dairy heifers in calf | 529 |
| Beef cows | 1,599 |
| Beef heifers in calf | 227 |
| Other cattle and calves | 6,857 |
| Total | 12,059 |
| *Sheep* | *'000 head* |
| Ewes | 16,760 |
| Shearlings | 3,650 |
| Lambs under 1 year old | 22,023 |
| Other sheep | 1,365 |
| Total | 43,798 |

*Source:* Ministry of Agriculture, Fisheries and Food

1

Figure 1.1 shows the pattern of land use from the official June census data for 1991. The grassland area is divided into three categories: under five years old, over five years old and rough grazing. Grassland under five years old – 1.6 million ha—includes newly sown swards and a range of swards from one to four years old. Some of these may have been deliberately sown to last only one to four years; others may be in the early years of an intended longer duration.

Short-term leys are based on Italian, hybrid and perennial ryegrasses (see Appendix 1 for Latin names of plants). Where seed mixtures are intended for medium-term or long-term duration, perennial ryegrass is the dominant constituent, often with the addition of other grasses, especially timothy, and including white clover. Seeds mixtures are discussed fully in Chapter 3.

Grassland over five years old—5.3 million ha—includes temporary grassland or leys which may be part of an arable/grass rotation, possibly up to ten or more years, and also permanent grassland which is old grassland that may originally have been sown or simply have been derived by management. The statistics do not indicate whether or not the grassland in the two age categories is part of a rotation with arable crops or land which will always be grassland. In indepen-

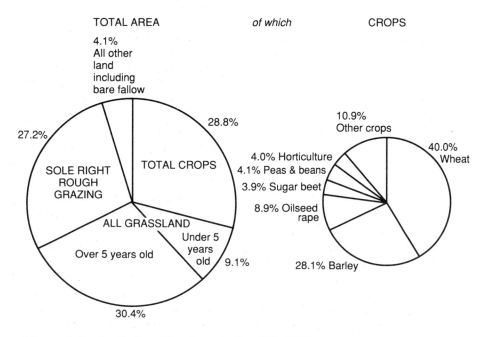

Figure 1.1   Agricultural land use in the UK: 1991
Source: Ministry of Agriculture, Fisheries and Food

dent grassland surveys, this latter category is regarded as permanent grassland regardless of its age at the time the survey information is recorded; the terms 'permanent grass' and 'old grass' have also been used to categorise grassland over twenty years old.

The third category—rough grazing—comprises 5.9 million ha, made up of 4.7 million ha of sole right and 1.2 million ha of common rough grazing. Usually unfenced and subject for many years to an extensive type of management, rough grazing is equivalent to open rangeland in many countries.

The percentage of agricultural land under grass increases from east to west and from south-east to north-west. This is the result of increasing rainfall, a generally decreasing soil fertility and the difficulties of arable cropping. Grassland farming is also predominant at higher altitudes, with an increase in rough grazing and a reduction in length of growing season on account of lower temperatures and poorer growing conditions. Most of the rough grazing—4 million ha—occurs on the hills and uplands of Scotland, with much of the remainder in the Pennines of northern England and the mountains of Wales.

## SWARD PRODUCTIVITY

The potential production of grass swards has been calculated to be 27 to 30 t/ha of herbage dry matter (DM). Production within this range has been achieved in very small experimental plots when water and nutrients were not limiting. (At this stage, suffice it to say that 1 t DM is equivalent to 67 grazing days for a 600 kg dairy cow with a 15 kg DM daily intake or 4 t of silage at 25 per cent DM content.) In experimental plots of 5 m² and upwards, production of 18 to 21 t/ha DM is more commonly obtained under high fertiliser nitrogen (N) use and infrequent cutting, which simulates conservation. Under more frequent defoliation, to simulate grazing, the best DM production has been 12 to 15 t/ha. In farm practice, herbage production is very variable because of numerous limiting factors, but maximum production seldom exceeds 10 to 13 t/ha DM.

For grass/white clover swards, a potential of 18 to 22 t/ha herbage DM has been calculated. Production as high as 15 t/ha has been achieved in experiments but 10 to 12 t/ha is more common, down to 6 to 8 t/ha in farm practice. However, production is very dependent on the content of clover, which fluctuates widely in practice, especially where the swards have not been managed to favour the clover.

An interesting classification made in the mid 1980s graded agricultural land by quality or potential. A set of physical criteria such as

3

altitude, gradient, climate, soil and drainage, and the extent to which these factors constrained agricultural usage, were evaluated on a scale graded from 1 to 5 (see Table 1.2). Grades 1 and 2 have few limitations for agricultural use, other than perhaps minor soil or drainage problems; such land is often under intensive arable cropping and horticulture. Grade 3 has moderate soil or even climatic limitations and forms a core of good arable land for flexible cropping and lowland grassland. The bulk of the hills and uplands fall into grades 4 and 5 and are regarded as mainly suitable for sheep and cattle rearing. Grade 4, mainly upland and permanent grassland, and grade 5, mainly rough grazing with limited areas suitable for improvement, have severe soil, climatic and altitude constraints. Some of the mountainous slopes and peaks can only support very poor rough grazing of sheep's fescue, bilberry, mosses, lichens and bog plants.

About 80 per cent of Scotland and Wales was in grades 4 and 5 in a ratio (grade 4:grade 5) of 1:7 in Scotland and 1:0.8 in Wales. England had lower proportions of grades 4 and 5 land—26 per cent—and a ratio of 1:0.7, while in Northern Ireland, with 55 per cent of grades 4 and 5 land, the ratio was 1:0.1. As Table 1.2 indicates, 9.9 million ha or 52 per cent of UK land was classed as grades 4 and 5, the area equating closely with the less favoured areas, comprising 9.8 million ha, specified by the EC.

**Table 1.2   Estimate of agricultural land by grade, 1984, Land Service Classification, MAFF ('000 ha)**

| Country | Grade | | | | | |
|---------|------|------|------|------|------|------|
|  | *1* | *2* | *3* | *4* | *5* | *Total* |
| England | 327 | 1,652 | 5,343 | 1,553 | 1,019 | 9,894 |
| Scotland | 20 | 155 | 880 | 660 | 4,765 | 6,480 |
| Wales | 3 | 39 | 295 | 745 | 604 | 1,686 |
| N. Ireland | – | 36 | 457 | 531 | 62 | 1,086 |
| UK | 350 | 1,882 | 6,975 | 3,489 | 6,450 | 19,146 |

*Source*: Burrell *et al*.

## SURVEY INFORMATION

Surveys carried out in different parts of the country have revealed interesting features associated with the different types of grassland. A major survey conducted on the permanent grasslands of England and Wales, published in 1980, showed that the presence of preferred species (mainly perennial ryegrass, timothy and white clover) declined steadily with age of sward and that non-preferred indigenous grasses increased (see Table 1.3). Meadow grasses were more frequent in the younger age group while bent grasses, Yorkshire fog and fine-leaved fescues increased in the older swards. Soft brome grass, which is an invasive weed grass, was abundant in frequently mown fields; docks occurred more frequently in younger rather than older swards while the converse was true for creeping thistle and buttercups. A good clover content, mainly white clover defined as well distributed and contributing more than 5 per cent of the ground cover, was commoner in the younger swards and on beef rather than dairy farms (see Table 1.4).

Table 1.3  Average content of preferred species in permanent grassland of differing ages on contrasting farm types (percentage contribution to total ground cover)

| Sward age (years) | Dairy farms | Suckler beef farms |
|---|---|---|
| 1–4 | 73 | 66 |
| 5–8 | 58 | 52 |
| 9–20 | 51 | 44 |
| Over 20 | 31 | 27 |
| All ages | 48 | 39 |

Source: Forbes et al.

Table 1.4  Proportion of swards with a good content of clover in permanent grassland of differing ages on contrasting farm types (percentage of farms in each group)

| Sward age (years) | Dairy farms | Suckler beef farms |
|---|---|---|
| 1–4 | 13 | 41 |
| 5–8 | 8 | 34 |
| 9–20 | 6 | 39 |
| Over 20 | 7 | 23 |
| All ages | 9 | 30 |

Source: Forbes et al.

A survey of upland permanent swards in England and Wales, in which 85 per cent of the land was above 120 m altitude, gave similar results to the survey of lowland permanent grasslands (see Table 1.5). It was also noted that perennial ryegrass, the dominant sown grass, was most abundant in swards receiving high fertiliser N inputs, particularly at lower altitudes and where mown rather than grazed, e.g. 16 per cent at nil N compared with 58 per cent where over 300 kg/ha N was applied. A good contribution of white clover, defined as 10 per cent or more ground cover, was associated with low N inputs, good drainage and grazed swards. Unsown fine-leaved fescues, especially red fescue, increased with sward age, particularly where no fertiliser was applied.

Table 1.5   Average botanical composition of upland permanent grassland of different ages in England and Wales (percentage of total ground cover)

| Sward age (years) | Sown species | | Unsown species | | | |
|---|---|---|---|---|---|---|
| | Perennial ryegrass | Other species | Bent grasses | Meadow grasses | Other grasses | Other species |
| 1–4 | 56 | 18 | 5 | 13 | 4 | 4 |
| 5–8 | 49 | 13 | 12 | 14 | 7 | 5 |
| 9–20 | 38 | 11 | 20 | 13 | 11 | 7 |
| 20–35 | 30 | 11 | 23 | 10 | 18 | 8 |
| Over 35 | 19 | 9 | 24 | 10 | 25 | 13 |
| All ages | 31 | 11 | 20 | 11 | 17 | 10 |

*Source:* Hopkins *et al.*

Another survey examined grassland on 'stock rearing with arable' farms in eastern Scotland. Two-thirds of the grass was rotational, under seven years old, and 30 per cent was ten or more years old. Species contribution was recorded on a herbage cover basis, bare ground being excluded. Sown-species content fell with age to 49 per cent in the oldest swards (see Table 1.6). Perennial ryegrass was the main sown constituent present, followed by cocksfoot and timothy. Half of all fields surveyed had under 15 per cent of white clover. Bent grasses, Yorkshire fog, meadow grasses and fescues were the main unsown grasses present.

6

Table 1.6   **Average content of sown species in rotational grassland of differing ages of stock rearing farms (percentage contribution to total herbage cover)**

| Sward age (years) | Sown species (%) |
|---|---|
| 1–2 | 93 |
| 3–4 | 86 |
| 5–6 | 83 |
| 7–9 | 74 |
| 10–20 | 66 |
| Over 20 | 49 |
| All ages | 77 |

*Source:* Swift *et al.*

# ROUGH GRAZING

Rough grazing is a generic name for semi-natural plant communities which have evolved under a host of influencing factors. Soil type and prevailing climate are predominant but the influence of man is over-riding. In the uplands, most areas below the tree line—which varies with the exposure of the land—were originally wooded. Except in a few surviving remnants and in modern afforestation, woodland has been unable to regenerate after clearance because of grazing and burning. The grazing management is chiefly extensive and open range and the sward species developed are resistant to grazing and burning. Purple moor grass, for example, with its well-protected growing points, thrives under burning every three to five years. Heather re-quires more infrequent burning, every eight to fifteen years, if it is to flourish. If overgrazed, heather is succeeded by grass heaths with vegetation related to the inherent land quality.

Plant communities are often dominated by only a few species. The growing season is short and highly seasonal with late spring flushes. Some of the subsidiary species are valuable at specific times of the year, when they become acceptable to stock either because of their stage of growth or because there is nothing else to graze. Sedges, heath rush and cotton grass all have grazing value in early spring, for instance, but they are neglected at other times of the year. Some species are rich in specific minerals; for example, young purple moor grass and cotton grass have high phosphorus contents while bilberry is rich in calcium. Variation in production capacity among vegetation types can be as high as twentyfold.

7

The main soil and sward types of the hills and uplands are sum-marised in Table 1.7. In addition to these, some chalk downland areas in southern England, dominated by fescue grasses, and sometimes upright brome, and herbs, are included in the rough grazing category.

Bent grass/fescue swards predominate on the least acid brown earth soils where these are freely drained. White clover and a range of herbs are also present in higher pH ranges. On the more acidic brown earths, fescues rather than bent grasses become dominant and there is less diversity of species present. Of all hill swards, bent/fescue com-munities are the most valuable for grazing and are preferentially grazed. In the open range situation these communities are visible as close-cropped areas within the mosaic of plant communities. Unfer-tilised swards produce 1.0 to 3.6 t/ha DM during the growing season, these levels rising to 3.5 to 4.5 t/ha under moderate fertilisation, but 6 to 9 t/ha DM has been obtained in experiments with heavy fertilising, using 350 to 450 kg/ha N and adequate phosphate ($P_2O_5$) and potash ($K_2O$).

Bracken is present on some of the deeper brown earths, while gorse and broom can be local colonisers on the shallower brown earths. On pooly drained gley soils with surface waterlogging, rushes and sedges become dominant over the bent and fescue grasses. Moor mat grass and tufted hair grass feature on the more acid but drier gleys. Moor mat grass may be grazed transiently at its early stage of growth but the physically coarse leaves of tufted hair grass are little grazed unless it is the most acceptable plant present, in which case it can be heavily grazed.

On freely drained podzols and peaty podzols with soil pHs of 4.0 to 4.5, grass or shrub heaths predominate. Grass heath is typically dominated by moor mat grass or wavy hair grass/sheep's fescue associations. Herbage production is only 1 to 2 t/ha DM. Ling and bell heathers together with bilberry plants are the main species on the drier shrub heaths, with ling heather usually predominating. Young heather provides valuable grazing for sheep at all seasons, and more especially during the winter, but periodic burning is necessary to encourage regeneration of young growth. The amount of green edible shoots ranges from 1.5 to 3.0 t/ha DM.

Purple moor grass dominates swards on poorly drained peaty gleys in high rainfall areas, often in association with ling heather. Its leaves are deciduous in autumn following strong growth in summer. (The leaf shedding gave rise to its other name, flying bent.) It is grazed by cattle in its green state, but not after it has matured—which it does quickly. Unimproved purple moor grass swards produce 1 to 2.5 t/ha DM, but where liming and fertilisation are carried out produc-tion of 7.5 t/ha DM has been recorded.

**Table 1.7  Summary of the main soil and vegetation types of the hills and uplands**

| Soil | pH | Vegetation type | Principal species |
|------|----|-----------------|--------------------|
| Brown earth freely drained | 5.3–6.0 | Bent/fescue grassland, high grade or species rich | Common bent grass<br>Red fescue<br>Sheep's fescue<br>Meadow grasses<br>White clover<br>Herbs abundant |
| Gleys poorly drained | 5.3–6.0 | As above with wetland species, sedges, rushes | Sedges<br>Rushes |
| Brown earth freely drained | 4.5–5.2 | Fescue/bent grassland, low grade or species poor | Bent grasses<br>Sheep's fescue<br>Bracken |
| Gleys | 4.5–5.2 | As above with moor mat grass and wetland species, sedges, rushes | Moor mat grass<br>Sedges<br>Rushes<br>Tufted hair grass |
| Podzols Peaty podzols freely drained | 4.0–4.5 | Moor mat grass or wavy hair grass/fescue grass heath Sheep's fescue<br>*or*<br>Heather shrub heath | Moor mat grass<br>Wavy hair grass<br><br><br>Heather<br>Bilberries<br>Bell heather |
| Peaty gleys poorly drained | 4.0–4.5 | Purple moor grass<br><br><br>*or*<br>Heather/purple moor grass | Purple moor grass<br>Sheep's fescue<br>Wavy hair grass<br>Ling heather |
| Deep blanket peat poorly drained | 3.5–4.0 | Deer's hair sedge/cotton grass/heather bog | Deer's hair sedge<br>Cotton grasses<br>Ling heath<br>Purple moor grass<br>Sphagnum mosses |

*Source:* Hill Farming Research Organisation

On blanket peats with a soil pH often below 4.0 and with depths of peat of 30 cm or more, the typical bog vegetation comprises mixtures of cotton grass, deer's hair sedge, ling and bell heathers, purple moor grass and mosses, especially sphagnum. The proportions of these constituents vary according to altitude, depth of peat and intensity of burning or grazing. The associations are very fragile and can be upset rapidly if environmental conditions alter or if management changes are made. Herbage production is poor, with green shoot material amounting to 1 to 3 t/ha DM annually. These communities are the least suitable of all hill swards for land improvement because of the substantial inputs needed to effect and sustain an increase in herbage production.

*Chapter 2*

# Characteristics of Grasses and Legumes

Next in importance to the profusion of water, light and air, those three great physical facts which render existence possible, may be reckoned the universal beneficence of grass.

J. J. Ingalls (1872)

## RYEGRASSES

### Perennial ryegrass

This is clearly the most important and widely used grass species in British seed mixtures, exhibiting rapid establishment, good tillering ability, excellent production response to fertiliser N and high acceptability to stock. At equivalent growth stages to other species, both perennial and Italian ryegrasses are more digestible, thus giving higher DM production at specific levels of digestibility. Their production at a digestibility of 67 D value can be at least a third more than that of cocksfoot, for example, which is a low digestibility grass species. Perennial ryegrass is persistent if soil fertility is high, particularly the intermediate and late heading varieties. It is also tolerant of intensive grazing and of cutting, regrowing quickly after defoliation, provided swards are not cut for repeated heavy crops of silage or for hay at a mature stage of growth, since these managements reduce tiller density.

There are numerous recommended varieties and seed is normally readily available, except perhaps the most recently recommended varieties. As the varieties exhibit a wide range of characteristics, e.g. in heading dates, earliness of growth, disease resistance and persistency, perennial ryegrass can fulfil the needs of most farming systems. It is also an important grass in other temperate regions of the world, such as New Zealand.

However, perennial ryegrass does not grow well under very dry conditions or on infertile soils, when the sward becomes weak and stemmy, lacking longevity and susceptible to invasion by indigenous grasses and weeds. It is susceptible to winter damage during cold periods, or frost damage on first spring growth, although this can be minimised by good management (see Chapter 18).

11

Distinctive varieties of the different ryegrass species have been bred with double the number of chromosomes of normal diploid varieties. These commercial tetraploid ryegrasses were introduced in the 1960s. Tetraploidy can also occur as 'sports' in natural populations. Some of these were developed, and cocksfoot grass and white clover, for example, are natural tetraploids. Timothy is a hexaploid, i.e. it has six sets of chromosomes. Plant breeders induce tetraploidy in ryegrasses by chemical, radiation or other treatment and they follow this with selection programmes. Compared with diploids, tetraploids have larger seed (two to four times heavier than corresponding diploids), tillers are fewer and larger and leaves are darker green.

Recent breeding and selection have aimed to improved tillering capacity. With higher levels of moisture and sugars and better digestibility, the herbage of tetraploids has good stock intake characteristics. The high moisture content (low dry matter) is often cited as a disadvantage in silage or hay making, where wilting and drying are required, but in fact the differences involved are small compared with fluctuations brought about by nitrogen fertilisation. The more open sward density of tetraploid ryegrasses compared with diploids makes them compatible with white clover.

## Italian ryegrass

This is the second most important grass species sown in Britain, although its popularity has declined with the increasing trend towards long-term swards in order to save reseeding costs. It is a biennial, that is, it completes its life cycle in two years, and is most productive in the first full harvest year after sowing. Italian ryegrass produces very well in early spring, especially if sown in late summer rather than in spring the previous year. Under intensive fertilising and cutting for conservation it gives good production of herbage of high digestibility. Early bite can be taken before cutting for silage since it is not an early heading grass, most varieties heading during the same period as intermediate perennial ryegrass. It is also productive in the year of sowing if sown in spring because of its rapid establishment and growth vigour. It can therefore be used as a catch crop, or as a pioneer constituent in mixtures with forage rape.

The main drawback of Italian ryegrass is its short-lived nature which is associated with a high degree of stem production; this lowers acceptability to grazing stock in mid summer if swards are not well grazed. Production from the second full harvest year is also substantially lower than in the first. Older swards of Italian ryegrass or swards which are allowed to enter winter with surplus growth are more susceptible to winter damage than grass swards based on the other main species. It is customary in Recommended Lists for each ryegrass species to be classified independently for winter hardiness—A (hardiest) to E (least hardy)—so the ratings for an Italian ryegrass variety represent lower degrees of winter hardiness than corresponding ratings for perennial ryegrass.

## Hybrid ryegrass

This is a hybrid bred by crossing Italian and perennial ryegrass, seeking to incorporate the best characteristics of both parents, for example, the rapid establishment and vigour of Italian ryegrass with the tillering ability and persistence of perennial ryegrass. Hybrids may resemble one or the other of the parents or be intermediate in appearance. The Italian types have awned seedheads, form an open sward and give good spring growth, whereas the perennial types have awnless or short awned seedheads, form a denser sward and have later spring growth. Most hybrids on Recommended Lists are tetraploids and the best hybrid ryegrasses are less productive than the best Italian ryegrasses but exhibit better ground cover and drought tolerance. The perennial types persist into a third harvest year, although with reduced productivity. Hybrids are best in two- to three-year mixtures for intensive production, cutting and grazing, on well-fertilised soils. In grazed swards, their midseason acceptability is usually higher than that of Italian ryegrass. The initial promise of hybrid ryegrasses has not been fully realised and their role, as with Italian ryegrass, has declined with an increasing emphasis on long-term swards.

## Westerwold ryegrass

This is a rapidly establishing annual species which gives high production in the season of sowing if seeded early and adequate moisture is available. It is therefore suitable for routine or emergency catch cropping. Both diploid and tetraploid varieties are available. Unlike Italian ryegrass, it heads profusely in the sowing year so defoliation, whether by cutting or grazing, must be geared to keeping swards leafy to prevent a decline in their digestibility with maturity.

## OTHER GRASS SPECIES

### Timothy

This grass is better suited to cutting than to intensive grazing. It is very acceptable to grazing animals and, when mixed with competitive grasses such as perennial ryegrass, its slow regrowth leads to its being grazed out if grazing is intensive.

Timothy is useful for the cool, wet conditions of northern and western Britain and is invaluable on wet, peaty and heavy textured soils. Being the most winter hardy of the main grass species, and also winter green, it produces valuable early growth in hill and upland areas, where it tolerates lower fertility than ryegrass. The seed is very small, 2 to 3 million per kg, slow to establish and should not be sown deeper than 10 to 15 mm.

Timothy herbage has a low mineral content. Its digestibility declines less rapidly with maturity than the other major grasses, though the decline starts at an earlier growth stage. Therefore, it should be cut well before ear emergence if high quality forage is required, since the mature herbage is very stemmy and extremely low in digestibility and crude protein content. In its present usage it is close to becoming a minor species, being increasingly supplanted in seed mixtures for grazing by tetraploid perennial ryegrass.

## Cocksfoot

The main reason for sowing cocksfoot is for its drought tolerance and, though a relatively minor species, it is valuable on light textured soils in dry areas. In general purpose mixtures it can provide useful early season and autumn production in upland areas. It is slow to establish but is conspicuous in swards due to a coarse tufted habit of growth, notably if the sward is understocked. If not kept leafy, its low digestibility results in poor herbage intake by stock. It is not recommended in seed mixtures for highly productive animals. A low water-soluble carbohydrate content and relatively low production capability make it unsuitable for silage making.

## Meadow fescue

This grass species grows best under cool, moist conditions, tolerating wet, even occasionally flooded soils, and is highly acceptable to grazing stock when kept leafy. Being slow to establish and non-aggressive, it contributes little in mixture with vigorous species such as perennial ryegrass. It was used in timothy/meadow fescue/white clover mixtures in the past because of its compatibility with clover, and also in hay mixtures, but weed ingress was sometimes a major problem.

Increased fertiliser N use and stocking rates have diminished the role of meadow fescue as it does not respond well to high N rates and lacks persistency under intensive stocking. Varietal choice is limited and it is now a minor species.

## Tall fescue

This is a very adaptable species, growing well in dry or wet conditions; it is also very winter hardy. Established tall fescue swards grow earlier in spring than Italian ryegrass, are very responsive to fertiliser N and have the potential for high production. Tall fescue can withstand the frequent mowing needed in dried grass production. It has several major drawbacks, however, being slow to establish, of poor acceptability to grazing animals and of lower digestibility than the other main grass species. Varietal choice is also limited. While little used in Britain, tall fescue is an important species in some regions of the world, e.g. in the southern United States of America and Argentina.

## Prairie grass

This minor grass, a species of brome grass, performs best in dry conditions and shows good autumn growth. It gives very high production on free-draining soils under mild southern English conditions where its susceptibility to frost is less of a problem than in northern conditions. Its digestibility is better than timothy or cocksfoot but not the ryegrasses.

As it produces an open sward with an upright growth habit it is more suited to cutting for conservation and rotational grazing than for continuous stocking. Both its leafy and stemmy growth are acceptable to grazing stock. In New Zealand, where it is a recognised drought-resistant species, swards have been rejuvenated by temporary cessation of grazing to allow seedhead development and natural reseeding. It has a very large awned seed with a 1,000-seed weight of about 13 g compared with 1.5 to 3.5 g for perennial ryegrasses.

## Sweet brome

This very minor species with similar attributes to prairie grass may have a limited role for conservation on light sandy soils in dry areas. It has shown some compatibility for forage legumes since it develops an open sward. It is unlikely to be suitable for widespread use due to its variable performance compared with other grasses.

# FORAGE LEGUMES

The importance of leguminous plants lies in their rhizobial N-fixation ability, soil improvement properties and their excellent feeding value for animal production. These advantages have been neglected in Europe in recent times, during the era of heavy fertiliser N use. High N was detrimental to forage legumes. However, with new agricultural policies aimed at environmentally friendlier production methods and sustainable lower input systems, there is growing interest in the potential of forage legumes. They are only briefly discussed here since more detailed information can be found in Chapter 22. White clover is by far the most important forage legume used in Britain, followed by red clover and then lucerne. Other species used occasionally include sainfoin and birdsfoot trefoil.

## White clover

White clover is a perennial legume which spreads through the sward by branching stolons. Finely rooted daughter plants develop from nodes on the stolons, eventually taking the place of the tap rooted mother plant which dies out. The amount of clover production is thus related to the amount of stolon present. The N-fixing bacteria of the root nodules are concentrated in the upper sward layers, releasing N to the associated grasses. In comparison with grass, clover is richer in protein and minerals but lower in fibre. It maintains a better digestibility over the season, is more acceptable to stock, and the digested nutrients are metabolised more efficiently. While best suited as a grazing plant it can also be successfully cropped for high quality silage.

There is a potential risk of cattle bloating on clover-rich swards, and it is undoubtedly a serious factor in New Zealand dairy farming which is largely based upon the utilisation of grass/white clover swards. However, various preventative and remedial measures are available. Experience from grazing trials in the UK and from those farms making significant use of white clover has so far not indicated bloat as a major problem, however.

## Red clover

This species accounts for 10 to 15 per cent of the forage legume seed used in the UK. It is primarily a plant for cutting rather than grazing, having an upright growth habit and with the shoots developing from a plant crown. Its use in general purpose seed mixtures has declined, but it can be used in special purpose, clover-dominant mixtures for silage. Clover rot from the fungus *Sclerotinia trifoliorum* and damage from stem eelworm (*Ditylenchus dipsaci*) are limiting factors, but resistant varieties are available. The main disadvantage of red clover is its lack of longevity, which does not fit well with the increasing use of long-duration swards for animal production.

## Lucerne

Lucerne is a special purpose plant primarily suited for conservation. It will produce well for three to four years under the better soil and climatic conditions found in southern Britain. It has a deep tap root requiring free-draining soil, high pH and adequate fertility for longevity. Varietal choice is limited but resistance to soil-borne *Verticillium* wilt and to stem eelworm are necessary for it to be grown successfully. Rhizobial inoculation of the seed is essential.

## Sainfoin

This minor perennial legume has similar attributes and agronomic requirements to lucerne, but with 20 to 30 per cent lower production. There have been no intensive breeding programmes, though several strains or ecotypes are available. The common or single cut type is more persistent than the giant or double cut type. The seed is best sown direct in spring at 50 kg/ha for milled seed, i.e. without the seed coat, and at double this rate for unmilled seed, perhaps mixed with a few kg of a non-aggressive companion grass such as meadow fescue. It does not appear to have any serious pest or disease problems, and is essentially a plant for cutting, but the aftermath can be grazed. The protein rich forage is highly acceptable to stock and is of high nutritive value. The naturally occurring tannins in the leaves limit degradation of the plant protein in the rumen. This improves the efficiency of protein use, since this protein can then be utilised more efficiently later in the digestive tract, giving sainfoin a valuable bloat-free characteristic. The potential of the species warrants more investigation and development than it has hitherto received.

## Birdsfoot trefoil

This species occurs naturally and widely in herb-rich hill and lowland swards, but it is rarely sown in Britain. It is used in Canada and the north-eastern United States of America, for example, where it is adapted to the humid conditions. It is more tolerant of lower levels of soil pH than white clover and has a potential place on soils where clover does not thrive. If sown in mixtures, a non-aggressive grass companion is required. The important trefoil varieties available vary in their tolerance to cutting and to grazing so the proposed use will determine varietal choice. Because of a high tannin content, birdsfoot trefoil does not cause bloat in cattle. The related species, marsh birdsfoot trefoil, has shown potential in wetlands but there are specific establishment and management needs. It has an extensive rhizome system, the development of which is a major key to persistence. It is used for wetland pastures in New Zealand and Australia and with agroforestry swards.

## SECONDARY GRASSES

Surveys have shown that many types of British grassland, especially long-term or permanent grassland, contain unsown grasses. These have variously been called natural, secondary, undesirable, non-preferred or even weed grasses. Bent grasses, fine-leaved fescues, Yorkshire fog, smooth- and rough-stalked meadow grasses, crested dogstail and sweet vernal are the most commonly found. A survey of permanent grassland in England and Wales, for example, showed that bent grasses occupied at least 25 per cent ground cover in 8 per cent of one- to four-year-old swards, increasing progressively to 57 per cent of swards more than 20 years old (see Chapter 1). Despite this, and their widespread occurrence in grassland of all types, including rough grazing where they fulfil a valuable role, secondary grasses have not received the attention they deserve from researchers.

Yet it must not be forgotten that in many areas of the UK a wide range of adverse soil, climatic and management conditions exist, resulting in environments unsuitable for optimal performance of perennial ryegrass and the other main or primary grasses. Several secondary grass species are adapted to withstand or even exploit these poor conditions. By ameliorating poor soil fertility and improving management, sown primary grasses can perform well. Witness the performance of hill land which is limed, fertilised and sown to winter hardy perennial ryegrass swards; unless soil improvement is maintained by lime and fertiliser, and grazing controlled by fencing, however, sown swards soon regress, perhaps to a poorer state than the original.

A predominantly ryegrass sward is the normal aim with lowland grassland, and for hill land as well provided the sward's requirements, especially for adequate soil fertility and good drainage, are met. Soil fertility surveys have shown that lowland fields are often deficient in lime, phosphorus or potassium. Hill swards are subject to these or even more severe limitations, e.g. competition from bracken or rushes, and periods of under- or over-grazing. It is not surprising, therefore, that primary grasses bred for better conditions lose their dominance in such swards after sowing.

It is fortunate there are grasses which can exploit poorer soil environments and tolerate adverse grazing practices. Bent/fescue associations are the best of the indigenous hill swards, often with crested dogstail, Yorkshire fog and sweet vernal interspersed. As noted earlier, bent grasses are also abundant in old lowland pastures. In the past, the bents have been condemned because they are associated with poor soil fertility and consequent low production. There has never been a high input of breeding and selection in the UK to improve such grass species for the farming niches they fill, although there has been limited work with red fescue. The argument against such work was the lack of a sizeable market.

There are many improved European varieties of secondary species selected for sports turf and amenity purposes, and it is possible that agricultural evaluation may identify one or two of these from the fescues or meadow grasses, albeit as a bonus to grassland farming and not as an objective. The value to agriculture may arise from criteria used in selecting amenity grasses such as good tillering capacity, leafiness, persistency, winter hardiness, winter greenness and tolerance of trampling. However, digestibility and nutritive value would need to be given prominence too.

The productivity and potential of secondary grasses have been investigated at Auchincruive, Ayr using differing fertiliser N rates, cutting managements and swards, with and without white clover, all compared with perennial ryegrass (see Figure 2.1).

Some grasses, notably Yorkshire fog, red fescue and creeping bent grass, were superior in herbage DM output to perennial ryegrass at low or moderate N rates (0 and 120 kg N/ha/year) but this superiority was not sustained as N rate increased. Nevertheless, Yorkshire fog, red fescue and smooth-stalked meadow fescue still gave high DM production at higher rates of N (360 and 480 kg N/ha/year). Crested dogstail and sweet vernal gave higher output than ryegrass only at nil N. The least productive grass was common bent.

However, the superiority of ryegrass in digestibility (D value) was unchallenged at all N levels. Consequently, in yield of digestible herbage, ryegrass was pre-eminent at all N rates save nil N where only Yorkshire fog and S59 red fescue performed better. Next to ryegrass in

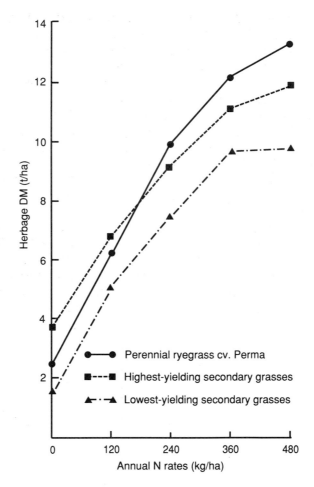

*Figure 2.1    Mean annual DM production over three harvest years at each fertiliser N rate (perennial ryegrass cv. Perma v. range of secondary grass species)*

digestibility were Yorkshire fog and crested dogstail. Red fescue, bent grass and smooth-stalked meadow grass were consistently low in digestibility. The lowest crude protein contents were associated with ryegrass and the most productive secondary grasses. Phosphorus, potassium and calcium contents were satisfactory for grazing stock in all the grasses, but magnesium levels were marginal.

Yorkshire fog, red fescue and sweet vernal were found to be more productive than ryegrass at nil N at the first (May) harvest. As N level increased, Yorkshire fog, red fescue and bent grass maintained a similar advantage over ryegrass. The same grasses outperformed ryegrass in midseason (June, July, August) at nil and 120 kg N/ha/year but the advantage switched to ryegrass at higher N rates. Later in the season, Yorkshire fog, red fescue and creeping bent grass outproduced or matched ryegrass DM yields at all N rates.

In other work, red fescue, smooth-stalked meadow grass and crested dogstail proved to be compatible with white clover and therefore have possible potential for use in grass/clover swards. The close-knit sward developed by Yorkshire fog and creeping bent was not conducive to clover stolon proliferation and development of clover growing points.

The above findings have been reinforced from work at English and Welsh sites, comparing permanent grass swards of mixed composition with reseeded perennial ryegrass. A range of fertiliser N rates was used and the sites covered a variety of environmental conditions. The permanent grassland was less productive than reseeded perennial ryegrass the first harvest year, partly because of extra N being available to the ryegrass swards from mineralisation in the soil resulting from cultivation. However, the advantage of the reseeded swards was only maintained in later harvest years at high rates of fertiliser N application—above 300 kg N/ha/year (see Figure 2.2).

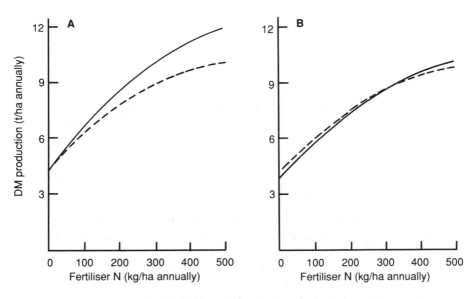

Figure 2.2   Herbage DM response from permanent swards (------) and reseeded perennial ryegrass (———) averaged over 16 sites cut at 4-weekly intervals: (A) 3-year mean and (B) 4-year mean, including establishment phase
Source: Hopkins et al.

These results indicate that certain secondary grasses, and some permanent swards, can match or outperform perennial ryegrass production at nil to moderate N application rates. It is apparent that some can also respond to high soil fertility and would repay increased N inputs on the farm. A few secondary grasses have a production advantage over ryegrass at certain times of year, which is a feature worth exploiting. Some fall down in their digestibility, but are not lacking in minerals essential for stock health.

It can be concluded that it is not always essential to reseed swards because they do not contain a good proportion of primary grasses, especially bearing in mind reseeding costs and the temporary loss of production during establishment. Performance and response to better management should be assessed first. Reseeding to a perennial ryegrass-dominant sward is thus not always warranted, particularly if the sward is to receive low to moderate N inputs.

The chief characteristics of selected secondary grasses are outlined below.

### Red fescue

This is suitable for inclusion in seed mixtures for land reclamation under harsh hill and upland environments and for infertile soils, because of good winter hardiness, winter greenness and sward density. Its main drawback is low digestibility and poor mineral concentrations. It is not highly acceptable to grazing animals because of its hard, needle-like leaves and early flowering habit if not well grazed. Many varieties have been bred for amenity grassland and sports turf, but few for agricultural grassland where it would be advantageous if its nutritive value was improved.

### Bent grasses

Bents are adapted to environments of low potential production. They can take up increased fertiliser N but are unable to metabolise the N for high DM production; they may therefore be rich in crude protein if soil fertility is high. They are generally low in digestibility, although high in mineral concentrations. Many varieties have been bred for amenity grassland but few have been tested for agriculture.

### Smooth-stalked meadow grass

Its notable feature is good DM response to fertiliser N so it has production potential and an ability to increase its proportion in the sward with increasing soil fertility. It is also highly acceptable to stock. It is included in seed mixtures for dry areas due to its drought resistance. Its disadvantages include poor digestibility and very slow establishment. A host of amenity cultivars are available but few agricultural varieties.

## Yorkshire fog

It would appear the productive ability of this grass has been underrated in the past, though some breeding effort has gone into it in New Zealand where there is a named variety, Massey Basyn. British trials show Yorkshire fog to be productive at both low and moderately high rates of fertiliser N, with good digestibility and mineral concentrations. Its main flaw is a lack of acceptability to stock if not kept at a leafy stage of growth, and a proneness to winter damage. Winter hardiness and a less hairy leaf type should be breeding aims.

## Crested dogstail

This species has poor production capacity, poor response to applied N and a propensity for producing stem. A variety bred in New Zealand is available. While crested dogstail has shown compatibility with white clover, there are better compatible grass species available.

## Sweet vernal

Overall this species has few advantages for agricultural grassland, although it can produce early in the season, and is relatively productive at low soil fertility. It has poor concentrations of minerals.

## Rough-stalked meadow grass

This species has given poor performance in Scottish trials, persisting for only one harvest year, although herbage quality was satisfactory. Even bred varieties from Denmark did not persist. The presence of rough-stalked meadow grass in sown swards should be regarded as undesirable although it may have some use in the reclamation of wet soils.

*Chapter 3*

# Herbage Variety Evaluation

But of all sorts of vegetation the grasses seem to be most neglected; neither the farmer nor the grazier seems to distinguish the annual from the perennial, the hardy from the tender, nor the succulent and nutritive from the dry and juiceless.

Gilbert White, *The Natural History of Selborne* (1788)

## CLASSIFICATION OF MATURITY GROUPS

Varieties within a grass species are classed into maturity groups based on date of ear emergence or heading. Base data are obtained from spaced plants grown at Crossnacreevy, Northern Ireland, the dates being recorded when ears become visible on 50 per cent of the reproductive tillers. For a given variety the date of 50 per cent ear emergence (EE) can vary at a location from year to year by up to two to three weeks, depending upon early season temperatures. However, the relative differences of heading date between varieties remain fairly constant. At the Scottish Agricultural College, Auchincruive, Ayr the date of 50 per cent EE for Aberystwyth S24 perennial ryegrass, one of the main reference varieties, has ranged from 13 May in an early year to 30 May in a late year. Heading date is progressively later the more northerly the latitude and the higher the altitude.

Knowledge of ear emergence is important in the prediction of cutting dates for silage or hay to attain herbage of target digestibility value. There is a strong relationship between the development of ear emergence and the pattern of digestibility, which falls as ear emergence and associated plant maturity increases. Earliness of heading date is closely related to earliness of spring growth, but each species has its own distinctive early growth pattern.

Varieties of perennial ryegrass are classed into five maturity groups: very early (VE), early (E), intermediate (I), late (L) and very late (VL) with most in the E, I and L groups. This species is exceptional in that the range of dates of 50 per cent EE among its varieties can spread over eight weeks during the May–June period. Varieties of Italian ryegrass and timothy are grouped into the three classes E, I and L with a spread of 50 per cent EE over two weeks for Italian ryegrass and

**Table 3.1    Grass maturity groups in relation to 50 per cent ear emergence (number of days after 1 May)**

| Species | Very early | Early | Intermediate | Late | Very late |
|---|---|---|---|---|---|
| Perennial ryegrass | Up to 26 | 27–34 | 35–45 | 46–57 | Over 57 |
| Italian and hybrid | | | | | |
| ryegrass | – | Up to 33 | 34–42 | Over 42 | – |
| Timothy | – | Up to 52 | 53–58 | Over 58 | – |
| Cocksfoot | – | Up to 31 | – | Over 31 | – |

*Source:* Scottish Agricultural College

three weeks for timothy. Heading dates and their maturity groups are specific for each species. For example, timothy is a late heading grass and early varieties head two to three weeks later than early perennial ryegrass (see Table 3.1). Only two groups, E and L, are assigned for cocksfoot, meadow fescue and tall fescue, with heading dates spread over approximately two weeks for cocksfoot, three for meadow fescue and four for tall fescue. The number of varieties commercially available in the latter two species is very small. Westerwold ryegrass has a profusion of heads in the seeding year when sown in spring but there are only small differences between varieties in ear emergence patterns.

White clover varieties are classified by leaflet size, using the length times breadth in mm² of the terminal leaflet as the unit of measurement. There are four groups: small (S), medium (M), large (L) and very large (VL), with most varieties falling into the first three categories. There is a good correlation between leaflet size and tolerance of defoliation, the smaller leaved varieties being more suitable for severe defoliation— typical of sheep grazing—and the larger leaved varieties more suitable for cattle grazing or silage cutting.

Red clover varieties are either early or late flowering, the range of flowering dates spreading over five weeks. All the lucerne varieties currently used in the UK are in a single maturity group, and flower within a week of one another.

## NATIONAL LIST VARIETAL EVALUATION

Grass and herbage legume varieties in EC countries must be on a National List before they can be sold in that country or listed in the EC Common Catalogue, entry to which can be gained after two years on a member country's National List. For National Listing the varieties must undergo comprehensive statutory assessment of varietal performance, termed 'value for cultivation and use' (VCU trials), and for 'distinctness, uniformity and stability' parameters (DUS trials). The latter trials, carried out at the Northern Ireland Plant Testing Station, Crossnacreevy on behalf of the UK, check that each variety is unique genetically so that protective Plant Breeders' Rights including royalties can be obtained for the breeder should the variety be successful commercially.

The extent and methodology of performance testing varies from country to country within the EC, but in the UK there are seven centres, using small plot experimentation to test the major species used: perennial, hybrid and Italian ryegrasses, timothy and white clover. In recent years a total of 60 to 70 varieties have been entered annually for testing. About one in four herbage varieties submitted is eventually National Listed. Cocksfoot, meadow fescue, tall fescue, Westerwolds ryegrass and red clover varieties, together with some other species, were formerly statutorily evaluated but now occupy a minor position in seed mixtures. Approximately 70 to 80 per cent of the varieties now tested are perennial ryegrass and 15 to 20 per cent are Italian ryegrass varieties. Government and plant breeders share the costs of assessment. Currently about 200 varieties of agricultural grasses and legumes are National Listed. The organisations which are responsible for the continuity of National Listed varieties are known as 'maintainers' who may be UK agents of foreign firms.

In the current National List testing procedure, perennial ryegrass varieties are sown for two consecutive years and each sowing harvested for two full years under a simulated grazing regime (7 to 9 cuts per annum) and a silage regime (4 to 5 cuts per annum). Varieties are compared with long-standing standard or control varieties. High level fertilisation is used with adequate $P_2O_5$ and $K_2O$ application and a high annual rate of fertiliser N—350 kg/ha for the silage regime and 360 kg/ha for the frequent cutting regime. Timothy varieties were formerly assessed in the same manner as perennial ryegrass but now only the silage regime is used. Italian and hybrid ryegrasses are managed together and productivity measurements start in the establishment year, when up to 5 cuts can be taken. Fertiliser N is applied at 50 kg/ha per cut. In the next two full harvest years, 5 to 7 cuts are taken: an 'early bite' harvest (1.1 t/ha DM target), 2 conservation cuts and then monthly cuts until the end of the season. Total annual N application is 400 kg/ha.

Assessment of the grass varieties is judged on seasonal and annual DM yield and on quality, i.e. digestibility as D value (digestible organic matter as a percentage of the dry matter), crude protein and water-soluble carbohydrate contents measured in the first two silage cuts of the first harvest year. First silage cuts are taken at a target D value of 67. This is achieved by monitoring the heading date of an early heading reference variety, whose ear emergence–digestibility relationship is well documented, and then relating the cutting dates of later maturing groups to a specific number of days after the early group. Survival or persistence is judged by ground cover. Tolerance to drought, diseases or pests is also recorded. In addition, varieties are exposed to specific diseases such as crown rust, Drechslera, Rhyncosporium and ryegrass mosaic virus in small-scale tests using disease inoculation techniques and infected sites to measure their resistance. Winter hardiness of ryegrass varieties is separately assessed at an upland site (300 m above sea level) in northern Scotland, where an autumn management of high fertiliser N input and late cutting is given to intensify any susceptibility to winter damage.

White clover varieties are tested in mixtures with a single intermediate variety of perennial ryegrass and cut 5 to 7 times per year, each at a target DM production of 1.5 t/ha. Production is only measured in the second and third harvest years, to allow full development of the clover, following sowings in two successive years. An annual fertiliser N rate of 200 kg/ha is applied together with adequate $P_2O_5$ and $K_2O$. The clovers are also sown at sites infected with Sclerotinia to assess resistance. In parallel sowings to assess persistence, the varieties are subjected to heavy defoliation pressure by cutting every 10 to 15 days to simulate intensive grazing, no production measurements being taken. National Listing is based on seasonal and annual DM production, quality (digestibility and crude protein content) and clover persistence as measured by survival and ground cover.

## RECOMMENDED LISTS

An important benefit derived from National List testing is the compilation of non-statutory Recommended Lists of the best varieties for use in seed mixtures. National List data are combined with information from supplementary testing on farms, including acceptability to grazing stock, spring growth, ground cover, disease resistance and other practical information. Plant breeders or agents for the variety can also submit further information when varieties are being considered for Recommended Lists. Older recommended varieties are retested periodically and their status reviewed alongside newer varieties. Table 3.2 shows a sample of the information given in the National Institute of Agricultural Botany (NIAB) Recommended List for England and Wales.

**Table 3.2 Example of information given in NIAB leaflet on recommended**

## LATE HEADING PERENNIAL RYEGRASS

First cuts of late perennial ryegrass will reach 67D about the 4th June in a normal season in central England.

### Recommended List of Late Perennial Ryegrass Varieties 1991/92

| | | Annual yield of dry matter as % Talbot | | Ground cover 1–9 good | Year first listed |
|---|---|---|---|---|---|
| | | Grazing | Conservation | | |
| G | Contender | 102 | 100 | 7½ | 1982 |
| G | Borvi | 104 | 104 | 6½ | 1978 |
| G | Parcour | 104 | 105 | 8 | 1979 |
| G | Condesa (Tet) | 108 | 108 | 7 | 1983 |
| G | Hercules | 103 | 102 | 7½ | 1987 |
| G | Trani | 104 | 102 | 7½ | 1977 |
| G | Belfort (Tet) | 104 | 103 | 6½ | 1982 |
| PG | Profit | 105 | 103 | 8 | 1988 |
| PG | Tyrone | 104 | 103 | 7 | 1987 |
| PG | Portstewart | 106 | 106 | 7 | 1989 |
| PG | Jumbo | 110 | 103 | 6½ | 1988 |
| PG | Tivoli (Tet) | 107 | 108 | 6 | 1989 |
| PG | Carrick | 104 | 102 | 7½ | 1988 |
| O | Meltra (Tet) | 102 | 106 | 6 | 1978 |
| O | Antrim | 104 | 102 | 7 | 1985 |
| O | Melle | 102 | 99 | 7½ | 1968 |

### RECOMMENDED FOR GENERAL USE

**Contender**   Average yields combined with good ground cover.

Nickerson, UK.

**Borvi**   Good yields under both managements with relatively good growth in early spring. Good resistance to crown rust. Below average midseason digestibility and ground cover for a diploid variety.

Danish Plant Breeding, Denmark.

**Parcour**   Good ground cover combined with high annual yields and relatively good spring growth. Good resistance to crown rust.   Petersen, Germany.

**Condesa** (Tet)   High yields under both managements especially in the second year combined with good ground cover for a tetraploid under both cutting and farm grazing. Good winter hardiness. Poor growth in early spring but good growth in mid summer and autumn. Good digestibility in the first conservation cut allows delayed cutting to achieve high yields at 67D.

Van der Have, Netherlands.

**Hercules**   Very good midseason digestibility and good ground cover combined with good resistance to crown rust.   Zelder, Netherlands.

**Trani**   High yields in the simulated grazing management combined with relatively good early spring growth. Good ground cover, winter hardiness and resistance to crown rust.   Danish Plant Breeding, Denmark.

**Belfort** (Tet)   Good simulated grazing yields and resistance to leaf diseases for a tetraploid.                                                          Mommersteeg, Netherlands.

## PROVISIONALLY RECOMMENDED FOR GENERAL USE

**Profit**   High yields under the simulated grazing management and in the conservation management in the second harvest year. Very high midseason digestibility. Good ground cover and winter hardiness.      Van Engelen, Netherlands.

**Tyrone**   High yields in the simulated grazing management and under first harvest year conservation. Good midseason digestibility.
                                                                         N.I. Plant Breeding Station, UK.

**Portstewart**   High yields under both managements combined with very good midseason digestibility. Good yields at 67D in the first conservation cut.
                                                                         N.I. Plant Breeding Station, UK.

**Jumbo**   Very high simulated grazing yields, especially in the second harvest year, despite below average ground cover for a diploid variety. Good conservation yields in the second harvest year and good late summer and autumn growth. Good resistance to crown rust. Low midseason digestibility.
                                                                              Green Genetics, Netherlands.

**Tivoli** (Tet)   High yields under both managements with above average midseason digestibility and relatively good early spring growth. Good winter hardiness. Good resistance to leaf diseases for a tetraploid.
                                                                            Danish Plant Breeding, Denmark.

**Carrick**   High yields under the simulated grazing management. The relatively good early spring growth is combined with a very late heading date. Good digestibility at the first conservation cut allows delayed cutting to achieve high yields at 67D. Good ground cover and above average midseason digestibility. Moderately susceptible to crown rust.         N.I. Plant Breeding Station, UK.

## BECOMING OUTCLASSED

**Meltra** (Tet)   High yields in the conservation management with good winter hardiness. Moderately susceptible to crown rust and midseason digestibility has been well below average.                              RvP, Belgium.

**Antrim**   High yield in simulated grazing management with good resistance to crown rust. Below average midseason digestibility.     N.I. Plant Breeding Station, UK.

**Melle**   Good ground cover but below average yields in the conservation management, and in early spring.                              RvP, Belgium.

## RYEGRASS VARIETIES WITH HIGH MAGNESIUM CONTENT

Two ryegrass varieties which have high magnesium contents, Magnet Italian ryegrass and Ramore early perennial ryegrass have been considered for addition to the Recommended List. Neither variety was considered to have sufficient overall merit for Recommendation.

*Source:* National Institute of Agricultural Botany

There are currently 35 perennial ryegrasses, 15 Italian ryegrasses, 5 hybrid ryegrasses, 11 timothies, 8 cocksfoots, 12 white clovers, 6 red clovers and 4 lucernes on the National Institute of Agricultural Botany (NIAB) Recommended List for England and Wales. Table 3.2 shows a sample of the typical information given. Since some herbage varieties in the EC Common Catalogue may not have been tested in any way in the UK it is obviously a calculated risk to use them, nor can they ever be put on to UK Recommended Lists unless they have been tested. Recommended List testing is now funded by the farming industry through a voluntary levy on retailed herbage seed. Logo labels of the testing authorities—NIAB in England and Wales and the Scottish Agricultural College (SAC) in Scotland—are displayed on merchants' seed catalogues and seed bags. The seed merchants receive results of varietal performance trials and advice from the testing bodies, information which they can utilise in formulating their seed mixtures and pass on to customers.

It is unfortunate that farmers do not pay the same attention to choice of herbage varieties as they do to cereal varietal choice, for example. This is because of the much larger numbers of recommended grass varieties available, because grass is not a cash crop, and also because varieties are marketed as constituents of seed mixtures. However, there is an advantage to be gained from matching varietal capabilities as far as possible to farm requirements. In addition to production performance ratings, all other listed characteristics such as earliness of spring growth, persistency, winter hardiness and disease resistance should be considered before selecting a particular variety. Seed mixture formulation is discussed in Chapter 4.

# Chapter 4

# Seed Mixtures

---

To break a pasture makes a man
To make a pasture breaks a man

Adage

A field requires three things; fair weather, sound seed and a good husbandman.

Anon.

---

There has been much folklore and mystique involved in the art of formulating seed mixtures, but they are now based on more scientific evidence. The main reason is the detailed information available for selection from the comprehensive UK-wide programme of varietal evaluation. A fund of knowledge has also built up from research concerning factors such as species compatibility, reaction of blends to different management systems and tolerance of species and varieties to special soil or climatic conditions. At the end of the day, the types of sward favoured and the varieties selected for the seed mixtures are matters of a farmer's choice. However, this choice is rarely made independently and individually. Most farmers choose seed mixtures via sales representatives and seed merchants' catalogues.

The merchants have a close association with varietal testing authorities and advisory services and have access to most of the information gathered. The species and varieties on offer should therefore be the best available, if full advantage is taken of the links. The advisory services also have computer programmes which make use of a database of detailed varietal characteristics to compile and generate appropriate seed mixtures for various situations.

The main change in seed mixtures in recent years has been the trend towards simplification, from the old-time 'blunderbuss' mixtures through the 'shotgun', and now to the 'rifle' mixtures, at their simplest for intensive grassland use. The rationale is that modern varieties have been bred and tailored to meet specific requirements and have highly distinctive growth, production and quality features, together with specific tolerance to environmental factors, pests and diseases. If their potential is to be exploited fully, they must be sown in simple seed

31

mixtures, either straight or with one or two other varieties with compatible characteristics. To do otherwise is to dilute or lose the advantages for which the variety or varietal combination was first chosen. A counter-trend has recently started with the development of grassland to be used extensively and the desire to create general purpose, multi-species swards. Here a wider range of grasses (both primary and secondary), legumes, forage herbs and wild flowers need to be considered. Precise formulation of such mixtures is at an early stage of development since there is a lack of research. Many of the existing commercial mixtures contain too many constituents, for example, without the certainty that they will make substantial contributions (see Chapter 23).

# CHOOSING SEED MIXTURES

Factors to be considered when choosing or making up a seed mixtures are shown in Table 4.1.

**Table 4.1  Key factors in choice of seed mixture**

- Intended duration of the sward, short or long term

- Management of the sward, grazing or cutting or both

- Intensity of management in relation to fertiliser use, stocking rate, class of stock and cutting for conservation

- Special conditions which may prevail such as disease, drought or winter kill

- Compatible heading dates of grass species and varieties in relation to cutting target quality silage

- Compatibility of grasses and clovers in relation to both components producing optimum production

For short duration swards, single or multiple varieties of the short-lived Italian or hybrid ryegrasses are recommended. Alternatively, simple blends of the two species with features complementing one another can be devised. Westerwold or Italian ryegrass or combinations of these make suitable one-year catch crops. It is a mistake to incorporate these aggressive species in long-duration mixtures for the sake of their speed of establishment and short-term production advantage. They are highly competitive and their vigorous growth reduces the establishment of longer lived, but slower establishing companion species. For 2-to-3-year leys, it is necessary to add some early or intermediate perennial ryegrasses, with early growth characteristics, to the Italian or hybrid ryegrasses. For 3-to-6-year leys, perennial ryegrass is the most important constituent, especially for intensive management. Varieties with similar heading dates should be blended for leafy, precision quality silage. A succession of first cut silage crops of target digestibility can be assured by sowing different varieties or complementary combinations to match the anticipated spread of silage making time.

Because of the growth vigour of the large seeded tetraploid ryegrasses it is not normally necessary to increase customary seed rates either when sowing straight or when replacing a proportion of the diploids, though an increase of 15 to 20 per cent may be required when sowing the extremely large seeded Italian or hybrid ryegrasses. At Ayr, one diploid and two tetraploid Italian ryegrass varieties (2.45 to 5.46 g per 1,000-seed weight) were sown at 9 seed rates, 10 to 90 kg/ha. Herbage production and quality parameters were similar regardless of seed rate, and it was concluded that 20 to 30 kg/ha was an adequate seed rate in practice, provided attention to detail was given during seedbed preparation, sowing and fertilisation.

On account of their openness of sward density, tetraploid perennial ryegrasses were formerly included in mixtures to replace only up to a quarter of the diploids, but the newer tetraploid varieties, which are more densely tillering, can be included to at least a half by weight, or up to three-quarters on land not liable to poaching.

For long-term or permanent swards, increasing emphasis is placed on the intermediate and late heading perennial ryegrasses as the basis of mixtures, because of their tillering ability and persistence. Timothy may be added for heavy or wet, peaty soil conditions while, conversely, cocksfoot may be added for light, drought-prone soils. These two species are also more suited than ryegrass for less intensive use in more difficult upland conditions; all three species are often incorporated in a general purpose upland mixture. In other specialised situations timothy can be sown as the dominant constituent for a long-term hay meadow, or cocksfoot where drought is endemic. The use of these and other non-ryegrass species for special soil and climatic conditions could increase in the future, as in New Zealand where varieties of cocksfoot, brome grass and tall fescue have been developed specifically for drought situations.

The amount of varietal testing on the hills and uplands has been limited in the UK. Reliance has mainly been placed on seed mixtures using the best varieties from lowland site testing but with characteristics suited to upland conditions, notably winter hardiness and persistency. Perennial ryegrass is again the favoured dominant constituent, with other species added for special conditions, as outlined above.

White clover is included in most seed mixtures intended for medium- or long-term duration, but red clover is included only in general multi-purpose mixtures or in special purpose red clover-dominant silage mixtures. Lucerne is mainly sown in monoculture or with a non-aggressive grass. There is increasing interest in using legume-based swards for animal production, so the role and management of grass/white clover, red clover and lucerne swards are discussed in detail in Chapter 22.

## SPECIMEN MIXTURES

### Short-term to long-term

Specimen mixtures using the above principles are outlined in Tables 4.2 to 4.4. The mixtures are given purely as examples which can be modified to suit individual circumstances, especially in relation to varietal choice. The seed rates are typical of those used in Scotland; they tend to be lower in England and Wales. There is some flexibility in the durations shown for medium- and long-term mixtures. The seed rates are suitable for sowing by broadcasting and should be reduced by 15 to 20 per cent for drilling. However, seed rates could also be reduced by 15 to 20 per cent when broadcasting if seedbed conditions are good, or increased by the same proportion for poor seedbeds, although there is a limit to the extent to which increased rates can compensate for ill-prepared seedbeds. In selecting white clover varieties, it is best to choose by leaf size to match the intended class of stock and management.

A full seed rate is not required for all situations, e.g. renovating deteriorated swards by oversowing or direct drilling (see renovation section, Chapter 7). A half or two-thirds seed rate will usually suffice provided there is a nucleus of existing sward.

### Set-aside land

Economy of seed use and therefore cost is necessary when sowing out land set aside from arable cropping. A dense vegetation cover is desirable to prevent soil erosion and leaching of soil N. Colonisation by weed seed from the soil or invasive seed is also prevented and the land is kept in a good agricultural state. Natural revegetation does not fulfil most of these objectives and can be unsightly to boot.

**Table 4.2   Short-term seed mixtures**

|  | Sowing year | | 1 to 2 years | |
| --- | --- | --- | --- | --- |
| *Mixture no:* | *1* | *2* | *3* | *4* |
| Westerwold ryegrass | 35 | 25 | | |
| Italian ryegrass | | 10 | 32 *or* | 16 *or* |
| Hybrid ryegrass | | | 32 | 16 |
| Perennial ryegrass | | | | |
|   Early | | | | 16 *or* |
|   Intermediate | | | | 16 |
| Total (kg/ha) | 35 | 35 | 32 | 32 |

**Table 4.3   Medium-term seed mixtures**

|  | 3 to 6 years | | | | |
| --- | --- | --- | --- | --- | --- |
| *Mixture no:* | *5* | *6* | *7* | *8* | *9* |
| Perennial ryegrass | | | | | |
|   Early | | 18 | | | |
|   Intermediate | 30 | 12 | 24 | 16 | 9 |
|   Late | | | 6 | 8 | 9 |
| Timothy | | | | 6 | 6 |
| Cocksfoot | | | | | 6 |
| White clover | 2 | 2 | 2 | 2 | 2 |
| Total (kg/ha) | 32 | 32 | 32 | 32 | 32 |

**Table 4.4   Long-term seed mixtures**

|  | 7 years plus | | | | |
| --- | --- | --- | --- | --- | --- |
| *Mixture no:* | *10* | *11* | *12* | *13* | *14* |
| Perennial ryegrass | | | | | |
|   Intermediate | | 6 | 24 *or* | 18 *or* | |
|   Late | 30 | 18 | 24 | 18 | 14 |
| Timothy | | 6 | 6 | 6 | 6 |
| Cocksfoot | | | | 8 | |
| Red fescue | | | | | 12 |
| White clover | 3 | 3 | 3 | 3 | 3 |
| Total (kg/ha) | 33 | 33 | 33 | 35 | 35 |

For rotational (1-year) set-aside, fast establishment is necessary, and this aim is met by a mixture of Italian ryegrass and red clover (see Table 4.5). A mixture of the more persistent and densely tillering perennial ryegrass together with white clover is more suitable for the full-term (5-year) set-aside option. The inclusion of a forage legume improves the soil N status for subsequent cropping. The mixture for the more permanent set-aside can be modified to take account of special situations such as the partial replacement of perennial ryegrass by cocksfoot for dry conditions or by timothy for wet, heavy soils.

Table 4.5   Specimen seed mixtures for set-aside fallow and for horse paddocks

|  | Set-aside | | Horse paddocks | |
| --- | --- | --- | --- | --- |
| *Mixture no:* | *15* | *16* | *17* | *18* |
| Perennial ryegrass | | | | |
| Intermediate | | | 5 | |
| Late | | 15 | 15 | 20 |
| Italian ryegrass | 16 | | | |
| Timothy | | | 5 | 4 |
| Red fescue | | | | 4 |
| Smooth-stalked meadow grass | | | 5 | 4 |
| White clover | | 1 | 2 | 2 |
| Red clover | 4 | | | |
| Total (kg/ha) | 20 | 16 | 32 | 34 |

Economy of cost must be maintained from the seed mixture to establishment through to maintenance of the vegetative cover by the officially specified cutting regime, yet still choosing the most suitable and effective method for establishment for a particular farm circumstance. Lime may be applied on set-aside land to improve soil pH, but not inorganic fertilisers containing nitrogen, phosphate or potash. Application of organic manures is only allowable in special circumstances, e.g. to aid the good establishment of a cover crop. The grass/white clover sward may require more cuts than specified to maintain it in a satisfactory condition because the improved soil fertility resulting from clover inclusion will lead to higher herbage production than from a grass sward.

## Horse paddocks

In recent years there has been an upsurge in the numbers of horses and ponies kept for recreational riding, often in semi-rural areas. Paddocks are frequently overstocked and often have to function as exercise grounds as well as providing forage. Thus, seeds mixtures have to contain persistent grass species resistant to trampling. While perennial ryegrass forms the core of mixtures, other grass species can be included for good acceptability, e.g. timothy, and for turf-forming ability, e.g. smooth-stalked meadow grass and red fescue (see Table 4.5). In some situations, as in the south-east of the United States of America, tall fescue fulfils a valuable role in thoroughbred horse pastures; it is much less popular in Britain. Small leaved white clover or a blend of small leaved and medium leaved white clover is included to enhance soil fertility and improve the nutritive value of the herbage. A blend of forage herbs at 2 to 3 kg/ha can also be usefully included.

*Chapter 5*

# Herbage Plant Breeding

---

Every blade of grass is a study; and to produce two where there was but one is both a profit and a pleasure.

Abraham Lincoln (1809–65)

But three-leaved grass (clover) soon yield a three-fold profit. Three volumes may be writ in praises of it.

Andrew Yarranton (1663)

---

In crop breeding the objective is often directed towards the annual harvest of a single part of the plant, such as the seed or root. In contrast, the production of forage grasses and legumes is based on sequential harvests, either frequently, if grazed, or infrequently, when cut for conservation. Permutations of harvest modes are possible, and lasting over several years for perennial species. The method, pattern and intensity of harvesting herbage at any time can influence the regrowth, especially under grazing, which is an irregular harvesting process, influenced by prevailing soil and weather conditions and by management decisions. Production of herbage is not an end in itself, as herbage must be grazed or fed to stock and converted into animal products. Thus the herbage plant breeder is faced with the complex task of breeding for feeding value to livestock with a variety of needs in addition to breeding for plant production and persistence. The management of grassland is the vital link between plant and animal.

## SELECTION AND GENETIC MANIPULATION

The breeder can select for required characteristics within a fund of genetic variation in indigenous species or existing varieties. He can also widen the genetic base by bringing in new material from plant collection abroad or from overseas germplasm stores. Hybridising of introduced species with indigenous plant material can produce new varieties with desirable characteristics. Growth at low tempera-tures and light intensities to extend the grazing season is a case in point. By hybridising Swiss white clover with indigenous material, varieties with early spring growth are being developed. Similarly, grasses have been developed with the same characteristic from cross-breeding indigenous and Mediterranean plant material. The early growth characteristic is often accompanied by reduced cold hardiness, so concurrent selection is necessary for resistance to cold as well as disease resistance, good seed production and other desirable characteristics.

Seed production can be a problem with interspecific and intergeneric breeding crosses, but there are ways of overcoming difficulties. Induced doubling of the grass chromosome numbers to achieve hybrid fertility is one such way. More recent biotechnological techniques for plant breeding include tissue culture and genetic manipulation. Specific genes for new characteristics can be inserted into the existing gene complement of the plant, for example.

The process of plant breeding can take up to 15 years or more from the initial cross to the fully tested variety reaching the farm in a seed mixture. The technique involves many thousands of spaced plants with cycles of detailed observations of characteristics leading to selection of elite mother plants to form the basis of a variety. The variety is then entered into DUS and VCU trials, as previously outlined in Chapter 3, and, if it successfully completes these trials, it is included in the National List. If it is of sufficient agronomic merit it will eventually attain Recommended List status. Figure 5.1 shows the typical sequence and time-scale in conventional plant breeding.

The modern plant breeder is supported by scientists from many disciplines: agronomy, plant physiology, chemistry and animal nutrition and, increasingly, cell biology and molecular genetics. All feature prominently during the process of producing a 'winner'. At the development stage, and again at the Recommended List stage, varieties are sometimes evaluated under practical conditions using grazing animals; this is not possible at the National List testing stage, however, due to the large number of test varieties. Nevertheless, many of the characteristics can be adequately assessed under simulated management conditions and by using selected compositional analysis and laboratory predictors of performance. Only a few of the best varieties will ever reach animal production trials, so the ultimate test is generally performance under commercial farming, which is not easily measured. For novel varieties that are synthesised from parental material for which background information is limited, or for varieties created by genetic manipulation, evaluation by animals should be introduced at as early a stage as possible in the plant breeding programme.

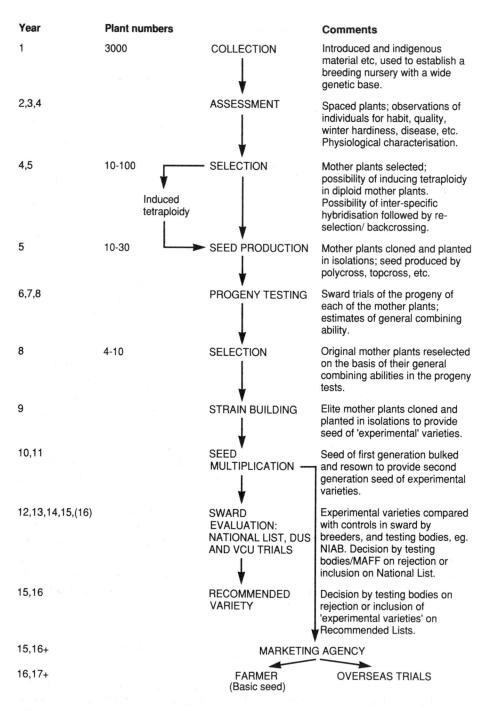

| Year | Plant numbers | | Comments |
|------|---------------|--|----------|
| 1 | 3000 | COLLECTION | Introduced and indigenous material etc, used to establish a breeding nursery with a wide genetic base. |
| 2,3,4 | | ASSESSMENT | Spaced plants; observations of individuals for habit, quality, winter hardiness, disease, etc. Physiological characterisation. |
| 4,5 | 10-100 | SELECTION | Mother plants selected; possibility of inducing tetraploidy in diploid mother plants. Possibility of inter-specific hybridisation followed by re-selection/ backcrossing. |
| | | Induced tetraploidy | |
| 5 | 10-30 | SEED PRODUCTION | Mother plants cloned and planted in isolations; seed produced by polycross, topcross, etc. |
| 6,7,8 | | PROGENY TESTING | Sward trials of the progeny of each of the mother plants; estimates of general combining ability. |
| 8 | 4-10 | SELECTION | Original mother plants reselected on the basis of their general combining abilities in the progeny tests. |
| 9 | | STRAIN BUILDING | Elite mother plants cloned and planted in isolations to provide seed of 'experimental' varieties. |
| 10,11 | | SEED MULTIPLICATION | Seed of first generation bulked and resown to provide second generation seed of experimental varieties. |
| 12,13,14,15,(16) | | SWARD EVALUATION: NATIONAL LIST, DUS AND VCU TRIALS | Experimental varieties compared with controls in sward by breeders, and testing bodies, eg. NIAB. Decision by testing bodies/MAFF on rejection or inclusion on National List. |
| 15,16 | | RECOMMENDED VARIETY | Decision by testing bodies on rejection or inclusion of 'experimental varieties' on Recommended Lists. |
| 15,16+ | | MARKETING AGENCY | |
| 16,17+ | | FARMER (Basic seed)          OVERSEAS TRIALS | |

Figure 5.1   *Sequence in conventional grass breeding*
Source: Thomson and Wright

# DEVELOPMENT OF HERBAGE PLANT BREEDING

Formal plant breeding in Britain started late in the nineteenth century leading to the release of varieties by commercial firms early this century. Before this, sowing land to grass was a high risk affair. Land was often simply allowed to convert to grassland naturally from existing seed in the soil—'tumble down'—after arable cropping. Commercial ecotypes or land races were sometimes sown, although these were often very stemmy types, or just seed sweepings from hay barns.

The greatest impetus to herbage plant breeding came from 1919 onwards with the setting up of the Welsh Plant Breeding Station (WPBS) under the direction of R. G. Stapledon. Genotypes were selected from the best fattening pastures and a series of leafy varieties of several grass species developed. Releases of these varieties, with S prefixes, started in the early 1930s, Aberystwyth S23 perennial ryegrass being the best known. These varieties eventually formed the backbone of ley farming during and after the Second World War. Imported seed supplies were scarce during the war and home-produced seed had to be substituted so that a seed production industry sprang up. At the same time a stimulus was provided for a seed certification scheme to ensure that seed of high quality was available. Legumes were also bred, mainly white and red clovers.

In recent years British herbage plant breeding has contracted considerably, but with WPBS still the main source of British-bred varieties. Analysis of the grass and legume varieties recommended for England and Wales by NIAB for 1991–2 indicates that 81 per cent of the grass varieties were developed abroad, 45 per cent coming from The Netherlands. The position for British varieties of legume is better, with 41 per cent being home bred, mainly white clovers from WPBS, with the remaining 59 per cent of the recommended white or red clovers and lucerne coming from overseas, mainly from Belgium and France. It is unfortunate that scarcity of resources has limited the number of herbage species being bred in the UK.

# PRESENT BREEDING OBJECTIVES

Currently at WPBS, the breeding programme is concentrated on perennial, Italian and hybrid ryegrasses, white clover and intergeneric crosses between ryegrasses and fescues. A major aim with perennial ryegrass breeding is to improve herbage quality by increasing the

41

water-soluble carbohydrate (sugars) content. This results in better acceptability and digestibility of the herbage, with increased animal intake and consequently more efficient animal production. High sugars in ensiled grass encourage a lactic acid fermentation which is necessary to obtain a silage of good feeding value.

Emphasis on herbage quality has to take place against the background of adequate or improved herbage production, whether annual or seasonal. NIAB data indicate a steady improvement in annual dry matter yields by approximately 0.6 per cent per year from new ryegrass varieties coming on to its Recommended List. Varieties are also being bred to use soil nutrients, especially nitrogen, more efficiently. The end result of this programme will mean that less N is required to achieve a given level of production while reducing N losses to the environment and improving cost effectiveness of animal production. Improved disease resistance against crown rust, mildew and Rhyncosporium is a constant aim. Adequate seed producing ability is essential in any new variety if it is to be exploited fully in farming practice.

A major objective with Italian ryegrass is to increase the digestibility of the stem in order to improve the overall attractiveness of the plant to animals, and also to slow down the plant's reduction in digestibility as it matures. These improvements benefit silage and hay crops since they allow more flexibility in cutting times and increased bulk at high levels of digestibility. Some success has been achieved in selecting Italian ryegrass varieties with enhanced magnesium content, with the objective of alleviating magnesium deficiency in grazing stock. (This can cause hypomagnesaemic staggers.) A similar improvement has been made in perennial ryegrasses at the Northern Ireland Plant Breeding Station. Mineral concentrations in herbage species and varieties could well receive more attention from plant breeders in the future. Resistance to ryegrass mosaic virus, mildew and Rhyncosporium diseases are other selection objectives.

Hybrids between Italian and perennial ryegrasses have been produced for some time, combining some of the valuable characteristics of each. The original intention was to provide longer lived varieties than Italian ryegrass, with wider flexibility of use and hybrid vigour. The creation of hybrid tetraploids has been particularly successful. By doubling the number of gene-carrying chromosomes through colchicine treatment, improved quality, better production and enhanced persistence have been achieved. Ryegrasses are the most susceptible species to damage from winter cold so improved winter hardiness is an important breeding objective.

If projected changes in world climate materialise, particularly global warming, drought resistance could assume more importance than in the past. This is partly catered for by the programme to develop intergeneric ryegrass × fescue hybrids which is at an advanced stage, although as yet there is no variety recommended for use. The broad objective of the programme is to link the herbage quality and establishment vigour of the ryegrass species with the persistent growth under environmental stresses of the fescues, e.g. drought or cold, and low soil fertility.

The objectives outlined are common to many breeding centres elsewhere, though their emphasis may differ. Some of the innovative work of plant breeders at the New Zealand Pastoral Research Institute provides a good example, particularly since a number of their varieties have proved suitable for conditions in Britain. To overcome the deficiencies of perennial ryegrass growth during severe drought conditions, breeders in New Zealand have developed alternative drought-resistant varieties of cocksfoot, tall fescue and phalaris. These can be sown with the drought-resistant herb, chicory, and with white clover, to create low input swards capable of high productivity. Varieties of Yorkshire fog and common bent have been developed for infertile soil conditions.

Another notable innovation in New Zealand concerns the endophyte fungus which may be present in perennial ryegrass swards. The natural presence of this fungus confers resistance against the Argentine stem weevil and other pests which can kill out ryegrass, but unfortunately it can cause an adverse metabolic condition in animals known locally as ryegrass staggers. Consequent financial losses in the livestock industry can be enormous. However, a safe strain of the endophyte has been isolated and, when inoculated into perennial ryegrass seed, causes the resultant sward to be resistant to the weevil pest and eliminates ryegrass staggers.


## FORAGE LEGUME BREEDING

There is tremendous scope for the improvement of white clover by breeding because of the range of material with differing characteristics which exists in many countries. At WPBS, breeding programmes have led to several new and superior varieties appearing in recent years, to suit differing intensities of grazing management and for improved output, persistence and disease resistance. A major current aim is better clover production and reliability by matching clover to compatible grass companions, which then benefit from rhizobially-fixed clover nitrogen. The work is mainly with perennial ryegrass but may be extended to other grass species. Varieties are also being developed from Swiss genetic material, incorporating characteristics of early spring growth and cold resistance; previously, these two attributes were not usually found together in clover plants. Lack of early clover growth at low spring temperatures meant that associated grasses could seize a competitive advantage over clover and reduce its spread later in the season. Good spring growth is desirable in swards for sheep farms on the hills and uplands. Genetic variation in white clover is also being explored for resistance to diseases and pests, particularly clover rot and stem eelworm.

Different types of clover are being developed for potentially differing intensities of defoliation from extensive, continuous stocking by sheep to rotational grazing with cattle and/or silage cutting. Bloat has always been a potential hazard when cattle graze clover-rich swards. Apart from specific anti-bloat measures, such as dosing with an anti-foaming agent, there are genetic means of altering the protein make-up of the plant in order to reduce bloat incidence. Alternatively, intergeneric hybridisation is a possibility. Birdsfoot trefoil, with its high tannin content, does not cause bloat and avenues are being explored of hybridising it with white clover using genetic engineering.

Improving nitrogen fixation ability is another key objective for white clover and advances are possible through matching superior nitrogen-fixing strains of *Rhizobium* with specific clover varieties; these effective strains must be maintained in the soil where they have to compete against existing, less effective rhizobia. Genetic manipulation is also used to try to introduce nitrogen-fixing capability into grasses.

Adequate seed producing ability is a requisite of a new clover variety but must also be a satisfactory means of multiplication, otherwise its use in practice will clearly be severely limited. Denmark is a major producer of clover seed in Europe, but supplies have fluctuated markedly as more reliable and profitable arable crops have been substituted. Climatic conditions for clover seed production in the UK are not ideal so multiplication abroad, in places such as New Zealand or the western USA, have been used for some new varieties. A European Community project is evaluating new areas in Europe which might prove suitable for clover seed production.

The variety Grasslands Huia from New Zealand has long dominated the world market, accounting for about 90 per cent of the seed sold; around 3,000 tonnes are exported to Europe annually, with almost a third coming to the UK. Many of the newer British and European varieties have superior characteristics to Huia but lack of seed has prevented them from ousting it from its premier UK position. In New

Zealand a number of white clover varieties have superseded Huia for specific situations, including improved large-leaved varieties with autumn and winter growing ability, though lacking cold hardiness for many European conditions. There is also a small-leaved prostrate growing variety specifically bred for sheep grazing in marginal land conditions.

New Zealand breeders have developed a marsh birdsfoot trefoil variety for acid wetland pastures, stabilisation of slopes and for soils low in phosphate. In addition there is innovative breeding of drought-resistant, mineral rich forage herbs, chicory, yarrow and ribwort plantain being notable examples.

# Chapter 6

# Seed Production and Certification

---

### On a Seed

This was the goal of the leaf and the root
For this did the blossom burn its hour
This little grain is the ultimate fruit
This is the awesome vessel of power

For this is the source of the root and the bud . . .
World unto world unto world remolded
This is the seed, compact of God
Wherein all mystery is enfolded.

George Starbuck Galbraith, *New York Times* (6 May 1960)

Copyright © 1960 by The New York Times Company. Reprinted by permission.

---

In addition to the VCU and DUS aspects discussed in Chapter 3, there are two other major links in the chain of legislation which ensures that high quality herbage seed reaches the consumer. The first is the warranty that varieties have been maintained true to genetic type and varietal purity, which is achieved by field inspection of seed crops and by maintaining verification plots of seed stocks. The second main aspect is the standardising of percentage purity and germination in seed lots for sowing out. To receive the seals of authenticity, growing seed crops and the harvested resultant seed issuing (from the breeder and/or maintainer of the variety or from the grower) must meet the statutory requirements laid down by the official Seeds Regulations. Full requirements of the United Kingdom Seed Certification Scheme—Certification of Seed of Grasses and Herbage Legumes—are published by the Ministry of Agriculture, Fisheries and Food of England and Departments of Agriculture for Scotland, Wales and Ireland.

## SEED GRADES AND STANDARDS

Breeders' seed is multiplied to produce pre-basic seed which in turn produces basic seed, made available to farmer growers under contract for multiplication to produce certified seed. This is sold by seed merchants to farmers and cannot be used for further multiplication. Certification of approval under the statutory scheme is indicated by different coloured labels tagged to the bags of seed: violet for breeders' seed, white with a violet stripe for pre-basic seed, white for basic seed and blue for certified seed. If certified seed meets specified quality standards above the EC minimum standards officially required for seed analytical purity and weed seed presence, it can be labelled as Higher Voluntary Standard (HVS) and marked with the special HVS symbol, a category available for most of the agriculturally important grass and legume species. Mixtures sold to farmers for grass production have a green label and, as a selling point, some seed merchants offer seed with standards higher than HVS. Imported seed must have gone through equivalent certification schemes, whether in EC or non-EC countries, before sale in the UK. Farmers should only sow authentic certified seed if they wish to exploit fully the potential of the varieties or mixtures they choose. The only way to do this is by purchasing seed from reputable merchants.

Seed purity and germination standards for certified seed of the main species are shown in Table 6.1. Certain standards are required in connection with seed impurities. No wild oat or dodder species are permitted in either the minimum or HVS categories for grass and clover seed and only limited numbers of dock species. Also, set percentages by weight of black grass and couch grass are laid down for grass seed in the minimum grade and set numbers in the HVS grade. Set percentages by weight are also laid down for the presence of undesirable melilot species in legume seed and maximum contents of other grass contaminants are specified; for example, no more than 0.4 per cent by weight of annual meadow grass or 0.3 per cent by weight of rough-stalked meadow grass are permitted in the HVS seed of perennial ryegrass. In HVS seed of white clover no more than 0.5 per cent by weight of any one other plant species is allowed and the certified seed must be healthy; red clover seed must be fumigated by an approved method before certification if stem eelworm in the crop is detected during field inspection. The analysis also details the presence of inert material such as dirt, chaff or broken seed.

Well-ripened, well-harvested herbage seed will usually have a high and uniform germination and the standards will be easily reached. The 1,000-seed weight is an indication of how well seed has ripened and is assessed against standards. There are limits to the amounts of hard seed permissible in forage legume seed. Hard seeds do not germinate during laboratory tests but may do so in the field should the seed coat become abraded, allowing moisture uptake.

Table 6.1 **Standards for certified seeds under the Seeds Regulations (bracketed figures for legumes are maximum hard seed content as per cent by number of pure seed)**

| Herbage species | Standard | Minimum analytical purity (%) | Minimum germination (%) | Maximum content of seeds of other plant species (% by weight) |
|---|---|---|---|---|
| Perennial ryegrass and timothy | Minimum | 96 | 80 | 1.5 |
| | HVS | 98 | 80 | 1.5 |
| Italian and hybrid ryegrasses | Minimum | 96 | 75 | 1.5 |
| | HVS | 98 | 75 | 1.5 |
| Cocksfoot | Minimum | 90 | 80 | 1.5 |
| | HVS | 90 | 80 | 1.5 |
| Meadow and tall fescues | Minimum | 95 | 80 | 1.5 |
| | HVS | 98 | 80 | 1.5 |
| Red fescue | Minimum | 90 | 75 | 1.5 |
| | HVS | 95 | 75 | 1.5 |
| White clover | Minimum | 97 | 80(40) | 1.5 |
| | HVS | 98 | 80(40) | 1.5 |
| Red clover | Minimum | 97 | 80(20) | 1.5 |
| | HVS | 98 | 80(20) | 1.5 |
| Lucerne | Minimum | 97 | 80(40) | 1.5 |
| | HVS | 98 | 80(40) | 1.5 |
| Sainfoin | Minimum | 95 | 75(20) | 2.5 |
| | HVS | 98 | 75(20) | 1.5 |

*Source:* Ministry of Agriculture, Fisheries and Food *et al.*

## SEED PRODUCTION AND USAGE

Herbage seed is produced in many areas of the UK but of necessity is mainly concentrated in the drier areas of southern and eastern England. Sunny dry weather is required during the period of seed ripening and harvesting. Scotland has only a small seed industry almost entirely based on some 500 ha of the local 'ecotype' variety, Scots timothy, in central Scotland. The county of Kent produces the white clover, Kent Wild White, but the vast proportion of white clover seed sown in the UK is imported because weather conditions are rarely ideal for its production. It is important for seed growers to regard seed production as a deliberate enterprise rather than one of occasional opportunity. To be carried out properly, considerable attention must be given to the details of management and to developing the necessary expertise. Adequate harvesting, cleaning, drying and storing facilities are required since there is no market for seed crops which fail to meet the standards for certification.

Table 6.2 shows the weights of seed harvested for individual species. About 80 per cent of the grass seed produced is perennial ryegrass and 13 per cent Italian ryegrass. Common vetch features markedly within the legume species, being an annual widely used at 5 to 10 per cent by weight in sand mixtures for arable silage. Varieties of turf-type perennial ryegrass make up nearly 90 per cent of the weight

**Table 6.2   Weights of seed certified for 1990 harvest from 1 July 1990 to 30 June 1991, England and Wales**

| *Mainly agricultural* | *Tonnes* |
|---|---|
| Perennial ryegrass | |
|    Early | 2,938.5 |
|    Intermediate | 3,431.4 |
|    Late | 4,072.2 |
| Italian ryegrass | 1,677.2 |
| Hybrid ryegrass | 655.1 |
| Westerwold ryegrass | 15.3 |
| Cocksfoot | 249.3 |
| Timothy | 55.3 |
| Meadow fescue | 18.5 |
| Tall fescue | 11.7 |
| White clover | 74.5 |
| Red clover | 20.0 |
| Common vetch | 273.7 |
| Total | 13,492.7 |
| *Mainly amenity* | *Tonnes* |
|    Perennial ryegrass | 1,556.1 |
|    Red fescue | 378.4 |
|    Smooth-stalked meadow grass | 1.8 |
|    Brown top | 0.1 |
| Total | 1,936.4 |

*Source:* National Institute of Agricultural Botany

of amenity seed produced. In addition to these weights of seed from England and Wales, Scotland produces about 200 tonnes of timothy and a small quantity of perennial ryegrass.

The UK has a thriving seed export trade, particularly in both fodder and amenity varieties of perennial ryegrasses, and also in Italian and hybrid ryegrasses. Nevertheless, it is heavily dependent on imports to meet the annual requirements for sowing, especially perennial ryegrass, timothy, red fescue and white clover. The estimated tonnages of herbage seed delivered for use by seedsmen are shown in Table 6.3.

## SEED GROWING RULES AND REGULATIONS

Previous cropping must satisfy certain requirements. For certified seed, four clear cropping years must elapse before another variety of the same species, whether of grasses or legumes, can be sown. If the same variety is to be sown for seed, there must be one intervening cropping year. For the ryegrasses there must be three clear cropping years if cocksfoot, meadow or tall fescue was previously sown. Again, to take a cocksfoot seed crop as an example, a minimum of four years must follow ryegrasses or three years after meadow or tall fescue.

Land free from likely weed infestation should be chosen and special precautions taken to avoid volunteer plants of closely related species to the seed crop or other agricultural species. Clean equipment should always be used. Potentially damaging perennial weeds are best controlled at the pre-establishment stage by crop rotation or the use of herbicides. Further weed control may be necessary in the seed crop, by spraying or in some cases by rogueing of contaminant plants, measures which are especially important where stringent standards have to be met in the certification scheme.

When the weed seeds are of similar size to those of the seed crop, seed cleaning is made more difficult and costly. Wild oats have seed sizes similar to the large seeded grasses such as tetraploid ryegrasses and diploid Italian ryegrass. Seeds of couch grass, black grass, rough-stalked meadow grass and Yorkshire fog are other problem weeds in grass seed crops. Heavily contaminated headlands or other areas may have to be discarded. Effective control of docks is required to avoid contaminating clover seed.

Apart from the seed purity aspect, weed control is necessary so that there is good growth and seed production, and to avoid hampering harvesting and drying. Good weed control can also extend the number

**Table 6.3  Estimate of certified herbage seed delivered for use by seedsmen in the UK during the year ended 31 May 1991**

| *Mainly agricultural grasses* | *Tonnes* |
|---|---|
| Perennial ryegrass | |
|     Early and very early | 3,486 |
|     Intermediate | 5,662 |
|     Late and very late | 8,182 |
| Italian ryegrass | 2,122 |
| Hybrid ryegrass | 745 |
| Westerwold ryegrass | 170 |
| Timothy | 1,283 |
| Cocksfoot | 163 |
| Meadow fescue | 93 |
| Tall fescue | 13 |
| Total | 21,919 |
| | |
| *Legumes* | |
| White clover | 753 |
| Red clover | 79 |
| Lucerne | 25 |
| Alsike clover | 26 |
| Trefoil | 2 |
| Vetch | 232 |
| Total | 1,117 |
| | |
| *Mainly amenity grasses* | |
| Perennial ryegrass | 2,478 |
| Red fescue | 2,822 |
| Chewings fescue | 1,440 |
| Fine-leaved, sheep's and hard | |
|     fescues | 76 |
| Smooth-stalked meadow grass | 491 |
| Rough-stalked meadow grass | 1 |
| Brown top | 463 |
| Creeping bent | 19 |
| Timothy (*Phleum bertolonii*) | 54 |
| Others | 46 |
| Total | 7,890 |

*Source:* Ministry of Agriculture, Fisheries and Food

of years a seed crop can be harvested in perennial species such as perennial ryegrass or timothy.

Most grass species are cross-pollinated, that is, pollen from the stamens on one flower must be transported to the stigmas of another flower, not necessarily on different plants. The pollen is normally wind borne. To avoid self-fertilisation, species have self-incompatibility mechanisms; for example, stamens of individual flowers may release pollen before or after the stigmas on the same flower open. When insects are involved in pollination, such as honey bees in the case of white clover, the structure of the flowers is designed to prevent self-fertilisation. Siting of the seed crop is important from the point of view of pollination. Wind-pollinated crops are best sited where the wind blows in a favourable direction to spread the pollen among the crop. In the case of bumble bees which pollinate red clover, nearby rough ground is needed for their nests.

It is necessary to grow the seed crop at certain minimum distances from crops of other varieties which could cross-pollinate it. The isolation distance is greater for pre-basic and basic than for certified seed, but the latter must still be 100 m away for seed crops up to 2 ha in size and 50 m for fields over 2 ha. In addition, a 2 m isolation strip or a physical barrier such as a hedge is needed between the seed crop and any other crop likely to cause contamination. Effective isolation is essential during flowering to avoid cross-pollination from related species in other fields. If two varieties of the same species are grown on the same holding there have to be at least seven days between their times of ear emergence. Separate harvesting and handling equipment and facilities are also necessary in order to avoid admixture.

For crops of Italian ryegrass or Italian-type hybrid ryegrass, certified seed can only be produced in the first harvest year. This is because shed seed would become second generation plants following germination, and these could mix with plants producing first generation certified seed—though sometimes a second seed crop can be taken in this first harvest year. For perennial herbage species it is possible to produce certified seed each year for several years, so long as the swards remain vigorous and are free from contaminant plants.

Official field inspection is made after ear emergence but before pollination for grasses, and during flowering for legumes. Apart from checking that isolation requirements have been met, the inspector checks on the cultural condition of the field, plant health, varietal or species purity and weed contamination. The seed crop must meet specified plant purity standards, e.g. for certified seed a maximum of one 'off-type' per 10 m$^2$ of the same grass species. The grower should be familiar with the required standards so that the right production measures are taken.

## MANAGEMENT OF SEED CROPS

### Grasses

In establishing swards intended for herbage seed crops the principles and practices outlined in Chapter 7 should be followed to ensure a vigorous and reliable establishment. Direct sowing is a better option for slow establishing grass species such as cocksfoot or timothy. Specimen seed rates of the main grasses sown for seed production are 12 to 20 kg/ha for the ryegrasses, 6 to 8 kg/ha for cocksfoot, 6 to 12 kg/ha for timothy and 10 to 14 kg/ha for red fescue, lower rates being more suited to drilling than broadcasting. It is customary if drilling cocksfoot and timothy to sow out in wider drills than normal—up to 50 or 60 cm wide—rather than the 10 to 20 cm used for the ryegrasses. The timing of sowing and whether or not a cover crop is used will be determined by the particular pattern of cropping on the farm.

Other matters to consider are the potential problems likely to arise from weeds, diseases and pests. Spring sown grass provides good opportunity for weed control prior to the first full harvest year, but there are advantages in sowing Italian and hybrid ryegrasses in autumn since there is less time for mite-transmitted ryegrass mosaic virus (RMV) infection to build up before the harvest year.

Liming and fertilisation at establishment and during the seed harvesting years will vary with seed crop needs and associated utilisation of the sward using periodic soil analysis as a guide. Lime should be applied to achieve and maintain a soil pH of 6.0 while at moderate P and K status (ADAS index 2), a typical recommendation would be 50 kg/ha each of phosphate and potash. Nitrogen fertilisation will vary according to the intensity of management being imposed. In Italian and hybrid ryegrass seed crops, vigorous early spring growth can be grazed or cut for silage, provided it is done by late April. Typical N dressings would be 50 to 75 kg/ha before grazing or silage and a similar amount afterwards for the seed crop. If the sward is to be unutilised in the spring, the N dressing can range from 75 to 125 kg/ha. The purpose of the early utilisation is to reduce lodging and its associated loss of seed. The lodged thatch of Italian and hybrid ryegrasses is more open than in perennial ryegrass crops, so the potential loss of seed from lodging is greater. The spring defoliation should be as even as possible so that the seed crop will be uniform in growth and ripening. After the harvesting of the seed crop in July, the field can be set up for a second seed crop or it can be grazed. A seed crop the following year is not permissible.

It is not usual to defoliate perennial ryegrass or cocksfoot seed crops in spring. The fertiliser N application can be in the range 100 to 150 kg/ha. The harvesting period will depend on the maturity of the variety sown but will normally be within the late July/early August period for perennial ryegrasses and early to mid July for cocksfoot. The field can then be fertilised according to the intensity of use planned and grazed in the autumn. In Scotland, fields for timothy seed crops are often hard grazed by sheep or cattle during winter and given 80 to 120 kg/ha N in spring; harvesting is usually in August.

Red fescue is unusual in that 70 to 90 kg/ha N is applied in the autumn prior to the seed harvest year. This is because application of fertiliser N in the spring can result in an excessive bulk of vegetative material and little benefit to seed yield. Spring nitrogen application is, however, the normal procedure for red fescue seed production in the major grass seed producing areas of the north-western United States of America.

## Clovers

Clover growth is more sensitive than grass growth to soil acidity or low soil P and K status so any deficiencies must be remedied, though fertiliser N application is unnecessary. The white clover seed crop is taken as a by-product in sheep farming systems, mainly from mixed perennial ryegrass/white clover swards in Kent, or as a break crop in arable cropping, when it is sown alone by broadcasting at 2 to 4 kg/ha. Seed yields from the monoculture swards are generally higher than from the mixed sward. The spring growth is grazed by sheep until about mid May before closing up the field for seed harvesting in August. Vegetative growth can be cut and removed if grazing is not possible, but the timing and height of cut should avoid defoliating the first crop of flowering heads, which are the highest seed yielding. Defoliation allows light to penetrate the sward canopy and this stimulates flowerhead buds from the ground-hugging stolon network. The spring defoliation should be as even as possible, to encourage uniform flowering.

Red clover requires similar soil fertility conditions to white clover. It is usually sown at 10 to 15 kg/ha in monoculture by broadcasting, or at lower rates when drilled. To avoid excessive growth and possible lodging at harvesting the vigorous early season growth of early flowering red clover is cut and removed before the end of May. The red clover seed crop is harvested in September from the regrowth. For late flowering red clover the first growth is allowed to mature to the seedhead stage for harvesting in September.

## SEED HARVESTING

**Grasses**

The objective when harvesting any herbage seed crop is to achieve satisfactory yields of seed of high germination capacity. Stage of maturity plays an important role and crops should be harvested before there is potential loss of seed by shedding. Yet if harvested too soon, in order to avoid shedding, the seed crop may suffer damage during threshing because of its high moisture content, and subsequent germination may be adversely affected. This is a particular hazard when direct combining crops, especially Italian and hybrid ryegrasses, timothy and cocksfoot. Cutting and swathing before threshing can help to overcome this problem and the crop can then be cut at a higher seed moisture content. However, swathing increases the field time and the risk of bad weather affecting the crop. Ryegrass seed crops can be swathed at 400 to 450 g/kg moisture content and combined at 300 to 350 g/kg, or direct combined at 350 to 400 g/kg moisture content. It is advisable to use an infra-red moisture meter to determine the moisture content of seeds but, for a reliable guide, this has to be done on seed of representative field samples of seedheads. Moisture contents fall on average by 10 to 20 g/kg daily.

The setting of the combine harvester, particularly the drum speed, and the drying process prior to seed storage, are important factors requiring skill in order to prevent loss of germination capacity. A satisfactory target moisture content for safe storage is 130 to 140 g/kg. Final drying has to be done without delay after harvesting, and uniformly throughout the mass of seed to avoid potential heating, mould growth or infestations of seed-eating insects. Floor drying systems using large capacity fans are the best means of handling large quantities of seed, but in-sack driers are satisfactory for small quantities. Continued vigilance is necessary to avoid contamination of the seed crop by seed of other species throughout the drying, cleaning and storage phases. Careful identification and labelling of individual seed lots is necessary. Seed yields are generally higher for the ryegrass species, e.g. 1 to 2 t/ha, in comparison with other grass species, e.g. 0.5 t/ha to 1.0 t/ha for cocksfoot and 0.25 to 0.5 t/ha for timothy.

A major problem with the seed production of forage legumes is the extended flowering period, because it means that time of ripening is not uniform. Time of harvesting is therefore a compromise between obtaining a sufficient proportion of ripe seed yet avoiding loss of overripe seed by shedding. Better synchronisation of plant flowering within a variety is a major aim of the breeder, but the crop must be managed to increase the number of seed producing inflorescences. Because of possible wet weather at pollination, ripening or harvesting, producing clover seed in the UK is a high risk business and this is the reason why only a small percentage of the requirement for white clover seed is met from home-grown sources.

## Clovers

The white clover crop is usually cut and swathed before threshing, or it may be direct combined when the majority of the flowerheads are brown and the seed is hard and light yellow in colour. Diquat herbicide may be applied as a desiccant before cutting in order to reduce the field drying time. Yields can reach 500 to 600 kg/ha but can be as low as 100 to 200 kg/ha. Autumn regrowth is grazed off. Subsequent drying of the seed should be as rapid as possible to ensure maintenance of germination capacity and safe storage. Natural drying or forced-draught warm air systems can be employed. For long-term storage the seed should be dried to 100 g/kg moisture content or less. White clover seed, being small and dense, requires highly specialised equipment and skills for cleaning.

Red clover is harvested in September when the majority of the seedheads are brown and the seeds are hard and purplish in colour. Early flowering varieties are usually only harvested for one year, but late flowering varieties can be harvested for two or more years. The crop can be direct combined or threshed after cutting and swathing. A desiccant such as diquat may be used two to three days before the anticipated combining date. The seed requires drying to the same moisture content as white clover.

*Chapter 7*

# Sward Establishment and Renovation

---

Behold a sower went forth to sow;

And when he sowed, some seeds fell by the wayside,
and the fowls came and devoured them up:

Some fell upon stony places, where they had not
much earth: and forthwith they sprung up, because
they had no deepness of earth:

And when the sun was up, they were scorched; and
because they had no root, they withered away.

And some fell among thorns; and the thorns sprung
up, and choked them:

But others fell into good ground, and brought forth
fruit, some an hundredfold, some sixtyfold, some
thirtyfold.

*Matthew* 13: 3 to 8

---

The foundation of a productive sward is laid in its seeding year, so as little as possible should be left to chance in the preparations and actual operations. The aim is to provide the right conditions for seed germination in the soil, for seedling shoot and root growth, and finally for the development of a dense sward. From the moment of sowing, which involves 15 to 25 million individual seeds per ha, depending upon the make-up of the seed mixture and the rate of sowing, there is competition among seedlings in a race for survival. Every 1 per cent survival of seedlings is equivalent to 150,000 to 250,000 plants per ha. In a good establishment only 15 to 25 per cent of the seedlings will have survived after a few months, possibly declining to as little as 10 to 15 per cent after a year. At the start of the first full harvest year, a successfully established sward should have 5,000 to 7,000 tillers per square metre; a sward of this density will resist weed invasion, be less prone to poaching and will form a sound basis for future use.

## THE SEEDBED

The field drainage system should be operating effectively, and if it is not, this should be rectified before cultivation. The end objective of the cultivations—which include ploughing, cultivating, discing, harrowing and rolling—is a fine, firm seedbed. A fine tilth is needed since grass and clover seeds are small and firmness guarantees contact between sown seed and moisture-supplying soil. A level seedbed will ensure that the shallow, uniform depth of sowing required will be attained; the use of a soil leveller is often neglected. If direct reseeding (that is, sowing from grass back to grass), the old sward should be well grazed down so that ploughing can turn the sod over completely, and good contact, moisture movement and continuity from the unploughed soil layer can be maintained. Secondary cultivation operations should be carried out promptly, yet not rushed, aiming first to obtain a crumbly soil structure and second to consolidate the seedbed. In this way, there is minimum drying out of the seedbed yet allowance is made for air and moisture movement, and seedling root development. Loose or puffy seedbeds should be avoided at all costs.

The soil should always be cultivated from the surface down; otherwise subsurface clods, stones and subsoil may be brought to the surface unnecessarily and the operations prolonged. A good choice of equipment is available whether for handling light-textured soils where consolidation is all-important, or wet, heavy-textured soils where the avoidance of overcompaction, cloddiness or surface smearing is necessary. Another important point is tractor power. Grassland, more than any other crop, is associated with sloping and even steep land. The draught power of a tractor which is suitable for level land may not be up to the task when faced with slopes. The final consolidation is the key element, and here the Cambridge roller with its ridged compressive rings reigns supreme whether just prior to drilling or broadcasting, and also after both these sowing methods.

### Drilling versus broadcasting

Drilling in an even seedbed can place the seed at the correct sowing depth (10 to 20 mm for grass), and in contact with soil moisture, so this method is superior to broadcasting for light soils and in dry areas; the seed rate can be reduced to two-thirds of that needed for broadcasting, though rate of working is slower for optimal efficiency. Intra-row competition is also greater and there is more space for weed growth between rows. To ensure uniform seed distribution, the seed should be drilled in two directions. Broadcasting results in better ground cover of sown species but surface moisture is critical in preventing death of the germinating seeds and young seedlings; harrowing and rolling to cover the seeds are key operations. An ideal sowing method to ensure adequate clover establishment in a grass/white clover sward is to drill the grass, but broadcast the clover to ensure shallow sowing of its smaller seed.

Sowing without a cover crop guarantees a better establishment than sowing with, since the faster growing cover crop competes strongly for light, water and plant nutrients. The main reason for a cover crop is to boost production per unit area in the sowing year during the slow initial grass establishment phase of two to three months. If a spring cereal is sown for a grain cover crop, an early maturing variety with short stiff straw should be chosen and its seed rate reduced to three-quarters of the normal. The grass seed mixture should be sown immediately after the cereal. Smother of the sow-out by a laid cereal crop is a major cause of grass establishment failures.

A better plan is to sow an arable silage crop, e.g. barley or a barley/forage pea mixture, which can be harvested earlier in the season thus allowing better tiller, leaf and root development of the grass sow-out before winter and possible early frosts. This is especially important for the white clover component in a grass/clover seeding. Sowing out after harvesting a winter cereal is another possibility, provided there is time for sufficient plant development. Where forage rape is sown as a cover crop to provide grazing for sheep, no more than 1 to 2 kg/ha should be sown with the grass seed mixture.

## Season of sowing

Spring (March to May) and late summer (July to August) are the two main sowing options, when temperature and moisture are most favourable; shortage of moisture is often a limiting factor between these two periods. A spring direct seeding will have a greater potential weed problem than a late summer seeding and, since it is in its establishing phase, may not be able to take full advantage, production-wise, of the good early season growing conditions. Local conditions and experience will be a guide to the best sowing time, but cold winds and frosts can check or kill young seedlings if it is done too early. There is usually a lesser weed problem with late summer or early autumn sowing, but if weed spraying is necessary, it can be difficult should grass and clover plants not have developed sufficiently to tolerate the herbicide.

## Direct drilling

Reseeding of grassland is sometimes done by direct drilling after chemical desiccation, e.g. by the systemic herbicide, glyphosate, or split application of paraquat, particularly when ploughing is not possible or is difficult. The main advantages are the short period the land is out of production, as seed drilling can be done 10 to 14 days after herbicide spraying, and also the fact that the fertile upper soil layers are not disturbed and deep rooted perennial weeds are killed as well as other weeds. However, it has proved a high risk operation and the technique is not widely used. Causes of partial failure can include: incomplete desiccation of the old sward; surface trash and the toxic products of its decay, because of an excessive amount of herbage present at spraying; heavily matted swards; poorly drained compacted swards; and poor placement of the drilled seed in relation to soil moisture as a result of drill design. However, guidelines have emerged with trial and experience which, allied to a new generation of drills, have markedly increased the chances of successful establishment. The major points are: adequate desiccation of the old grass sward; minimising the sward trash; timing the operation to ensure adequate soil moisture is available yet avoid soil smearing; drill penetration of the uncultivated soil and good sowing depth control; rolling; pellets to deter slug attack and damage; and the use of pesticide against frit fly, if drilling at the time of year vulnerable to infestation.

Sowing grass seed by direct drilling or after minimal cultivations into cereal stubbles in autumn is fairly reliable provided there is not excessive trash and it is early enough in the autumn to allow good plant development before winter, e.g. after winter barley has been harvested in August. Wheel rutting of the soil surface by the cereal husbandry and harvesting operations can seriously interfere with the effectiveness of direct drilling. The intermingling of cereal stubble and young grass growth can present some problems in efficient utilisation of the grass, especially by lambs and sheep.

## FERTILISER APPLICATION

The best guide to fertiliser requirements is a prior representative soil analysis of the area to be seeded. The normal pH target for grassland is 5.8 to 6.0, and if the amount of ground limestone needed to attain this is over 7 t/ha, it is usual to apply the lime in split dressings, part to be ploughed in, and part to be worked into the soil during the final seedbed cultivations, or one to two years later. The aim is to have a uniform distribution of lime in the upper soil profile from the surface down, so that setbacks to establishment due to acidic conditions can be avoided, especially during the vulnerable early phase. (Soil acidity severely restricts the development of seedling roots.) Excessive local concentrations of lime due to uneven distribution can also hinder germination and growth.

A typical fertiliser requirement (in kg/ha) for direct sowing into a soil of moderate fertility would be 60 N, 75 $P_2O_5$ and 60 $K_2O$. Nitrogen stimulates early growth of plants, but some reduction in N is advisable for late season direct sowing to avoid excessive soft growth before winter; reduction in N is also needed where clover is expected to play an important role in the future sward. For a cereal grain cover crop the $P_2O_5$ and $K_2O$ would be adjusted upwards by 40 kg/ha each but the N downwards by about 25 kg/ha. The key nutrient for grass and clover seedlings is phosphorus, to stimulate root growth; the major part of the applied $P_2O_5$ should be in a readily available, water-soluble form.


## POST-SOWING MANAGEMENT

The aim of post-sowing management, irrespective of method or time of sowing, is to achieve a densely tillered, leafy sward canopy. This can best be done by grazing periodically from a sward surface height of 8 to 12 cm down to 3 to 6 cm in short grazing periods. Removal of the herbage allows light to reach and stimulate grass tiller buds and clover growing points. Sheep or young cattle are less likely to poach the developing sward than heavier adult cattle, but all stock must still be kept off in wet conditions when trampling damage is likely. The danger is greatest with late season sowings, because of increasing rainfall then. Nevertheless, the benefits of stock treading to consolidate light dry soils have long been recognised, as witness the phrase 'golden hoof'. Consolidation of newly established swards benefits root contact with soil moisture, improves the anchorage of the seedling roots and inhibits the growth of annual weeds. The post-sowing grazings will be effective against many of the weeds which emerge, especially annuals; topping, or forage harvesting and removal of the material, can also be done if necessary.


## WEEDS

The moral here is: prevention of weed invasion is better than cure.

Faulty seedbed preparation can result in failures or poor takes of sown seed and there is then scope for invasion by the vast burden of weed seed present in most soils. While grazing and topping will control most of the weeds which spring up in grassland, especially tall growing annuals, herbicides may be required for more difficult ground-hugging weeds such as clumps of chickweed or infestations of annual meadow grass which colonise thin or patchy takes; the problem is exacerbated in mild wet autumns.

The choice of herbicide will depend on the weeds present and whether or not clover safety is a factor; time of spraying is governed by stage of growth of the sown species. For example, certain benazolin- or bentazone-based products would be chosen to control chickweed when clover presence is an objective, and mecoprop products where it is not. As a rule, grass seedlings should have two to three leaves and white clovers at least one trifoliate leaf to ensure tolerance to the herbicide. Hormone-based herbicides, e.g. mecoprop, MCPA, MCPB, 2,4-D and 2,4-DB, should not be used to control perennial weeds before a direct reseed since herbicide residues in the soil may interfere with the establishment of grasses and particularly clovers.

If obstinate perennial weeds such as docks or creeping thistles are present before sowing, the mature plants are best sprayed out before ploughing. Glyphosate, which is translocated throughout plants, is highly suitable for controlling persistent weeds such as couch grass or docks which have considerable underground food reserves. Best results from any spray are obtained when weeds have adequate leaf area to absorb the herbicide and the plants are actively growing; herbicide manufacturers' approved label recommendations of the stages of plant growth for spraying and application rates should be followed.

## PESTS

A vigorously growing sward can often tolerate and recover from attacks by pests such as leatherjackets or slugs, but it may be very vulnerable at the seedling stage. Direct reseeds are more at risk than sow-outs after arable cropping since higher populations will have been maintained in the old grassland. While thorough seedbed cultivation will reduce pest numbers, there can still be considerable survival. Recovery from attack is more likely when the damage occurs early in the season, because of subsequent good growing conditions. If the damage occurs late in the season, the plants have less opportunity to recover during winter and may even succumb to winter kill. Ironically undersowing, which in principle is a less certain method of grass establishment, may better enable grass seedlings to escape attack since the cover crop provides an alternative source of food.

It is possible to forecast the onset of infestation or measure the actual population of certain pests such as leatherjackets, predicting potential damage and hence taking suitable cost-effective control measures, e.g. treatment by chlorpyrifos pesticide. On-farm do-it-yourself kits are available, which include liquid expellant (modified St Ives fluid) to bring larvae to the surface of soil cores for counting.

The damaging effects of frit fly larvae on grass shoots in establishing swards are well documented but can be avoided or ameliorated by sowing when the risk of infestation is low, i.e. by early rather than late spring, using resistant grass varieties if possible, and chemical control through seed treatment by fonofos or on the sward by chlorpyrifos application, for example, if an attack occurs. Slugs can also be a problem but chemical control is available, e.g. methiocarb. While certain root feeding nematodes are known to affect grass and clover plants adversely, their effects are rarely of practical importance. Table 7.1 sums up the key guidelines for establishment.

**Table 7.1 Key points for sward establishment**

- Soil pH and soil P status should be satisfactory
- Apply water-soluble $P_2O_5$ at sowing
- Seedbed should be fine and firm
- Direct sowing gives the best results
- If undersowing, preferably cut cover crops for arable silage.
- Avoid midsummer sowing since soil moisture will be limiting
- If sowing in late season, adhere to mid August deadline
- Control any problem weeds as soon as possible
- Graze down to 3 to 6 cm at intervals during early establishment phase
- Grazing is better than cutting in the year following sowing
- Aim for about 6,000 tillers/m² in spring of year after sowing

## SWARD RENOVATION

There are situations where an existing sward may require rejuvenation or renovation. Both are usually cheaper than conventional reseeding, but the rate of improvement is slower. The objective of rejuvenation is to reverse deterioration and make a gradual improvement of the existing sward by dealing with the factors causing deterioration which are amenable to correction. Corrective measures include: ameliorating poor drainage conditions: alleviating soil acidity or plant nutrient shortage; weed or pest control; better grazing procedures to avoid overgrazing, especially in spring, or conversely to avoid undergrazing which creates an open sward; minimising poaching of grazed swards, or reducing field tracking of silage or hay fields.

In some years, perhaps due to adverse weather conditions, it is not possible to prevent some degree of sward deterioration; in other cases, the deterioration can definitely be laid at the door of bad management practices. If the sward sole is too open, with many bare patches, and/or there is a scarcity of productive preferred species, a more drastic form of sward renovation is called for with a choice of partial or complete reseeding (see Table 7.2).

**Table 7.2  Strategies for sward renovation**

|  | *Rejuvenation* | *Partial reseeding* | *Total reseeding* | |
|---|---|---|---|---|
|  |  |  | *Surface methods* | *Cultivation methods* |
| Old sward | Retained | Partially replaced | Completely replaced | Completely replaced |
| Destruction of old sward | None | None or partial | Chemical | Physical |
| Soil cultivation | None or surface | None or minimum | None or surface | Ploughed |
| Herbicide | Selective | None or grass suppressants | Total sward destruction | Weed control, total sward destruction |
| Methods | Improved overall management | Oversowing to direct drilling techniques | Oversowing to direct drilling techniques | Cultivations and drilling or broadcasting |
| Seed | None | Reduced rates | Full rate | Full rate |

Increasing disturbance of sward and soil
Decreasing competition from old sward
Increasing inputs and costs

*Source:* Tiley and Frame

Sward renovation would include some of the above outlined remedial actions where necessary but, in addition, the introduction of grass and/or clover seed by various partial reseeding techniques. On a very open sward simple oversowing or surface seeding is possible; this technique is also suited to wet environments when the old sward can be cut down, severely grazed or burned in some situations to remove the existing vegetative cover and hence reduce competition to the oversown seed. 'Mob stocking' can be used to tread the seed and improve seed–soil moisture content. Light spike rotavation, light discing or heavy harrowing can be used to provide some tilth for a partial seeding operation. Rotavation is a good choice where there is matting at the base of the old sward; alternatively some form of pioneer cropping can be used for this purpose prior to seeding.

An Irish technique of renovating thinned-out Italian ryegrass swards has made use of oversowing followed by slurry application or, alternatively, mixing the seeds into the slurry before application. Production was boosted, although not to the high level of production associated with Italian ryegrass in its first harvest year following conventional seeding.

In recent decades direct drilling has become an option for the purpose of partial reseeding. Drills of differing design have come and gone as their shortcomings were exposed. However, a new generation of drills such as the New Zealand Aitchison Seedmatic and Hunters Rotary Strip Seeder have been introduced. The Seedmatic has proved highly effective in lowland conditions on uniform land, its winged coulters producing an inverted 'T'-shaped slit into which seed and fertiliser are placed. The Hunter has been successful in various situations, including difficult hill and upland country. It has a series of small rotavators, independently mounted to follow ground contours, which create 75 mm cultivated slots at 225 mm centres; drill depth control is by an articulated skid. Both drills can be fitted with a chemical applicator to band spray astride the slots; the chemical desiccation of existing herbage adjacent to the slots reduces competition to drilled species. Direct drilling is in the main a contractor operation, since the needs of an individual farmer could rarely justify the cost of the equipment.

It has to be remembered that all forms of partial seeding fall short of the ideal of conventional cultivations and seedings, but may have to be resorted to if conditions are against the usually reliable conventional techniques. Ploughing may be undesirable or impossible because it could lead to soil erosion, to nutrient leaching, or to inversion of a fertile topsoil—as on peat, for instance. The site may be inaccessible or too steep for machinery, or it may be too shallow, stony or wet.

Ploughing and cultivations also mean incurring considerable costs and a temporary loss of production and grazing days from the land, both of which may not be affordable.

In partial reseeding techniques, sown seeds are introduced into a hostile environment where competition from the existing sward can be intense. Seed germination, establishment, and subsequent development and survival all have to be encouraged. It is a common experience that good initial germination and seedling emergence are followed by a rapid decrease in survival as various adverse factors come into play. Adequate soil moisture for germination is necessary, so time of sowing is critical; obviously known dry periods must be avoided. Early season and late summer/early autumn are two periods when soil moisture conditions are normally suitable. Machines with mechanisms which engender good soil–seed contact by covering the seed and pressing the disturbed soil have been the most successful. Some species such as timothy have small seed which is more easily covered than species with larger seed, such as perennial ryegrass, but the small-seeded species may lack rapidity of establishment and therefore are not so highly competitive.

Successful continued development of plants requires adequate moisture and nutrient supply, particularly available phosphate to stimulate root growth. Fertiliser placement close to the seed is better than broadcasting over the whole sward since it allows preferential uptake by the seedlings rather than stimulating competition from the existing sward. It is essential that light reaches the young leaves, so open conditions at the base of the sward are vital, achieved by grazing, cutting, burning or chemically suppressing the existing sward vegetation. The developing seedlings are also susceptible to the ravages of pests such as slugs or frit fly, but these are controllable by pesticides.

Post-seeding management must again favour the needs of the introduced plants at the expense of the old sward. The existing sward needs to be grazed yet, at the same time, the establishing plants must not be overgrazed. This is easier to manage with rotational grazing rather than continuous stocking. The sward should not be used initially for cutting a silage crop. To sustain the improvement it is necessary to maintain an enhanced soil fertility and overall management to match the needs of the sown species; otherwise the improvement will not last, especially where the starting point is an infertile soil and unproductive sward.

Interest in partial reseeding techniques has waxed and waned in recent decades and initial enthusiasm has often been tempered by failures. However, many of the past problems associated with choice of equipment or its design, and choice of sward type or condition to be improved are now more clearly understood. Consequently a number of management guidelines exist (see Table 7.3) which, if followed, will increase the chances of successful renovation.

**Table 7.3  Guidelines for sward renovation**

- Ensure that the sole of the old sward is as open as possible in order to reduce competition to sown seeds. Do this by pre-sowing management if necessary

- Defoliate old sward severely before seeding. Achieve this by cutting, grazing, burning or chemical treatment

- Control perennial weeds present prior to seeding

- Rectify soil fertility factors, such as acidity or inadequate nutrient supply, which limit plant growth

- Choose suitable soil type carefully. Avoid heavy-textured and stony soils which are the most difficult

- Make sure soil conditions at seeding are in a fit state for seeding operations

- Choose machinery most suitable for the particular sward and soil types and operate it correctly

- Sow at times of the year when soil moisture is not likely to limit fast establishment

- Control pests where these are a problem—slugs, for example

- Sow rapidly establishing and competitive species and varieties where possible

- Graze the sward hard after drilling to reduce competition from the old sward but balance the grazing with rest intervals to avoid overgrazing the developing introduced species

*Chapter 8*

# Controlling Weeds

---

The even mead, that erst brought sweetly forth
The freckled cowslip, burnet and green clover
Wanting the scythe, all uncorrected, rank, ·
Conceives by idleness, and nothing teems
But hateful docks, rough thistles, keksies, burs,
Losing both beauty and utility.

William Shakespeare (1564–1616), *Henry V*, V. ii

---

Weeds have always been tolerated in grassland to a greater extent than in arable crops, and herbicide control is often used as a last resort. Many weeds can be effectively controlled by cultural or mechanical measures, especially when dealt with at the correct stage in their life cycle. There is likely to be a renaissance of these measures, together with likely advances in biological control—since research in this field has increased—as extensification policies take effect and public concern about pesticide use in general continues. Nevertheless, a host of effective herbicides is available to control the serious weeds which invade grassland. In many cases herbicide application is necessary and can be used to economic advantage, but its use for cosmetic reasons should be shunned.

The use of herbicides, and also fungicides, insecticides and other crop protection chemicals, is covered by government legislation and all users should be familiar with the requirements. Various official Codes of Practice are also available. Only trained and qualified personnel should handle and apply agrochemicals on the farm. Users must always read and follow manufacturers' label instructions carefully before use.

This chapter deals with methods of controlling the most widespread and undesirable weeds of grassland, both lowland and upland.

## DOCKS

Docks are a constant problem in intensively used, highly fertilised grassland. The commonest species are the broad-leaved dock and the narrow-leaved curled dock. Along with creeping thistle, spear thistle and ragwort, these two docks are scheduled as injurious weeds and by law should not be allowed to spread or seed. Plant form and vigorous growth make docks formidable competitors to grass for light, water and nutrients, and severe infestations reduce grass production markedly. Cattle graze them in the leafy state but not when the plants are stemmy and mature unless grass is scarce. The plants are perennial and develop ever stronger shoots as they age. The upper root section and crown can initiate new shoots after cutting down or following damage to the plant. They are also prolific seed producers and many swards have large reservoirs of seed in the soil, which can remain dormant for many years and then germinate if gaps appear in the grass.

Unlike other perennial grassland weeds, docks thrive in high fertility situations, so improving fertility further will not help control them. However, preventing sward damage by poaching, tracking and uneven slurry application is important to maintain dense vigorous swards and prevent colonisation or recolonisation after control.

In addition, plants which do emerge should be controlled and not allowed to produce seed. If they do flower they should be removed and burned, as used to be done when making hay. Cutting alone will not control docks since new shoots will regenerate from the tap root. This was why labour-intensive spudding rather than cutting was once common practice. If using herbicides it is easiest to tackle docks at the susceptible seedling growth stage when cheaper herbicides are successful. Spraying established docks with herbicides rarely gives a total kill and a follow-up spraying of the survivors is usually necessary. Dock control is not a once-for-all operation: there must be constant vigilance because infestations can build up quickly from the seed burden in the soil, and initial localised infestations can quickly spread.

Herbicides such as MCPA, 2,4-D or mecoprop can all control dock seedlings effectively in the establishing sward, but they kill clover too. Thus if clover is to be encouraged it is necessary to use clover-safe herbicides such as those based on 2,4-DB, MCPB or benazolin. Docks in established grassland can be controlled by herbicides based on chemicals such as dicamba, fluroxypyr, triclopyr or mecoprop, or mixtures of them. Where clover is important in a sward, overall spraying by asulam operates selectively. For small infestations or scattered plants these herbicides or glyphosate can also be used for spot treatment. Glyphosate applied by rope-wick applicator can selectively control dock plants standing above the level of the grazed sward. Alternatively it can be sprayed to destroy grass swards prior to reseeding when the swards are severely infested with docks and other perennial weeds such as thistles and nettles. The weeds should be in active growth but not yet at flowering stage. Swards can be grazed from five days onwards after glyphosate spraying and the reseeding operation then begun. For silage or hay the crop can be sprayed and then cut down after the minimum five-day period. This technique of pre-grazing or pre-cutting application minimises the period of time the sward is out of production.

## THISTLES

With their erect and spiny habit of growth, plus their conspicuous autumn thistledown, thistles are highly noticeable weeds. They occur in long-term and permanent grassland and spread especially when soil fertility is low to moderate and where extensive grazing permits rank growing, under-utilised swards in summer. Both creeping thistle and spear thistle are officially classed as injurious weeds, so they should be controlled and not allowed to spread or to seed. The perennial creeping thistle is more difficult to control than the biennial spear thistle. Creeping thistle spreads mainly by underground roots which can survive in a dormant state for some years before pushing up shoots into poor growing, open swards.

Twice- or thrice-yearly cutting can control creeping thistle to a certain extent, but it is apt to survive and invade again from its underground root system. Suitable and effective herbicides include 2,4-D, MCPA, mecoprop and clopyralid. Overall spraying, spot application and rope-wick application are all possible methods, depending upon the level of infestation. Stock must be kept out of sprayed fields since senescing material may be eaten, causing digestive upsets. As with other perennial weeds of grassland, control will only be effective in the long term if management can be improved to create and maintain a dense, vigorous and competitive sward.

## RAGWORT

Common ragwort, which is a scheduled and injurious weed, is widespread on less intensively used grassland. Another species, the marsh ragwort, is found on heavier soils and poorly drained land. Spread of ragwort is encouraged by lack of vigour in sward growth and bare spaces created by overgrazing or trampling damage. Ragwort is a biennial plant which develops a prostrate rosette of leaves in its first year of growth and then flowers and grows to maturity in its second year. Wind-borne seeds are produced profusely and a sward which appears free of the weed may have a considerable burden of seeds in the soil, awaiting the opportunity to germinate.

Unlike many grassland weeds ragwort is poisonous to livestock, particularly cattle and horses, although they do not readily graze it if there is plenty of grass available. It does, however, become attractive to grazing stock following spraying or cutting because of chemical changes in the dying material, but it is still poisonous. Similarly, dried ragwort will be consumed readily if it is contained in hay. Its presence in silage is particularly dangerous since the poisonous alkaloids can spread from ragwort plants through the silage pit. The health and performance of stock will suffer when ragwort is ingested by grazing or from conserved grass because it causes progressive liver damage; death can also occur once a lethal dose has been consumed.

Control by cutting, which was once commonly practised, is not satisfactory since it can keep the plant vegetative and prolong its lifespan, encouraging it to become perennial. Ploughing up and re-seeding is effective but reinfestation may occur from seed if grass establishment is poor or if the sward cover becomes open. Sheep are less susceptible than cattle to ragwort poisoning; they can be used to graze out ragwort in winter and early spring, but only if light infestations are present, otherwise their health may also be adversely affected. Close grazing by sheep in early season will also encourage the development of a well-tillered, dense grass sward resistant to ragwort invasion.

Spraying 2,4-D or MCPA can be highly effective in controlling ragwort infestations, providing grassland management can be changed to prevent the problem returning. A number of other herbicides based on mecoprop or triclopyr are effective, but as they are more expensive, should only be used when other problem weeds such as docks are also present. The optimum time to spray grazing fields is in late spring during active growth but while plants are still at the rosette stage, before flowering. Emerging seedlings will also be controlled then. Stock should not be put into a sprayed field until at least three to four weeks after spraying, to allow the ragwort to senesce and become unavailable. Autumn spraying is best for fields intended for silage or hay the following year. Timing is less critical than weather conditions; application during a mild settled spell will give the best results. Following the eradication of ragwort, measures such as improving soil fertility in order to encourage a vigorously growing sward, and management aimed at promoting sward density are necessary to prevent reinvasion of the weed.

## COMMON CHICKWEED

Common chickweed occurs frequently in both establishing and established swards. It is an annual plant and infestations arise from seed previously shed and present in the soil. As with many weeds, open ground during the grass establishment phase or bare spaces in older swards give chickweed seeds the chance to germinate. Invasion is most common in autumn reseeds or in undersown grass following the removal of a cereal cover crop in autumn. With a rapid rate of growth chickweed becomes very competitive, especially in wet conditions on fertile soils when the dense spreading clumps shade out sown species. Its growth is soft and low in dry matter content and while it is not highly acceptable to grazing animals it can be grazed out by heavy cattle or sheep stocking, preferably by adult animals since large intakes by young stock may result in digestive problems.

It is worthwhile controlling a heavy infestation in a young sward in autumn; otherwise grass production can be severely depressed the following year because of poor development of the sown species. Chickweed also has the capacity to grow slowly during the low temperatures of winter and it can flourish at the end of the grazing season because of this ability, especially where thinning out or damage to the sward has occurred. Winter kill or grass smother from slurry applications in winter will also permit chickweed to spread. Its high moisture content causes difficulties when trying to wilt chickweed-infested herbage for silage or to dry it during hay making and prolongation of the drying process in the field increases the risk of high losses of dry matter and nutritive value in the conservation crops. Excessive chickweed also interferes with the silage fermentation process, adversely affecting silage feeding value.

While many effective herbicides are available for the eradication of chickweed, choice will depend on the specific situations prevailing. Examples range from clover-safe herbicides such as benazolin or bentazone to those based on mecoprop which damage or kill clover. Since clover is invariably sown in seed mixtures for swards intended to last over three years, clover-safe herbicides are the most appropriate in the establishment phase. Both the sown grasses and clovers must be at an adequate stage of growth, two to three leaves for grass and at least one trifoliate leaf for clover. In many older swards clover will have become a minority species if present at all, so the choice of herbicide will be more a matter of cost and the spectrum of other weeds requiring control too. Mecoprop and fluroxypyr will control quite large chickweed but best results are achieved from summer application.

72

## NETTLES

Stinging nettle is a perennial with a creeping rootstock from which new shoots emerge at short intervals; it can also spread by seed, but these need bare ground to germinate. It frequently occurs in clumps on highly fertile grassland or on disturbed areas around stock handling facilities. Avoiding overgrazing in early spring and utilising other management measures to maintain a vigorously growing sward will limit colonisation by nettles. Clumps can be controlled by cutting twice- or thrice-yearly for several years in succession, with the first cut preferably just before flowering stage. Spraying with hormone herbicides such as MCPA, 2,4-D or mecoprop will kill young nettles whereas triclopyr-based herbicides are more effective against established clumps. The extent of nettle infestation is rarely sufficient to warrant overall spraying of a field and spot treatment will usually suffice.

## ANNUAL MEADOW GRASS

Annual meadow grass is a widespread opportunistic weed, readily colonising bare spaces in both establishing and established grass swards. The growth, development and production of sown grasses and clovers can be reduced by its competitive effects. Seeds are produced prolifically and these germinate readily in spring or autumn. Swards established by drilling the seed directly are more prone to invasion than swards sown by broadcasting or undersowing due to the inter-drill bare spaces. By following the tenets of good establishment (see Chapter 7) the problem can be minimised in new sowings.

In established swards the spread of annual meadow grass can be prevented by maintaining good growing conditions and avoiding, as far as possible, soil compaction, poaching and other factors which reduce the vigour of sown grasses and clovers or create open spaces in the sward; Its seed does not germinate in dense, vigorously growing swards. Selective control in establishing and established swards can be achieved by the application of residual herbicides based on ethofumesate during mild periods in autumn and winter. The success of this treatment will depend on a high content and even distribution of sown grasses in the sward; otherwise weed grasses will tend to reinvade. Clover will be severely damaged. Pre-emergence treatment of autumn sowings with this chemical is also possible where there is a high risk of annual meadow grass invasion.

## RUSHES

Rushes are mainly a problem with rough grazings and on long-term or permanent pastures. The most widely distributed species, and the greatest problem, is the soft or common rush. It is distinguishable from other species of rush by the small tufts of brown flowers high up on the green flowering stems and by the evidence of a continuous white pith when the flowering stem is split open. It is most frequently found on wet soils and shallow peats at altitudes up to 800 m and is a perennial plant which forms clumps by vegetative spread of short creeping rhizomes. Its spongy root mass can impede the flow of water to field drains. Once established the clumps become very persistent and resistant to damage or to drought. The soft rush is a profuse seed producer and the tiny seed can be widely spread by wind, water and animals.

Many soils contain a considerable burden of dormant rush seeds awaiting the opportunity to germinate, and this opportunity arises when the sole of the grass sward is broken up by poaching or overgrazing and the seeds are exposed to light. Newly established swards on marginal land are prone to this damage, especially if new drainage systems are not yet functioning at full efficiency. Soft rush seedlings are susceptible at this stage to trampling by livestock or competition from the grass sward, but if they survive they soon develop into resistant clumps.

Control measures should aim both to eradicate the rushes present and to prevent new invasions. Reseeding after drainage, deep ploughing and cultivations are options on suitable land, if it is economically justified. Larger clumps may require destruction before ploughing in order to prevent regeneration. Control can be done cheaply with the selective hormone herbicides MCPA, 2,4,-D or dicamba/mecoprop combinations. However, hormone herbicide residues may damage the newly establishing grasses and clovers if there is an insufficient time interval between spraying and reseeding. It is safest to spray rushes in the summer of the season before reseeding. Alternatively, the complete vegetative cover can be killed by spraying with glyphosate before ploughing.

When full renovation by reseeding is not possible, an infested sward can be rejuvenated by overall spraying with a selective herbicide. Eventually a repeat spraying will become necessary, unless the conditions which encouraged the initial invasion and spread of the rushes are rectified. Spraying is most effective in spring and early summer when the plants are actively growing, just before flowering. Sward rejuvenation will be aided if the killed rush material is removed.

74

Control of old established clumps of rushes is improved if they are cut and the litter removed before spraying; the young regrowth will be at a susceptible stage about a month or so later. Scattered soft rush infestations can be controlled by the application of glyphosate from rope-wick applicators. Control of other rush species is more difficult and may not be justified, given the quality of land they grow on: the jointed rush is found on waterlogged upland areas of little value, for example.

Improved sward management is essential after eradication of the rushes if reinvasion is not to occur from buried seed. Drainage must be effective and soil fertility sufficient to support a dense vigorous sward, which should be kept in that condition by good grazing management. While cattle grazing can discourage young rush growth, poaching has to be avoided at all costs; cattle should only be grazed when the risk of poaching is negligible.


## SOFT BROME

There has been an increasing awareness of the encroachment of soft brome or lop grass into hay fields in western Scotland and northern England. Brome may be either annual or biennial and is a prolific seed producer: following spring germination, plants mature and shed their seed by early summer. Plants which establish in autumn from freshly shed seed or from seed already present in the soil seed bank will overwinter and reach maturity by late spring or early summer. The seed heads will then shed their seed before the hay crop is cut, adding further to the seed burden in the soil. The open sward sole typical of hay fields, particularly those which are cut late, encourages the rapid spread of brome. If stock are grazed on the hay fields during winter, poaching damage can result in further bare spaces which are ideal niches for brome colonisation. When the hay crop is cut, soft brome is at an advanced stage of maturity so the feeding value of the hay is reduced while the hairiness and awned seed of the brome reduce the attractiveness of the hay to stock.

The spread of soft brome is clearly linked to the system of management. Changing to early cut silage followed by summer grazing will prevent brome seed production, so further invasion will be entirely dependent on existing seed in the soil. To prevent these seeds from germinating it is necessary to encourage density and vigour in the existing sward. The maintenance of good soil fertility, avoidance of overgrazing and poaching in winter, and earlier cutting for silage will all help.

Spraying during winter to early spring with the herbicide etho-fumesate is an additional option for control of brome. Established brome plants must have two to three leaves for the spray to be effective. This herbicide may also give some residual control of germinating seed for several weeks. Since the cost of control with herbicide is considerable, there is a question as to the degree of infestation at which it is advisable to spray. If the ground cover of the preferred grass species is only 50 to 60 per cent, it is better to plough and reseed the field anew, ensuring rapid establishment of a dense competitive sward. If 60 to 70 per cent of the ground cover is likely to be retained after spraying out the brome, it is necessary to patch up the sward by direct drilling with perennial ryegrass; other grasses and clover cannot be sown safely within eight weeks of spraying ethofumesate. With a subsequent remaining ground cover of over 70 per cent, additional seeding is unnecessary since the existing sward species will tiller out and fill in the gaps, particularly if encouraged by fertiliser nitrogen application and close grazing.

## BRACKEN

Approximately 360,000 ha of the British hills and uplands are currently affected by bracken, mainly in the sheltered valleys and on less exposed slopes below 500 m altitude, and the area is said to be increasing. Bracken occurs mainly on well-drained brown earth soils which support the valuable bent grass/fescue pastures, land which has the greatest grazing potential on upland farms but is then rendered valueless. The old Scottish adage 'Bracken is gold, gorse is silver, heather is copper' is indicative of underlying land quality. Historically, bracken has flourished because of its release from the shading of its original woodland habitat.

Bracken was once kept in check by frequent cutting, but this labour-intensive method of control has been superseded by more effective chemical control. However, chemical control has not been sufficiently widespread to slow down the expansion of bracken, chiefly because of the high cost of follow-up measures and other economic constraints on hill farms. A reduction in the trampling of young fronds by the declining numbers of cattle now kept on the hills and uplands has also contributed to the spread. Areas of bracken impede good shepherding and harbour parasites such as sheep ticks. Ingestion of the young fronds or rhizomes can cause disorders in livestock, including death by poisoning. More recently, concern has been voiced about carcinogenic properties in bracken affecting livestock and humans, including risk from spores in autumn. Bracken-covered land does not rate highly in terms of nature conservation and interesting wildlife, but the golden-brown colour in autumn and winter looks very attractive.

Bracken spreads vegetatively by underground rhizomes which serve as a store of food reserves. The frond canopy and its autumn die-back litter shade out the grass understorey where production can be reduced to zero where the canopy is dense. The litter layer also protects young emerging fronds from frost in the spring.

On suitable and accessible sites, ploughing and cultivations in midsummer followed by grass reseeding are ways of reclaiming braken-infested land. Selective chemical control is by the use of foliar absorbed asulam. This is useful where a good cover of grass exists below the bracken canopy, e.g. a moderate infestation of 30 to 40 bracken fronds per m², or less. If grass is too thin, or absent, reseeding should follow spraying to prevent invasion by weeds such as creeping soft-grass, foxglove, thistles or nettle. To be effective asulam must be sprayed at full frond development and before the start of autumn die-back, usually from mid July to late August. The chemical glyphosate is also highly effective, but will kill all vegetation present and is more costly than asulam; reseeding would be essential after its use. Extreme care is necessary to avoid spray drift of glyphosate onto neighbouring grass, trees or crops. A range of hand-held spraying equipment is available for spraying small areas of bracken or for spot application, including knapsack sprayers, mistblowers and controlled droplet applicators. Larger scale sprayers can be mounted on tractors or all-terrain vehicles. Contract spraying by helicopter is an option suitable for large and inaccessible areas of infestation.

Following effective control by whatever means, a watch must be kept for reinvasion, which may come from growing points on the underground rhizomes or from plants missed by the spray. Some of the small-scale hand-held equipment can be used to apply chemical to this regenerated bracken. Cattle grazing is another means of deterring new frond growth, while liming and fertilisation will improve the growth and vigour of the grass sward which remains after spraying. Alternatively, following the removal of a heavy infestation, cultivations appropriate to the terrain and soil can be followed by a full fertilising and reseeding programme.

## GORSE AND BROOM

Gorse and broom are leguminous plants capable of fixing nitrogen which colonise the shallow brown earth soils in the hills and uplands and other marginal or neglected land. Broom usually occurs on land of a lower potential than typical gorse land. The woody perennial growth makes access to the understorey of grass difficult and the grasses are suppressed. The underlying sward is usually a relatively valuable bent grass/fescue association. Gorse and broom bushes do offer some shelter to stock in adverse weather, however, and have some limited value for wildlife, while the yellow flowers are an attractive feature on hillsides.

A number of eradication measures are possible where economics permit. Invariably, follow-up control measures are necessary against regenerating plants. Cutting is an option, using manual or mechanical methods, and young bushes can be flailed and the cut material burned. Burning of standing bushes is also sometimes used as a control measure but will stimulate germination of a crop of seedlings, so follow-up chemical spraying may be necessary. Alternatively, actively growing bushes may be sprayed in summer using a number of effective herbicides based on the chemical triclopyr. High volume spraying is necessary to ensure that the bushes are adequately covered by spray, and spraying is more effective on younger bushes. Following control, regeneration from stumps or from the seed burden in the soil will require after-measures. A combination of cutting or burning followed by spraying is the most effective control strategy.

After the eradication of standing gorse and broom a dense vigorous sward must be maintained if reinfestation from buried seed is to be prevented. Liming, fertilisation and improved grazing procedures may all be required. Gorse and broom seedlings are susceptible to trampling by grazing cattle and when goats have been introduced to hill farms, they selectively browse and thus control gorse as well as other weeds such as rushes and thistles.

Rough grazings often contain areas of deciduous scrub woodland. Gorse, broom, bracken and other vegetation of little grazing value may be present in the understorey. Some grasses such as bent as well as forage herbs are present when the undergrowth is open enough to allow light to penetrate, but access by stock can be difficult. The scrub areas will contribute little to grazing unless some positive management is exerted in terms of clearing, thinning and in some cases drainage measures. Properly managed, they can be an asset not only for sheltered grazing but for amenity, timber and wildlife.

*Chapter 9*

# Effect of Soil Factors on Grass

Human vanity can best be served by a reminder that whatever his accomplishments, his sophistication, his artistic pretension, man owes his very existence to a six inch layer of top soil and the fact that it rains.

Anon.

There is nothing in the whole of nature which is more important than or deserves as much attention as the soil. Truly it is the soil which makes the world a friendly environment for mankind. It is the soil which nourishes and provides for the whole of nature; the whole of creation depends on the soil which is the ultimate foundation of our existence.

Friedrich Albert Fallou (1862)

The soil which supports grass swards has taken thousands of years to develop. The topsoil is one of the most important components in grassland and sustains the sward physically, chemically and biologically. A knowledge of the underlying soil on the farm is therefore essential to the understanding and correct management of grassland. Scientists often refer to the soil–sward–animal complex or interrelations to underline the fact that the proper functionings of all three are inextricably linked.

Farmers have or soon acquire an intimate knowledge of the behaviour of the soil in different areas of their own farm, and also within each field. This will be based initially on factors such as physical behaviour under different climatic and management conditions, trafficability, poaching, drainage, and handling properties during cultivations. Maintenance or installation of drains offers soil inspection opportunities, while observation pits may be dug to study soil profiles and subsurface conditions. For the healthy growth of grass the soil chemical nutrient status and pH will be determined by soil analyses of representative samples.

All these pieces of information can be allied to observing a sward's response to inputs such as fertilisers and how sward and soil have reacted to grazing and cutting management. The total fund of knowledge and experience built up will enable a high level of soil management expertise to be achieved, including prevention of potential problems or, when they do arise, their identification and solution. Incidences of grassland soil problems—both physical and chemical—have increased in number and variety in recent years, mainly because of the intensification of grazing and of silage cutting. Compaction by stock or field equipment, soil acidity and lack of specific nutrients are typical problems.

## SOIL FORMATION

The processes of soil formation in a given area often have a direct or indirect bearing on the soil's physical and chemical characteristics. Soil forming processes during geological times have resulted in the present day topsoil and subsoil. In Britain, glaciation in the Ice Ages had a major effect on geomorphology and subsequent soil formation. Extensive physical erosion of the rock surfaces took place, with gross physical transport of much of this material as glacial 'drift' in the form of gravels and tills, so that the soils which developed subsequently did not necessarily reflect the underlying parent material.

Glacial soil types such as glacial tills, water-worked tills and fluvoglacial sands and gravels are typical of many areas in the north and west. Glacial tills are the most fertile because of their accumulation of clay and silt particles, but drainage may be impeded. Fluvoglacial soils and coastal raised beach soils lost much of their finer clay and silt particles by the action of fast moving water, so leaching of nutrients occurred. These soils are also susceptible to drought and wind erosion. The water-worked tills are intermediate in these characteristics.

Other soils derived from transported material include alluvial soils, which were carried and deposited by rivers, and colluvial soils, which moved downhill due to water action or gravity. Their sandy or gravelly nature resulted in leaching of nutrients and a basic proneness to drought, except where the clay and silt particles were retained.

Non-transported or sedentary soils reflect the composition of their parent material and occur most commonly in central and southern Britain: they include chalk, limestone, sandstones, clay soils and shales. Heavy textured clays may have drainage problems which lead to poaching by grazing stock or low bearing capacity for grassland equipment. The lighter textured soils may exhibit typical nutrient and water shortage problems. Some limestone soils are very shallow.

The past history of soil formation at a particular location is displayed when the soil profile is examined. This exhibits layers or horizons which are characteristic for each soil type. The soil profile is a result of the interaction of soil-forming factors, the main ones being climate, especially rainfall and temperature, and the vegetation which existed on the parent rock material. Within the soil, mineral particles are washed down to the lower horizons, and nutrients leached from upper horizons to lower. Leaching of soluble iron compounds is apparent in colour bands or ochreous mottling; grey or grey-blue layers are indicative of anaerobic conditions caused by water stagnation. Dissolved iron and manganese may have cemented a clay layer into an impervious pan, impeding root penetration and drainage.

Farming operations have effects, whether beneficial or detrimental, and can be clearly seen and interpreted, especially in the soil's upper layers; admixture of the upper horizons may have taken place during cultivations if the horizons were shallow. Beneficial influences include liming which aids soil structure, application of manures which supply organic matter, or drainage installation. Detrimental influences include the formation of compacted pans at ploughing depth, destruction of the upper soil structure or accumulation of a mat of undecomposed organic matter because of failure to correct soil acidity or poor drainage.

The amount, degree of decomposition and incorporation of organic matter as humus plays an important role in the soil formation process. In addition to its beneficial effects on soil structure, humus has a high water-holding capacity and is a valuable source of readily available nutrients for grass growth. This availability is due to its high cation exchange capacity, where positively charged cations such as potassium, calcium, magnesium and ammonium are absorbed and made available for uptake by grass roots. In the initial stages of organic matter breakdown, nutrients are released from the tissues from the process of mineralisation by soil micro-organisms. Mineral soils have organic matter contents of 5 to 10 per cent, the higher levels being typical of long-term grassland and the lower of rotational grassland.

## SOIL COMPONENTS

Most soils are composed of mineral matter in different size grades. The major part of most soils is made up of different forms and sizes, ranging from fine clays and silts to gravel and stones, together with the minerals originated from parent rock material which has been subjected to many thousands of years of 'weathering'. In addition to wetting and drying, freezing and thawing, the leaching of water and soluble chemical substances has taken place. Plant roots branch through the spaces among the disintegrating rock material and assist its further decomposition into smaller rock particles.

An important action of water on rocks was the hydrolytic formation of clay minerals. Their colloidal properties control retention and availability of water and nutrients for vegetation growth and thus clay exerts a dominant influence on soil fertility. The rate of weathering is governed by the hardness of the original parent material. Igneous rocks such as granite weather more slowly than softer sedimentary chalks. Metamorphic rocks formed under heat and pressure, e.g. quartzite from sandstone parent material, decay at a slower rate than the original unmodified rock.

The remainder of the solid part of soil consists of organic matter derived from decaying or decomposed vegetation together with the remains and excreta of soil biota which range from micro-organisms and fauna living in the soil to large animals living on the surface.

The soil air exhibits higher carbon dioxide than the atmosphere because of root respiration. Soil water contains dissolved nutrients and interacts with the soil air in the pore spaces, the volume of pores being highest in heavy textured clay soils and lowest in light textured sandy soils. Impeded drainage and hence poor aeration is also highest in water-retentive clay soils.

## SOIL TEXTURE AND STRUCTURE

When classifying soils, emphasis is placed on their texture and structure since these criteria strongly influence soil properties. Texture is determined by the relative proportions of clay silt and sand minerals. In the texture triangle (see Figure 9.1) the co-ordinates of the clay, silt and sand percentages as determined by mechanical analyses are used to classify the texture of mineral soils. In the field it is possible to gauge soil texture by sight and by the feel of moist soil.

The properties of a specific texture class can be modified by the presence of gravel and stones. If the volume of pore space is reduced this has adverse effects on water retention, aeration and the space available for plant root ramification, and these can be serious drawbacks in sandy soils which already have limitations in these respects. However, stony soils warm up quickly in the spring, encouraging early season grass growth. The texture classes of mineral soils are amended to include the description 'organic' when the proportion of organic matter is between 10 and 25 per cent. At 25 to 50 per cent organic matter, soils are termed peat soils and above 50 per cent, peats. Some

▲ Experimental site at SAC, Auchincruive showing plots of grass and grass/white clover being evaluated under different fertiliser and cutting regimes   *Chapter 3*

White clover variety evaluation under cutting in experimental plots at SAC, Edinburgh
▼   (M. W. Morrison)   *Chapter 3*

White clover varietal assessment in sheep grazed experimental plots at WPBS,
▼   Aberystwyth   *Chapter 3*

▲   Set-aside experimental plots at SAC, Edinburgh showing the unsightly and weed infested consequences of not sowing a seed mixture   *Chapter 4*

Birdsfoot trefoil plants under assessment in a legume breeding programme
▼   at Paraná, Argentina   *Chapter 5*

◄ Cocksfoot selections being monitored in a grass breeding programme at Szarvas, Hungary  *Chapter 5*

▲ Aitchison Seedmatic direct drilling white clover into a perennial ryegrass sward previously cut for silage to make the base of the sward 'open'  (G. E. D. Tiley) *Chapter 7*

Vakuumat Slotter strip ► seeding at Graz, Austria. Trash is returned to the cultivated strip  (G. E. D. Tiley)  *Chapter 7*

▲   Silage sward showing heavy infestation of dock plants   *Chapter 8*

▲   Grazed sward showing infestation of ragwort plants   *Chapter 8*

▲   Hill sheep grazings in southern Scotland showing bracken infestation on the lower slopes   (G. E. D. Tiley) *Chapter 8*

◄   Ill-drained upland sward showing invasive rushes   *Chapter 8*

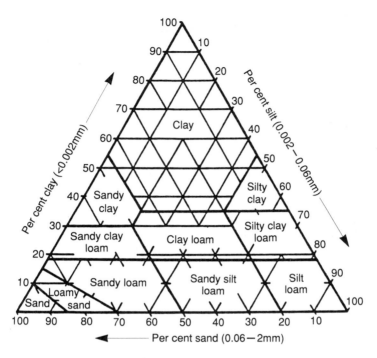

Figure 9.1 *Soil texture triangle*
Source: Avery

deep peats are 80 to 90 per cent organic matter, whereas there is little or no organic matter in very sandy soils.

Soil structure is determined by the extent of aggregation of the mineral particles in soil. This aggregation is the consequence of the weathering action, such as wetting and drying, and organic matter incorporation. If the organic matter is well decayed and humified it binds the clay particles and this has a stabilising effect on the small crumb and granular structural units. These small units are most abundant in the upper soil layers where biological activity is greatest. Grasslands, especially long-term grasslands with a history of dense rooting, provide ample organic matter for humus formation. If the soil is not acidic or waterlogged the biological activity of soil fauna and flora is at a high level leading to humification and the subsequent desirable development of a porous crumb and granular structure. Porosity encourages root proliferation, uptake of nutrients and good aeration for gaseous exchange in root respiration. Drainage is promoted and so too is the water-holding capacity.

83

The structural units become larger with depth, and from blocky and rounded in shape to angular. Rooting is inhibited since the soil is less porous and water and nutrients are less accessible to the rootlets. The reduction in pore volume leads to a lower available water capacity and poorer aeration, especially in heavy textured soils. These limiting characteristics to grass growth are intensified as the blockiness changes to columnar prisms lower down in heavy textured subsoil.

Soil structure can be altered for better or worse by farming operations, principally in the upper layers. For instance, liming grassland not only 'sweetens' the soil but improves the aggregation process and these effects encourage biological activity. Effective drainage systems improve the balance of air and water in the soil and this helps the maintenance of structure in the drained horizons. On the other hand, soil structure can be easily damaged or destroyed by periodic waterlogging through impeded drainage or compaction caused by grazing stock or wheeled equipment. Balanced inputs of fertiliser nutrients encourage vigorous sward growth which augments the supply of organic matter. Animal slurries and manures are other valuable sources of organic matter and plant nutrients.

## SOIL MAPS

For detailed information on soil types and groups reference can be made to official soil classification maps prepared from detailed soil surveys throughout the country. These maps are based on the study of soil development profiles from representative pits dug to about a metre in depth. Soils are classed by series, derived from specific types of parent material, and by soil group or subgroup together with their associated characteristics which have resulted from past soil formation processes and parent rock material. All soil types in the surveys have been sampled for laboratory analyses.

'Land capability' maps classify areas according to the limitations for agriculture imposed by climate, gradient, soil, wetness, erosion and vegetation factors. Emphasis is placed on limitations to cropping with regard to potential productivity, flexibility and ease of management. In Scotland, according to the classification of the Macaulay Land Use Research Institute, there are seven main classes, some with up to three subclasses. Land classes 1 to 3 produce a wide to moderate range of crops: classes 1 and 2 are intensively used for arable cropping, while most of the better grassland is on class 3 land. Hill and upland rough grazings are grouped in land classes 5 and 6, the land in class 5 being capable of improvement by various reclamation techniques and in class 6 largely unimprovable. Class 7, e.g. mountain tops, is of low agricultural value and capable only of supporting very poor rough grazing or for use as amenity land. In England and Wales the corresponding agricultural land capability system makes use of five main classes.

Soil survey information has also been used to make up special purpose maps which forecast the risk of specific problems, e.g. waterlogging, proneness to drought, poaching susceptibility, and predicting potential need for drainage or irrigation.

## STOCK TRAMPLING AND SWARD POACHING

Grazing animals trampling or treading the sward is part and parcel of the grazing process (see Chapter 16). Excessive trampling—poaching—modifies soil structure by altering the equilibrium between soil particles, air and water. Resulting compaction of the upper soil layers impedes grass root development and reduces root mass. Uptake of soil nutrients is then restricted, thus limiting grass growth rate and production. Other signs of poaching are poor water infiltration, puddling of the soil surface and reduced soil porosity. Subsequent cultivations are made more difficult.

In extreme cases there may be run-off of surface water and soil erosion on slopes. Abnormal aerial tillering may also occur in the sward and lead to sod pulling—uprooting of tillers, plants or clumps of grass including the roots. This is discussed in the next section.

Soil and plant damage is most likely in early season, before the soil has dried out, and in late season, as the soil is wetted and overwetted by increasing rainfall. In the west of the country, rainfall is greatest in September to December, intermediate in June to August (and January) and lowest in February to May. Even in low rainfall areas, or where rainfall is fairly evenly distributed over the year, low evapotranspiration from late autumn to early spring may result in soils remaining at field capacity during this period and hence swards are at risk from poaching damage. Finely textured, poorly structured clays, silts and soils with very high organic matter content, all of which have high retention of excess water, are the most susceptible to poaching when wet. Light sandy soils and friable loams, because of their free-draining characteristics, are the most tolerant of trampling.

Poaching can directly damage or destroy plant growing points, leaves, tillers and roots, and whole plants may be displaced or buried. Grass production is adversely affected in the short term since growth vigour and recovery rate after grazing are reduced. In the long term, the preferred sown grasses may not persist so that the botanical composition of the sward is adversely changed. High tillering dense swards such as develop under intensive continuous stocking and in long-term or permanent pastures are more resistant to poaching than the open soled swards which develop under heavily cropped silage or hay systems, or intensive rotational grazing. Trampled, soil-contaminated grass is obviously less acceptable to grazing stock, while fungal disease which can invade damaged plants also decreases grass acceptability.

Various measures can be taken to ameliorate the adverse effects of poaching, but cost and practicability must be considered. Any measure which reduces the distance animals have to walk on growing swards is obviously beneficial, so a well-planned field layout, with well-sited hard tracks, gateways and water troughs, is called for. Needless to say, good provision of an adequate supply of herbage for grazing will cut down the distances over which stock have to forage in order to satisfy their daily intake requirements. This means attention to the various factors, such as soil pH and soil fertility, which influence grass production, and the use of effective flexible systems of grazing which allow a reduction in stocking densities during periods of risk. Soil heaving by winter frosts and the activity of soil fauna are natural agents which aid the restoration of soil structure.

Seed mixtures with grass species and varieties which are densely tillering are obviously the most tolerant, while long-term or permanent swards develop a more resilient surface-bearing capacity than the more open, short-term swards. Recently established swards which have not developed a good load-bearing capacity are particularly susceptible. Fields with soils at high risk from poaching are thus best kept as long-term grassland and their drainage systems maintained in good working order. The use of sacrifice fields to bear the brunt of grazing in critical wet periods is a strategy which can be employed, provided there is a good reason for the sacrifice—the intention to plough and reseed a particular field, for example.

In extreme circumstances the stock may have to be removed from the field and fed indoors; high likelihood of poaching is one of the factors determining the cessation of grazing in autumn and housing of stock. Conversely, in spring, stock turn-out may be delayed because of potential poaching. Curtailment of the grazing season means increased reliance on conserved and/or purchased feed. Zero grazing has been advocated, and sometimes undertaken, early or late in the grazing season but the equipment for cutting and carrying the forage can also cause sward and soil damage by wheel tracking. This measure and other options, e.g. 'partial storage' feeding, are outlined in Chapter 17.

Table 9.1 lists important management factors which reduce poaching damage, improve sward tolerance or ameliorate the damaging effects of poaching.

**Table 9.1  Key points to minimise poaching**

- Maintain effective soil drainage systems
- Use tolerant grass species and varieties
- Make use of long-term rather than short-term swards
- Encourage development of dense, vigorous swards by grazing management and balanced fertiliser use
- Maintain adequate supply of grass in grazed fields
- Optimise winter grazing management of swards
- Use young cattle or sheep on establishing swards
- Reduce stock traffic by planned layout of fields and tracks

## SOD PULLING

Sod pulling, i.e. the uprooting of plants by grazing animals, is a matter of concern to farmers on occasion. It is most prevalent in late summer and early autumn on intensively N-fertilised and heavily grazed swards, but it occurs on extensively managed swards too. Older swards seem to be the most prone. Some sod pulling or tiller uprooting is inevitable in view of the way grass is prehended—torn off—by grazing animals, cattle in particular. The question is: how serious is it, or is it of visual concern only? It can be serious when large clumps of sward come away, first for the immediate loss of production and second for the long-term loss if the bare space is colonised by unproductive grasses and weeds. These effects happen when scattered small tufts are removed too, but to a lesser extent. Some of the causes are preventable or quickly curable if they are straightforward: the control of leatherjackets which feed on the sward roots in autumn and spring, reducing plant anchorage, for example. Other less obvious causes may require investigation and remedial action by a subsequent change in management.

High N application can be a major cause since it allows a greater intensity of sward use. Hoof trampling by stock and operations involving wheel traffic associated with high intensity of use can lead to surface compaction or the development of a subsurface compaction pan. Fertiliser N applied as nitrate or urea has a strong acidifying effect on the soil, especially in the upper layers. Compaction and/or the acidity prevent development of vigorous root growth and good anchorage, so plants are easily uprooted. In addition, the high level of N available at the surface encourages the spread of shallow soft roots and a low root:leaf ratio, especially in moist conditions when plants are prone to damage and not well anchored. Clearly, remedial action lies in the maintenance of an adequate soil pH by liming, minimising compaction as far as possible and if necessary using measures to alleviate it such as subsoiling. It will also help the overall vigour of root and shoot growth if the fertiliser programme is balanced in relation to nutrient supply.

On grazed swards, where distinct rejected patches appear, as around sites of decaying dung pats, adequate light may not reach the base of the sward so basal tiller bud formation is restricted and grass plants may respond by developing rooted tillers on stem nodes above ground level (so-called aerial tillers). Such tillers are easily uprooted and may be dropped by the grazing animals. Rotationally grazed swards defoliated too laxly soon develop a layer of vegetation above ground level which spawns aerial tillers. The solution to aerial tillering, when it occurs, is to graze such swards intensely and remove the layer of sward vegetation to a low level. While this will leave a thin sward, slow growing initially, the increase of light to the base will stimulate a new crop of ground-rooted basal tillers, far less prone to uprooting by stock.

## WHEEL-INDUCED EFFECTS ON SOIL

It is sometimes forgotten that as intensity of grassland use increases, swards have to bear more frequent wheel traffic from tractors and equipment for operations such as fertiliser distribution, slurry spreading, cutting, swath treatment and hauling. Machines have increased in size and weight over the years and multi-cut silage systems have become more popular. Wheels can damage sward tissues by bruising and crushing, growth and regrowth capabilities being subsequently impaired, and with serious wheel rutting, as around gateways, completely destroyed. Damaged herbage soon begins to decay, loses nutritive value and may be prone to disease. Recently reseeded swards which lack a well-developed shoot and root complex, and therefore have a low weight bearing capability, are at high risk from damage by wheel tracking and slipping.

In systems work at the Scottish Centre of Agricultural Engineering, Edinburgh, where farm-scale machinery and management were used, herbage production was enhanced with zero traffic or when larger tyres were employed, inflated to lower pressures than conventional tyres (see Table 9.2). More rapid and uniform grass growth, particularly before the first cut, was evident from the zero and reduced ground pressure systems.

**Table 9.2    Effect of wheel-traffic systems on relative DM and crude protein (CP) production from perennial ryegrass swards**

| Traffic system | Year 1 | | Year 2 | | Year 3 | | Year 4 | |
|---|---|---|---|---|---|---|---|---|
| | DM | CP | DM | CP | DM | CP | DM | CP |
| Conventional | 100.................................................................................100 | | | | | | | |
| Reduced ground pressure | 107 | 124 | 119 | 131 | 119 | 129 | 113 | 122 |
| Zero | 102 | 128 | 132 | 151 | 115 | 135 | 115 | 127 |

*Source:* Douglas and Crawford

Wheels can also cause serious surface and subsurface soil compaction. Heavy textured clays, clay loams or silts in high rainfall areas are most susceptible, whereas well-structured, lighter textured sandy soils in drier areas are most tolerant. Soil compaction, resulting in increased soil bulk density and reduced air pore volume, impedes root spread, soil aeration and water drainage; soil nutrient uptake is also restricted. Accordingly, herbage production and quality are adversely affected. These effects were evident in a soil compaction experiment at Edinburgh where a reseeded sward was subject to zero, small or large tyre/soil contact stress by two passes of tractor wheels when the large stress (72 k Pa from tractor rear wheels) was 2.4 times that of the small stress (30 k Pa); herbage production results are shown in Table 9.3.

On many farms the tractor power available, size of equipment and working rates exceed reasonable requirements. Any build-up of wheel rutting in silage fields makes it difficult to achieve target cutting levels consistently and to harvest herbage free from soil contamination. A major objective in minimising wheel traffic damage should be to keep the equipment size and weight to the minimum necessary for efficient field operations. Soil/tyre contact area should be increased and the pressure stress correspondingly reduced by the use of suitable tyre

**Table 9.3    Effect of grassland compaction on DM production from perennial ryegrass swards**

| | DM (t/ha) | | |
|---|---|---|---|
| Treatment | First cut | Second cut | Third cut |
| Zero treatment | 5.7 | 4.8 | 3.9 |
| Small compactive effort, spring | 5.0 | 4.4 | 3.5 |
| Large compactive effort, spring | 2.1 | 3.8 | 3.4 |
| Small compactive effort, first harvest | – | 4.4 | 3.5 |
| Large compactive effort, first harvest | – | 3.1 | 3.2 |

*Source:* Douglas and Crawford

types and inflation pressures, though this involves extra cost. During actual field operations, traffic frequency should be reduced to the minimum level possible, avoiding traffic movement over fields and unnecessary journeys. Good and well-planned farm roads, tracks and field layout are helpful in this respect.

*Chapter 10*

# Weather and Water Control: Irrigation, Drainage and Subsoiling

---

... as the tender grass springing out of the earth by clear shining after rain.

*2 Samuel* 23: 4

And Ahab said to Obadiah, go into the land, unto all fountains of water, and unto all brooks; peradventure we may find grass to save the horses and the mules alive, that we lose not all the beasts.

1 *Kings* 18: 5

---

Listening to or watching weather forecasts is of consuming interest to farmers—more than most people—for the simple reason that weather has such a powerful influence on grass production and utilisation, and on farming operations generally.

## TEMPERATURE

As light intensity and duration (solar energy or radiation) increase in the spring, soil and air temperatures begin to rise and grass growth becomes more noticeable. The threshold value at the soil surface or in the upper layers is around 5°C for grass and 9°C for white clover. Some growth does occur slowly during winter at temperatures lower than these, but not below freezing point.

As spring temperatures rise further the rate of grass growth accelerates. The dominating effect of temperature is the reason various temperature-related methods have been devised to predict dates of fertiliser nitrogen application in spring (see Chapter 18). The optimum growth of the temperate grasses and legumes used in the UK occurs at temperatures of 18 to 24°C but these temperatures are only sustained during parts of the summer. As temperatures fall in autumn, grass growth tails off correspondingly. These falling temperatures are often associated with dull cloudy weather and lower levels of solar energy which, combined with the physiological state of the grass being past the early season reproductive phase, results in lower growth generally than in the first part of the grazing season.

The growing season is longest in south-western coastal areas of England, where it may reach 300 to 350 days, while in the north-west of Scotland, mainly on account of the Gulf Stream modifying the effect of latitude, the growing season can be 250 to 300 days in length (see Figure 10.1). In colder eastern areas, the period is reduced further to 200 to 250 days.

The growing season is also shortened at both ends of the season with increasing altitude since temperature falls at approximately 1°C per 150 m, and with north-facing relative to south-facing aspects. It should be noted that the grazing season is 5 to 6 weeks shorter than the growing season since there has to be adequate grass on offer before grazing can begin. Grass production potential is greatest with long rather than short growing seasons, provided limiting factors such as drought do not intervene.


## RAINFALL

Rainfall is the aspect of weather to which farmers must necessarily pay most attention. Considering the average amounts of rainfall from spring to early autumn (April to September)—the main grass growing season in the UK—there are marked differences between regions (see Figure 10.2). Rainfall during April to September is clearly the most important for producing grass; nevertheless, sufficient winter rainfall ensures that soil water is plentiful at the start of spring. Water loss by drainage is highest during winter and lowest during the growing season, due to summer evapotranspiration.

Considering the importance of water supply for grass and crop growth, soil water management does not often receive the attention it deserves, even in drier regions. Yet management which optimises soil water retention and availability will pay dividends in better growth. The production of 1 t/ha of grass dry matter requires 200 to 250 t/ha of water, absorbed by roots and transpired through the grass foliage into the atmosphere. This corresponds to a depth of water of 25 mm/ha whether taken up from soil reserves or falling as rainfall. The transpiration stream provides soil nutrients in solution and keeps the grass plants turgid. This turgidity ensures optimal photosynthesis since leaves are well positioned to absorb solar radiation and leaf stomata are opened to allow free entry of carbon dioxide. Most transpiration is at a maximum during the day.

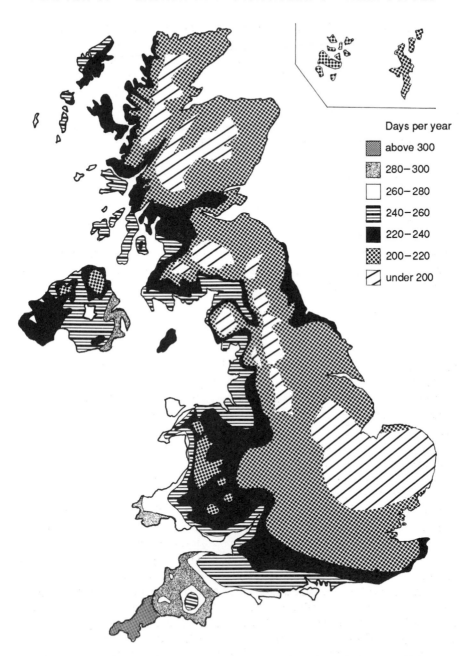

Figure 10.1   *Grass growing days – soil temperature adjusted for drought*
*factor and altitude*
Source: Down *et al.*

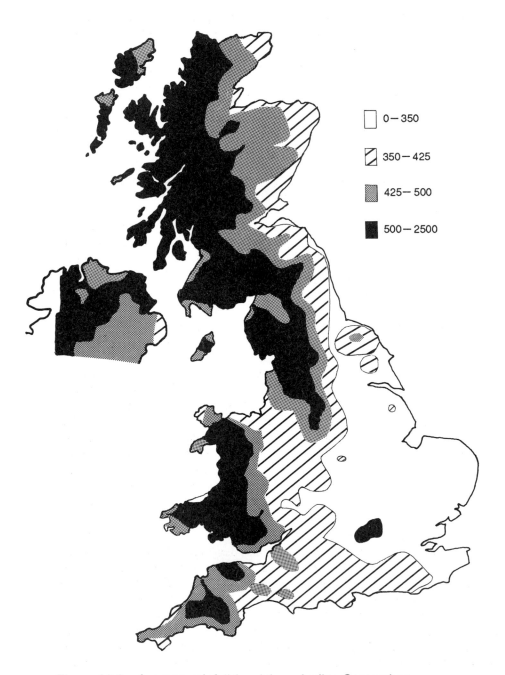

| | |
|---|---|
| ☐ | 0 – 350 |
| ▨ | 350 – 425 |
| ▨ | 425 – 500 |
| ■ | 500 – 2500 |

*Figure 10.2   Average rainfall (mm) from April to September*
Source: Thomas *et al.*

When rain falls, some may evaporate from sward foliage before it can percolate into the soil, especially in hot weather. This loss together with that from transpiration is referred to as 'evapotranspiration'. Excess rainfall may run off into ditches from steep land or from sloping, compacted land. Once infiltrated into the sward, a part of the rainfall replenishes the water supply held by the soil.

If the soil is not free draining, such as heavy textured, poorly structured soils or soils with an impermeable subsoil, it is soon saturated. This waterlogging leads to poor aeration, lack of oxygen and a build-up of toxic gases, and nitrogen is rendered unavailable. These problems produce the typical symptoms of poor drainage such as inhibited root growth and retarded grass production, suppression of desired herbage plants and their substitution by weeds and rushes.

When excess water has drained away under gravity, that held by soil particles and micropores is at field capacity of the soil. However, as far as grass growth is concerned, the water available to its roots over their rooting depth is the most significant and is termed the available water capacity (AWC) of soil. The remainder is unavailable and while this is highest in heavy textured clays and clay loams, these soils also have the greatest AWC. The AWC of soils is improved by an increase in their organic matter content, as a result of both its absorbtive properties and its benefits to soil structure.

Grass roots can grow and absorb water down to 0.75 to 1.0 m, or deeper if soil structure is good, although subsoil water is usually poorer in nutrients than topsoil water. During vigorous grass growth 25 to 30 mm/ha of water can be transpired in a week, so water deficit on light soils soon limits grass growth in low rainfall areas. As roots take up the available water from the matrix of soil pores, the soil gradually dries out from the surface downwards. Unless this water is replaced by rainfall, or upward capillary movement from water held lower down, the soil develops a 'soil moisture deficit' (SMD) which is the amount of water required to restore a soil to field capacity.

When the water deficit reaches a critical stage, transpiration by the sward is reduced below that when soil water is plentiful. Accordingly, grass growth slows down and production is reduced. Soil nutrient uptake is restricted and this also contributes to reduced growth rate and yield. Application of fertiliser does not alleviate the nutrient shortfall since there is little moisture other than dew to wash in the fertiliser or hold nutrients in soil solution. Eventually the grass foliage wilts and collapses to a state known as the 'permanent wilting point'. Plant growth virtually ceases and older leaves and tillers begin to die.

# IRRIGATION

Irrigation is used in low rainfall areas to correct the SMD when it has reached the critical stage for growth. As the noted scientist H. L. Penman once put it, 'The value of successful irrigation is that it provides the water of a wet summer in the sunshine of a fine one'. Irrigation has never achieved the level of popularity, even in dry regions, that it would seem to deserve—probably because of bother, expense and erratic need. Risk of drought and therefore reduced grass production is as high as seven or eight years out of ten in the south-east of England but decreases to a negligible risk in the west and north.

On average, drought occurrence is higher in the early summer, coinciding with the period of maximum grass growth. Irrigation can guarantee peak growth as well as a consistency of supply of grazing and conservation crops, which is important on livestock farms where the numbers of animals are relatively fixed. Grassland irrigation is only likely to be cost effective on intensive livestock enterprises and the economics must be checked before installing a system.

In grassland experiments in the south-east of England annual grass production increased on irrigated swards by 25 per cent compared with unirrigated swards. However, the increases ranged from 10 to 60 per cent due to fluctuations in summer rainfall. The effect of irrigation on seasonal grass production is shown in Figure 10.3. Irrigation is most beneficial on light textured soils with relatively low available water capacities. An adequate supply of available nutrients, especially nitrogen, is necessary if the growth stimulating effect of irrigation is to be fully exploited. To maintain grass output in dry weather the general recommendation is that irrigation water should be applied to restore field capacity when SMD reaches 25 mm on lighter soils and 40 mm on more water-retentive, heavier soils.

Irrigation scheduling has to be precise and a number of aids are available. The Meteorological Office calculates the potential weekly loss of water by evapotranspiration within 40 × 40 km grids covering the UK. A service, called the Meteorological Office Rainfall and Evaporation Calculation Systems (MORECS), gives information on rainfall, potential and actual evaporation, and soil moisture deficits. This information assists local assessments of how much irrigation water to apply after consideration of local conditions such as on-farm rainfall and available water capacities of the farm soils. Computerised scheduling systems with a high degree of accuracy are available from advisory services.

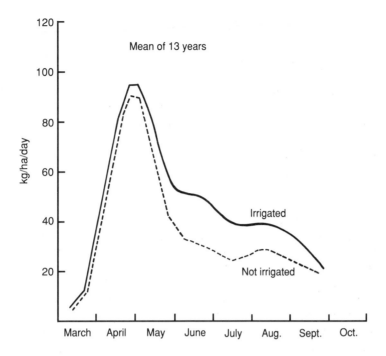

*Figure 10.3* *Seasonal pattern of herbage production from irrigated and unirrigated grassland (early perennial ryegrass)*
Source: Corrall

# DRAINAGE

Rapid removal of excess water from the soil is a key factor in soil management for productive grassland. This improves soil structure, air/water relationships, root development and nutrient uptake, while the chances of surface ponding or run-off are lessened. The natural drainage characteristics of different soil types were described in Chapter 9. Drainage problems usually become most evident in the winter months during periods of high rainfall. The soil may become temporarily saturated but with an effective underdrainage system it will soon drain to field capacity. Problems arise when blockages occur in the drainage pipes due to silt or iron ochre, or at the outflow. Soil compaction caused by grazing stock or grassland equipment and the consequent formation of surface or subsurface pans restrict water flow through the soil to the drains and the upper soil layers may become waterlogged.

There is no doubt that there has been a reduced investment in underdrainage in recent years, particularly due to decreased grant aid and reduced farm profitability. Maintenance of existing underdrainage systems has also declined, yet many systems were installed at about the turn of the century and require renewal if soil fertility is to be sustained. Regular checking and maintenance are also required for younger drainage systems. It is well worthwhile for every farm to have field plans of the positions of drainage pipes. These can sometimes be mapped during a dry summer when grass growth will be more vigorous above the drain lines.

The risk of drainage problems and potential remedies can be assessed from soil survey information but local knowledge, practical experience of individual field behaviour and soil profiles will also be necessary. The presence of grey soil colours predominating over brown indicates periods of summer dryness and winter waterlogging. The presence of pans, deteriorated soil structure and restricted root development at depth are other indicators of impeded drainage.

The planning and installation of underdrainage is a skilled job requiring survey of the topography, examination of the soil and decisions of spacing, depth and type of drain. Choice of drain is generally between traditional tile or plastic pipe, both of which are expected to last 60 to 70 years or more. Mole drains have a much shorter life span of 5 to 6 years, so repetitive moling is necessary. They are only suitable for heavy textured soils with 25 per cent or more clay content to retain their shape, though on lighter soils mole channels can be filled with gravel. Mole drainage has proved economic and effective when used in conjunction with a widely spaced collector system of drains which are backfilled with water-permeable gravel.

Grassland surveys have shown that a high proportion of permanent grassland is poorly or badly drained. This reduces potential grass production, the species persistence (see Table 10.1) and the length of grazing season. Sward conditions may be adverse at both the beginning and end of the season. On individual farms greater reliance may have to be placed on more expensive conserved grass unless there are some well-drained fields suitable for grazing. The benefits of good drainage are undeniable and are sustained over many years, provided the drainage system is well maintained. Nevertheless, there are very few long-term experiments which have actually quantified the benefits. Results in the short term have confirmed the benefits, although the improvements in grass yield have been less than expected. The results of drainage studies conducted in Devon on a clay loam soil slowly permeable to water under high rainfall conditions are shown in Table 10.2.

**Table 10.1** Average content of 'preferred' species in swards classified by age and drainage status (percentage contribution to total ground cover)

| Age (years) | Drainage status | | | |
|---|---|---|---|---|
| | Good | Imperfect | Poor | Bad |
| 1–4 | 73 | 72 | 65 | * |
| 5–8 | 63 | 60 | 53 | * |
| 9–20 | 53 | 52 | 45 | 35 |
| Over 20 | 36 | 35 | 30 | 20 |

* insufficient swards
*Source:* Forbes *et al.*

**Table 10.2** Effect of drainage and reseeding on herbage production and liveweight gain (LWG) by beef cattle (average of four years)

| Sward | Annual N (kg/ha) | Treatment | DM (t/ha) | Grazing days | | LWG (kg/ha) |
|---|---|---|---|---|---|---|
| | | | | To 31 May | Total for year | |
| Old grass | 200 | Undrained | 8.3 | 170 | 894 | 716 |
| | | Drained | 8.7 | 204 | 962 | 780 |
| | 400 | Undrained | 9.7 | 204 | 963 | 810 |
| | | Drained | 10.4 | 286 | 1,079 | 879 |
| Reseeds | 400 | Undrained | 9.5 | 193 | 1,200 | 792 |
| | | Drained | 10.3 | 251 | 1,254 | 990 |
| Average | | Undrained | 9.2 | 189 | 1,019 | 772 |
| | | Drained | 9.8 | 247 | 1,098 | 889 |

*Source:* Garwood

# SUBSOILING

Subsoiling at about 300 mm depth can make compacted soil more permeable to water by bursting the pan and creating fissures and cracks. It also improves the structure and drainage of subsoils with low water permeability and this assists the effectiveness of the existing underdrainage. The timing of subsoiling is important since the soil

99

must be dry to produce a shattering effect; summer is normally the best period. A mole plough can also be used.

Non-disruptive cultivation which both alleviates compaction and improves drainage may be an alternative to ploughing and reseeding pastures which have been badly poached. On soils susceptible to poaching the risk of damage is highest on reseeded swards which have not yet developed a sufficient root mass to strengthen the soil surface. Ground condition scoring on an 8-point scale can be used to gauge safety from and susceptibility to poaching (see Table 10.3). The risk of poaching increases progressively from score 4 upwards.

**Table 10.3   Ground condition scoring**

| Score | Condition | Score | Condition |
|-------|-----------|-------|-----------|
| 1 | Baked hard | 5 | Squelchy in patches |
| 2 | Dry | 6 | Squelchy throughout |
| 3 | Damp and firm | 7 | Very soft |
| 4 | Damp and soft | 8 | Waterlogged |

Source: Smith et al.

The effects of three soil loosening/secondary drainage cultivations—subsoiling, moling and paraplowing—on sward productivity on a soil compacted by poaching were monitored over three years at Ayr. Cultivations produced an average 16 per cent increase in early grass production and, by controlling surface waterlogging and lowering the soil water table, the season was extended during early spring and late autumn when the grass could be grazed without causing poaching damage. The beneficial effects from subsoiling and moling were more persistent than paraplowing. Increased grass output was attributed to better root development and nitrogen uptake following improved soil drainage and aeration, and reduced denitrification of applied nitrogen.

*Chapter 11*

# Soil Fertility and Grass Production: Nitrogen

And he gave it for his opinion, 'that whoever could make two ears of corn, or two blades of grass, to grow upon a spot of ground, where only one grew before would deserve better of mankind, and do more essential service to his country, than the whole race of politicians put together'.

Jonathan Swift, *Gulliver's Travels* (1726)

How lush and lusty the grass looks! how green!

William Shakespeare (1564–1616) *The Tempest*, II. I.52

Nitrogen is the most important nutrient influencing grass production and its supply is largely under the farmer's control. These facts led to the increased use of fertiliser N associated with grassland intensification in the past. The average annual usage of fertiliser N on grassland, excluding rough grazing, has been in the region of 115 to 135 kg/ha in recent years, but can range from 0 to 450 kg/ha. The highest rates are usually associated with intensive dairying and the lowest with extensive beef and sheep systems. Fields for conservation receive higher rates than grazing fields. Swards which receive little or no N from fertiliser or organic manures and contain few N-fixing legumes are not highly productive, so even if efficiently utilised, animal output will not be high.

At a nil rate of applied N, an all-grass sward produces 2 to 5 t DM/ha annually, depending on the available N in the soil. In contrast, a grass/white clover sward receiving nil N produces 6 to 9 t DM/ha annually, the higher levels being associated with 25 to 35 per cent of N-fixing clover. About half of the 1.5 million tonnes of fertiliser N used annually in Britain is applied to grassland. This contrasts with an annual usage of 10 to 20,000 tonnes of fertiliser N in New Zealand, where grassland production is sustained by N-fixation from white clover—an estimated one million tonnes of N fixed per annum (compared with 80,000 tonnes in the UK).

101

The grass roots take up soil N mainly in the form of nitrate, although they can also absorb ammonium. In the plant, the nitrate is converted to ammonium ions which are combined with carbohydrate to synthesise amino acids, used to form protein. This increases the size and area of the leaves which in turn results in more green tissues capable of photosynthesis, and further herbage production. Nitrogen surplus to the plant's requirements remains in the plant tissues as non-protein nitrogen, some of which may be nitrate. The concentration of N in grass DM typically ranges from 20 to 40 g/kg; thus N offtake is 20 to 40 kg/t DM and two cuts of silage amounting to 10 t/ha DM will remove 200 to 400 kg/ha.

The overall impression of nitrogen deficiency in a sward is one of stunted grass growth with the leaves ranging from pale green to yellow in colour. The older leaves show these colour changes first, starting at the tips. Root growth also becomes stunted and the roots then have decreased capability for seeking soil nutrients and moisture. The change to greener, larger leaves following application of fertiliser N can be visually dramatic, provided other essential nutrients are not limiting. Excessive nitrogen application is unnecessary for the sward, however; apart from being uneconomic, it increases the risk of atmospheric and water pollution. The leaves become blue-green in colour, and sappy, making them less attractive to grazing stock; and in silage or hay systems with long growing periods before cutting, the crops lodge, causing death and decay in the underlying sward. High moisture content also causes difficulties in drying the cut crop.

## SOIL NITROGEN STATUS

The amount of soil N available to plants cannot yet be measured precisely by scientific means, but it can be estimated. Soil N reserves are higher under grazed swards than regularly cut swards because of the cycling of N-rich excreta, although application of slurry or farmyard manure augments organic N reserves in cut swards. Only 1 to 2 per cent of the soil organic N may become available during the growing season by mineralisation from the action of microbes but the amounts range from 20 to 120 kg/ha depending on the soil reserves. Mineralisation is increased following ploughing up of a grass sward, e.g. in order to reseed.

Estimation of soil N status is particularly important when breaking up grassland for arable cropping so that the arable crop can be fertilised with some degree of precision. It is also important when establishing grassland following cropping rotations, which deplete soil

N, so that seedbed N application matches the needs of the establishing grass. For established grassland the rate of N application is governed mainly by the intended target level of herbage production, but the rate can be adjusted from knowledge of the estimated soil supply. Typical examples of how previous field history can be used to classify soil N status into estimated low, moderate and high categories are shown in Table 11.1.

**Table 11.1    Soil nitrogen status based on previous crop**

| Low | Moderate | High |
| --- | --- | --- |
| Leys (1–2 years) grazed semi-extensively with less than 125 kg/ha N per year | Leys (1–2 years) grazed semi-extensively with 125–250 kg/ha N per year | Leys (1–2 years) grazed intensively with more than 250 kg/ha N per year or a strong clover sward |
| Leys (1–2 years) cut | Oilseed rape | |
| | | Forage crop grazed |
| Forage crop removed | Potatoes – ware | |
| Cereals | Peas, beans | |
| Potatoes – seed | Field vegetables | |

*Source:* Dyson/Scottish Agricultural Colleges

Small amounts of nitrogen, around 12 kg/ha on average, are deposited from the atmosphere onto the land by rainfall and by dry deposition. There are also free-living soil micro-organisms capable of fixing N from the atmosphere though the amount fixed may only be a few kg per ha. These sources, together with symbiotic N-fixation by legumes, are significant in rough grazings which rarely receive N from fertilisers or organic manures.

## SYMBIOTIC N-FIXATION BY LEGUMES

The amount of N fixed by white clover in grass/clover swards ranges widely depending on clover content but an estimated 75 to 280 kg/ha in lowland swards and 100 to 150 kg/ha on hill and upland pastures have been reported from experiments. The amount declines with increasing fertiliser N rate and other factors which depress both clover content and N-fixing capability of the rhizobia. The role and potential of forage legumes in grassland, including the interactions between clover and nitrogen application, are discussed further in Chapter 22.

## NITROGEN IN ANIMAL EXCRETA

Between 75 and 90 per cent of the N in the herbage consumed by cattle and sheep is excreted, the proportion depending upon the type and class of animal. Beef store cattle utilise the least and dairy cows and most N from herbage. With an assumed DM intake of 3,000 kg/ha over the grazing season, a dairy cow may excrete 45 to 90 kg/ha N. At a stocking rate of 3 cows/ha, 135 to 270 kg/ha N is circulating or 'cycling'; this is not a net addition since the same nitrogen can be utilised more than once during the season and some is lost from the cycle by leaching of nitrate or volatilisation as ammonia. The components of the nitrogen cycle in grassland are shown in Table 11.2. About a quarter of the excreted N is in the dung and not immediately available, being mostly in organic form, while the remaining urinary N is readily available.

Grassland also receives N from animal manures collected over the winter period, mostly in the form of slurry. The amount of available N within the total N depends on the type of manure, rate of application, content of N and time of spreading. Total N application must not exceed 250 kg/ha according to the official Codes of Practice such as the MAFF Code for England and Wales. The N in slurry is used most effectively and with least loss if applied during the growing season. The use of slurry and other organic manures on grassland is discussed in Chapter 13, while excretal return of N during the grazing process is dealt with more fully in Chapter 16.

## TYPES OF NITROGEN FERTILISER

The most commonly used nitrogen fertiliser in Britain is ammonium nitrate, 34.5 per cent N, in the form of prilled granules. This fertiliser absorbs moisture rapidly and becomes rapidly available for uptake by the sward roots after application. Calcium ammonium nitrate with 21 and 26 per cent N is also marketed. The inclusion of calcium carbonate—40 and 26 per cent respectively—reduces the soil acidifying effect of the ammonium nitrate, but less so than the original mixture which had only 15 per cent N and 57 per cent calcium carbonate.

**Table 11.2    The main components of the nitrogen cycle**

*N inputs*

Nitrate and ammonium fertilisers (by application)
Organic manures (by application)
Symbiotic N-fixation (by legume rhizobial bacteria)
Mineralisation from soil organic matter (by soil micro-organisms)
Wet and dry deposition (from the atmosphere)
Non-symbiotic N-fixation (by free-living bacteria)

*N outputs*

Animal products, e.g. milk, meat, wool (from grazing the sward)
Conserved silage or hay (from cut swards)

*N cycled*

Urine and dung (from grazing animals)
Slurry and farmyard manure (from housed animals fed conserved grass and
    concentrates)
Unutilised herbage and root tissues (by senescence and soil organisms)
Nitrification of ammonium to nitrate (by nitrifying bacteria in the soil)

*N losses*

Volatilisation of ammonia (from urine)
Immobilisation of N in soil organic matter (from applied N inputs and cycled N)
Leaching of nitrate (by drainage water)
Denitrification of nitrate to nitrogen gases (by denitrifying bacteria in the soil)
Run-off of slurry (following application in unsuitable conditions)

Ammonium nitrate and urea are widely used as the N constituent of compound NPK fertilisers.

Fertiliser urea (46 per cent N) has increased in use in recent years because it is highly concentrated and cheaper per kg of N than ammonium nitrate. Care in application is necessary because its small, lighter prills give a narrower spreading width than ammonium nitrate. Before uptake by the roots, urea is converted to ammonium carbonate which then decomposes to ammonia and carbon dioxide. The ammonia dissolves in the soil moisture, and is converted to nitrate by nitrifying bacteria. There is thus opportunity for loss by volatilisation of ammonia into the atmosphere, though loss from denitrification or nitrate leaching is lower than with ammonium nitrate. The ammonia loss is encouraged by high temperatures, dry conditions and high soil pH, so the effectiveness of urea for herbage production has been more variable than with ammonium nitrate.

In trials at the Scottish Agricultural College in Ayr and Edinburgh comparing these two sources of N, annual herbage production from urea application was 3 to 6 per cent less, with reductions greater than this at individual cuts during dry conditions. The lower production from urea was associated with reduced uptake of N, resulting in lower amounts of herbage N and lower N contents. These results suggest urea is a suitable alternative to ammonium nitrate, especially for spring grass production and for silage, but effectiveness later in the season is dependent on adequate soil moisture or enough rain soon after application to wash in the urea. Some additional urea could compensate for its reduced efficiency at rates up to 100 to 120 kg/ha since its efficiency progressively decreases at higher N rates. The cost of the additional N needed would also reduce the price advantage of urea. Overall, the results suggest that urea is an economic alternative to ammonium nitrate, provided it is 10 to 20 per cent cheaper per kg N.

Aqueous ammonia is a liquid nitrogenous fertiliser containing gaseous ammonia dissolved in water. It is injected into the soil as a single high dressing in early spring which then persists throughout the summer, especially in a dry season and if applied in early spring. It is slow acting, though, and requires topping up with solid N fertiliser for early grass production.

Anhydrous ammonia with 82 per cent N is the most concentrated form of fertiliser N available. It is handled as a liquid under high pressure and after subsurface injection reverts to its normal gaseous state. It is used extensively in the United States of America but much less in Europe. Injection by applicator into slits cut in the sward must be 10 to 20 cm below the surface to prevent loss of gas and scorching of the sward. Sealing by press wheels of the slits cut by the injection apparatus is essential and slits must be spaced 30 to 35 cm apart since closer spacing would result in excessive slicing of the sward and possible damage from wheel slip. Wider placement would result in a typical banding effect of alternating dark green nitrogen-rich grass and light green nitrogen-deficient grass.

After injection the ammonia becomes dissolved in the soil moisture or adsorbed on the surfaces of clay and humus particles. The ammonia is largely converted to ammonium ions some of which are utilised directly by plant roots with most of the remaining ammonia being converted by nitrifying bacteria into nitrate and absorbed by the roots. Because of the high cost of application and its mode of action, a large single dressing of 250 to 350 kg/ha N in spring has been advocated. However, trials at Ayr in the past showed that such a dressing did not last the whole season. The herbage response to N was slower and annual herbage production was 15 per cent less than when solid fertiliser N was used in repetitive dressings to make up the same annual N total. Further development of anhydrous ammonia will depend on the economics of its usage in relation to other N fertilisers.

## REDUCING N LOSSES

Recently there has been a searching examination of the use of fertiliser nitrogen on grassland. Economic pressures demand optimal efficiency of use, for example, precision in timing and the amounts applied. There is also concern in the long term that fossil fuel reserves may be insufficient to sustain economic fertiliser N manufacture, a concern which has led to increasing reappraisal of the role of N-fixing forage legumes. Organic manures as a source of N are being used with greater effectiveness than in the past, and with greater care in minimising environmental pollution.

Public and governmental concern about the environment has focused particularly on reducing nitrate levels in drinking water in some areas where the water is derived from aquifers, e.g. eastern England. The EC limit of 50 mg/l is the maximum concentration allowed and trial Nitrogen Sensitive Areas have been designated where there are restrictions on the rates of nitrogen application and management measures to control the entry of nitrate into water. Gaseous losses, e.g. ammonia, are also under scrutiny because of their impact on atmospheric constituents, influencing the acidity of rainfall and contributing to global warming.

More research is thus being focused on the N cycle, particularly on how to reduce the losses or 'leakage' of N from animal production systems based on grassland. The drive towards better efficiency of N use includes Codes of Practice, voluntary agreement schemes, advisory service recommendations and selective legislation.

Ultimately, potential losses are dependent on total N supply of which fertiliser N is a major component both directly, e.g. as nitrate, and indirectly from its effects in increasing herbage production, consequent stocking rates and excretal N returns. The concept has been put forward of a critical 'breakpoint' corresponding to the N application rate at which losses rise steeply. Where environment considerations are pre-eminent, applications should be well below the breakpoint. Preliminary work in southern England indicated that the environmentally safe annual N rate there ranges from 150 to 200 kg/ha for grazing and 250 to 350 kg/ha for cut swards. Because of the complexity of the N cycle the safe rate is not easily determined for all the widely variable soil and climatic conditions in the UK. Nevertheless the answers are being sought with ongoing research.

Table 11.3 summarises various measures to reduce N losses, although not all may be practicable on individual farms.

**Table 11.3  Measures which reduce nitrogen losses**

- Precision of rate and timing of fertiliser N use
- Balanced fertiliser use
- Use of good quality fertilisers
- Accuracy of spreading
- Use of ammonium N in spring
- Application of slurry in spring/summer
- Emphasis on cutting in late season
- Use of grass/white clover swards
- Reliance on long-term grassland
- Reseeding leys in spring
- Soil management to increase ammonia retention

## RESPONSE OF GRASS SWARDS TO N

With increasing rates of fertiliser N applied to a grass sward, herbage production increases linearly in a response of around 15 to 25 kg DM per kg N up to an annual N rate of 250 to 350 kg/ha (see Figure 11.1). Both the amount of response and the N rate needed to maintain production linearly are influenced by soil characteristics, moisture supply and frequency of defoliation. As the annual N rates are further increased from 350 to 450 kg/ha, herbage DM response decreases to 5 to 15 kg DM per kg N, and an asymptotic point is reached where there may be no further herbage response, perhaps at an N rate between 450 and 600 kg/ha.

It is also important to realise that 80 to 90 per cent of the maximum herbage production is achieved with only 50 to 60 per cent of the N needed to attain the maximum production. This is the effect of the law of diminishing returns coming into play when herbage DM response per kg N steadily declines with each additional kg of N on the curvilinear part of the N response curve.

The N rate likely to be economic will vary with the type of

enterprise but will be below the amount of N needed to achieve maximum DM. Thus, a herbage response between 5 and 10 kg DM per kg N may be economic in a dairying enterprise but not in beef and sheep production systems where the economic animal output per kg DM utilised is lower. The efficiency with which the extra DM production is utilised and the productive ability of the stock utilising it are also important factors. Again, it may not necessarily be the farmer's objective to obtain the absolute the maximum DM theoretically possible.

## RESPONSE OF GRASS/WHITE CLOVER SWARDS TO N

When rates of fertiliser N are increased on a grass/white clover sward, total (grass plus clover) herbage production increases linearly up to N rates of 250 to 350 kg/ha, and then curvilinearly. The increases per kg N are lower than from a grass sward since the grass response to N in the mixed sward is accompanied by a decrease in clover contribution. The herbage response curve is therefore flatter (see Figure 11.1).

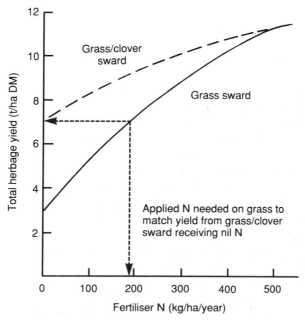

Figure 11.1  Effect of increasing fertiliser N application on grass and grass/clover swards

However, with a DM production of 6 to 9 t/ha at nil N a grass/clover sward starts at a higher level, although it can be seen from the graph that this production advantage decreases with increasing N rate. Eventually, as the mixed sward becomes all grass, its response to N is similar to an original all-grass sward. To match the production from a grass/clover sward receiving no N and with a vigorous clover content, a grass sward would require an annual N application of 180 to 200 kg/ha. This rate is referred to as the 'fertiliser N equivalent'.

## SEASONAL DISTRIBUTION OF NITROGEN

In rotationally grazed fields it is customary to split the annual N rate into five to seven applications; under continuous stocking, N may be applied to a quarter of the area every week so the principle of spreading N over the season is still maintained. Application timing ranges from early February in south-west Britain to mid April in the north, with local variations due to soil, and topographical and climatic factors. Nitrogen rates are normally reduced towards the end of the season to avoid unnecessarily high N contents in the herbage; N is taken up by the grass then but herbage production is lower than earlier in the season because of poorer growing conditions, leading to a correspondingly higher risk of leaching. General management guidelines for early, mid and late season grassland production are discussed in Chapter 18.

Table 11.4 shows the effect of time of season on average herbage DM responses to fertiliser N from trials on perennial ryegrass swards at Ayr. Within the monthly responses, the higher levels were associated with less frequent defoliation.

Alteration in the amounts of N applied at each dressing have little effect on annual herbage production but may affect seasonal production

**Table 11.4   Effect of season on herbage DM response to fertiliser nitrogen at Ayr**

| Month | kg DM/kg N |
|---|---|
| April | 5–15 |
| May | 15–25 |
| June | 30–40 |
| July | 20–30 |
| August | 15–25 |
| September | 10–20 |
| October | 5–15 |

Source: Frame et al.

considerably. Using results from perennial ryegrass swards cut five times annually, prediction models of seasonal production for two contrasting N application schedules, both supplying 335 kg/ha, are shown in Figure 11.2. Schedule A, with five equal N dressings, accentuates the normal pattern of herbage production with a peak in June. Schedule B, with heavier dressings supplied for the early cuts—particularly the first—and none in later season, is appropriate for silage making, 77 per cent of the annual production being obtained by late July.

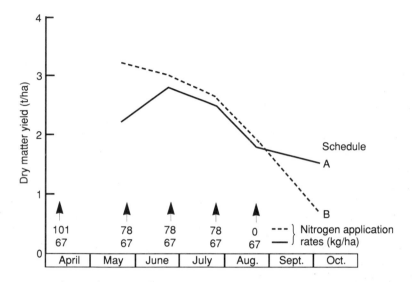

*Figure 11.2   Seasonal patterns of dry matter production from two nitrogen application schedules on S23 perennial ryegrass*
Source: Reid

On farms, the intensity of use governs the seasonal and annual rates of fertiliser applied. Rates recommended by the SAC Advisory Service for high, moderate and low intensities of use are shown in Table 11.5.

Potential herbage DM production and stocking rates are illustrated in Table 11.6. Analysis of the results of recording schemes, e.g. the Meat and Livestock Commision 'Flockplan' or various dairy herd costing services, invariably emphasises the powerful influences of stocking rate in achieving optimum physical and financial performance.

## NITROGEN AND INTENSITY OF DEFOLIATION

Cutting close to ground level in silage systems or grazing close in

**Table 11.5    Fertiliser nitrogen recommendations for grazing (kg/ha)**

|  | *Soil N status* | | |
| --- | --- | --- | --- |
| *N applications* | *Low* | *Moderate* | *High* |
| *High N policy* | | | |
| 1st | 105 | 80 | 55 |
| 2nd | 90 | 65 | 40 |
| 3rd | 90 | 65 | 40 |
| 4th | 90 | 65 | 40 |
| 5th (not later than mid August) | 75 | 50 | 25 |
| *Moderate N policy* | | | |
| 1st | 95 | 70 | 45 |
| 2nd | 75 | 50 | 25 |
| 3rd | 75 | 50 | 25 |
| 4th | 65 | 40 | 20 |
| *Low N policy* | | | |
| For early grass | 85 | 60 | 35 |
| For maintenance (mid season) | 25 | 0 | 0 |

*Source:* Younie *et al.*/Scottish Agricultural Colleges

**Table 11.6    Potential herbage DM production and stocking rates per ha over 180 to 200 day grazing season for dairy cows, beef cattle (200 kg live weight at turn out) or sheep**

| | | *Stocking rate per ha* | | |
| --- | --- | --- | --- | --- |
| *Intensity of use* | *Herbage DM (t/ha)* | *Dairy cows* | *Fattening beef cattle* | *Ewes with lambs* |
| High | 11–12 | 3.5–4.0 | 7–8 | 15–18 |
| Moderate | 9–10 | 3.0–3.5 | 6–7 | 12–15 |
| Low | 7–8 | 2.5–3.0 | 5–6 | 9–12 |

rotational systems increases herbage DM production compared with less severe defoliation over a range of fertiliser N rates and sward types. Perennial ryegrass swards cut regularly to 20 to 30 mm over the season produce 20 to 30 per cent more herbage DM than swards cut at 50 to 60 mm. The reasons for this include inhibition of stem and flower development, with resultant stimulation of tiller and leaf, and the efficiency of utilisation, i.e. herbage removed as a proportion of the amount present. On occasion, less output is obtained from severe

112

compared with lax defoliation when rest intervals between defoliations are very short, or during drought.

There are, however, practical issues involved which sometimes override the advantage of higher DM to be gained from close defoliation. Conservation fields may have surface irregularities and overclose cutting may scalp swards, damage equipment and cause soil contamination of cut herbage. Drying of a cut swath overlying a short stubble is retarded due to restricted air circulation below the swath. Herbage quality and its effect on animal production also require consideration, since close defoliation results in utilisation of the less leafy, less nutritious herbage, typical of the lower sward horizons. Some sward types, such as hybrid and Italian ryegrass, do not tolerate repeated close cutting, e.g. at 3 to 4 cm, to the same extent as perennial ryegrass, and stubble heights of 5 to 6 cm are more suitable. Severity of defoliation is thus a compromise between efficiency of utilisation and practical requirements on individual farms.

Frequency of defoliation has been widely investigated, particularly in cutting trials to simulate grazing or silage management, using a range of fertiliser N rates. Several treatment variations have been used, e.g. time interval, stage of growth or sward height at cutting. In general, frequent cutting reduces herbage production compared with infrequent cutting. This is illustrated in Figure 11.3, derived from trial data at Ayr over three seasons including one dry year. Similar effects have been found elsewhere, although because of differing soil and climatic conditions the levels of herbage production and herbage responses to fertiliser N differed. Figure 11.3 shows that when a perennial ryegrass sward was cut four times annually at approximately 50-day intervals, DM production increased almost linearly with N rate from 0 to 360 kg/ha, each kg of N giving an extra 25 kg of DM. The response from 360 to 480 kg/ha decreased to 10 kg DM per kg N, declining to negligible responses above this rate. In contrast, when cut eight times at approximately 25-day intervals annually, the production response was approximately linear up to 480 kg/ha with an extra 11 kg DM for each kg of applied N. It decreased to 7 kg DM per kg N up to the 600 kg/ha annual rate but became negligible at N rates above this.

The increased herbage production from infrequent compared with frequent defoliation was greatest during the first half of the season. This is because the bulk—60 to 70 per cent—of annual production is produced then, due to better growing conditions and to high growth rates during the period when reproductive tillers are produced. Production differences between the different frequencies were least in late season. Evidence elsewhere has shown that even when rest intervals were lengthened in the second half of the season compared with the

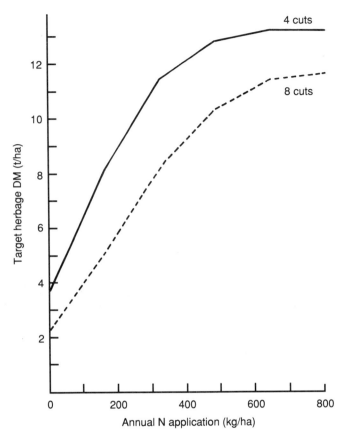

*Figure 11.3   Herbage response to fertiliser nitrogen under differing cutting
systems*
Source: Frame et al.

first, there was little effect on herbage accumulation since leaf senescence also increased.

It can be deduced from Figure 11.3 that a high N rate is needed to compensate for the output-reducing effect of frequent cutting. Thus in a multi-cut silage system, when the objective is high digestibility silage, high N rates and more frequent cuts are needed in order to attain specified production targets. The converse objective of higher bulk of herbage with lower digestibility is achieved by fewer cuts and longer rest intervals. Factors, including cutting management,which affect digestibility are discussed in Chapter 14. The rates of N required to achieve specified target DM yields at the two cutting frequencies in the Ayr trial are shown in Table 11.7.

▲    Severely poached sward with water accumulating in the hoofmarks of grazing livestock because of underlying soil compaction    (D. Howat)
*Chapter 9*

Silage fields showing tracking damage caused by equipment during
▼    harvesting    (D. Howat)    *Chapter 9*

▲    Different intensities of wheel tracking being imposed on experimental grass plots, SAC, Auchincruive    *Chapter 9*

Experimental grass plots ▶ exhibiting different growth vigour and production in response to differing wheel track intensities—untracked control plot in foreground— at SAC Auchincruive.
*Chapter 9*

Winged subsoiler in use during summer to alleviate soil compaction    (C. Smith)
▼    *Chapter 9*

Mole plough for soil ▶
loosening and drainage
(C. Smith) *Chapter 9*

▲ Trench showing mole
plough slit and channel to
improve soil drainage
(C. Smith) *Chapter 9*

Mole plough and hopper ▶
which delivers gravel to the
channel to prevent it collapsing
(C. Smith) *Chapter 9*

▲ Paraplowing to ameliorate subsurface soil compaction    (C. Smith)
*Chapter 9*

▲ Large-scale irrigation of Italian ryegrass silage crop (*above*) and
(*below*) of the sward immediately following removal of the crop, east
▼ Germany   *Chapter 10*

**Table 11.7    Estimated annual fertiliser nitrogen (kg/ha) required to achieve specified DM yield levels**

| Target DM yield (t/ha) | 4-cut system | 8-cut system |
|:---:|:---:|:---:|
| 4 | 5 | 45 |
| 6 | 10 | 165 |
| 8 | 90 | 315 |
| 10 | 245 | 520 |
| 12 | 475 | 770 |

*Source:* Frame *et al.*

# PREDICTION OF N REQUIREMENT

A national series of fertiliser N trials, cut monthly on sites differing in soil and climatic conditions, confirmed the importance for herbage production of water supply during the growing season and the available water capacity (AWC) of the soil (see Chapters 9 and 10). The concept of site class and expected herbage production emerged (see Tables 11.8 and 11.9) and this has proved valuable in developing models of grassland utilisation and expected animal performance which can then be validated under practical conditions. However, the cutting regime on the series was rigid and the database limited so the figures are only indicative of potentialities; it is difficult to extrapolate to individual farms which often have fields of differing output potential. Rainfall and its distribution in any one year also affect output

**Table 11.8    Site classes**

| Soil texture | Average April–September rainfall | | | |
| | More than 500 mm | 425– 500 mm | 350– 425 mm | Less than 350 mm |
|---|:---:|:---:|:---:|:---:|
| All soils except shallow soils over chalk or rock and gravelly and coarse sandy soils | 1 | 2 | 2 | 3 |
| Shallow soils over chalk or rock and gravelly and coarse sandy soils | 2 | 3 | 4 | 5 |

*Source:* Thomas *et al.*

**Table 11.9** **Probable yields of grass when cut for conservation or monthly to simulate grazing (based on medium N status and optimal N per cut)**

| | Annual DM production (t/ha) | | |
| Site class | Conserved (2 cuts at 61 D value)* | Conserved (3 cuts at 68 D value)† | Grazed |
| --- | --- | --- | --- |
| 1 | 16.0 | 15.4 | 14.3 |
| 2 | 15.4 | 14.4 | 12.8 |
| 3 | 14.3 | 13.4 | 11.4 |
| 4 | 13.4 | 12.6 | 10.5 |
| 5 | 12.6 | 11.7 | 9.6 |

* Cuts taken on 10 June, 12 August followed by grazing
† Cuts taken on 18 May, 22 June, 27 July followed by grazing

*Source:* Thomas *et al.*

significantly, e.g. by as much as plus or minus 20 to 25 per cent. Consequently, outputs from commercial farms are usually less than those predicted from small-scale trials, especially at higher N rates. Losses associated with utilisation, whether from grazing or cutting, and the large-scale nature of farm operations reduce actual outputs. There are also influencing factors on the farm of excretal N return at grazing and slurry N application on cutting fields.

## NITROGEN AND CUTTING SYSTEMS

The interactions between N rate and practical cutting regimes have been examined in relation to target herbage production and quality for silage making. The date of the first or primary cut has an important influence on annual herbage output. When this is delayed there is increased production, though at the expense of digestibility and the rate of regrowth. In dry areas or where grazing follows the first cut, the slow regrowth may be a significant problem. These effects are of practical significance since it is important to make enough silage—and with minimum losses—for the winter feeding period, particularly since silage intake by stock is greater at high than at low D value. On-farm consideration must weigh the needs of the livestock system, and the cost of quality silage against the cost of alternative feeds. Some of the above effects are illustrated in Table 11.10 from cutting systems work at Ayr. Intensive frequent cutting (System C) was successful in achieving high quality herbage for silage, at D values consistently over 70. However, this system could not sustain the silage

**Table 11.10**   Herbage production and quality from three silage systems, each at three rates of nitrogen application, on perennial ryegrass swards (S = silage cut; G = simulated grazing)

| System | Date of first cut | Cutting interval (days) | Annual N rate (kg/ha) | DM (t/ha/year) | | Average D value of silage cuts |
| | | | | Silage cuts (S) | Silage plus grazing (S+G) | |
|---|---|---|---|---|---|---|
| (A) 2S+1G | 7 June | 50 | 150 | 8.5 | 9.4 | 68 |
| | | | 250 | 10.6 | 12.4 | 66 |
| | | | 350 | 12.2 | 14.6 | 65 |
| (B) 3S+1G | 28 May | 50 | 250 | 11.6 | 12.4 | 69 |
| | | | 350 | 12.6 | 13.6 | 68 |
| | | | 450 | 13.2 | 14.3 | 67 |
| (C) 4S+1G | 18 May | 40 | 250 | 9.6 | 10.2 | 72 |
| | | | 350 | 11.0 | 12.0 | 72 |
| | | | 450 | 12.4 | 13.4 | 71 |

*Source:* Harkess and Frame

or annual herbage production of the more traditional 2-cut System A without additional fertiliser N. System B (3 cuts) resulted in increased herbage production and quality of silage over System A (2 cuts).

Timing of spring application, as previously discussed for grazing fields, ranges from mid February onwards, depending on region. Several guides, e.g. the 'T-sum 200' system (see Chapter 18), are available to determine the optimum timing for the first spring application but need to be tempered by local soil conditions and current weather. However, fertiliser N and slurry should be applied at least six to eight weeks before the expected date of first cut, and if the fertiliser N application is split, the second part should be applied no less than six weeks before the expected date of cut. It is a cardinal rule for second and later cuts (and also on rotationally grazed swards) to apply fertiliser N as soon as possible after the herbage from the previous cut is removed to give adequate time for an optimum response. Also, the number of growing days in a season is limited and prompt application of N encourages rapid rates of regrowth. Delay of a few days on each occasion N is applied results in 5 to 10 per cent less annual herbage output.

Prolonged winter grazing on silage fields or their use for an early grazing, e.g. by sheep, reduces herbage output from first cuts of silage. The value of the grazing for the sheep system has to be weighed

117

against the depression in silage output. None the less, there is good cause to be aware of the adverse effects of previous grazing where high silage production is an aim. Table 11.11 shows the effect of defoliations during February to early May on a silage cut on 17 June.

Table 11.11    Effect of simulated spring grazing on herbage production at a first silage cut on 17 June

| Silage system | | Grazing plus silage system | | |
| | | N rate (kg/ha) | | |
| N rate (kg/ha) | Herbage DM (t/ha) | Grazing | Silage | Herbage DM (t/ha) |
|---|---|---|---|---|
| 60 | 7.5 | 30 | 60 | 4.3 |
| 100 | 8.0 | 50 | 100 | 5.0 |
| 140 | 9.1 | 70 | 140 | 5.5 |

Source: Harkess

Rates of fertiliser N recommended by the SAC Advisory Service for silage and hay systems are shown in Table 11.12. The N rates are adjusted in accordance with the soil N status and should also be modified to take account of the available N supplies from organic manures (see Chapter 13).

Table 11.12    Fertiliser nitrogen recommendations for conservation (kg/ha)

| | Soil N status | | |
| | Low | Moderate | High |
|---|---|---|---|
| Silage: 2 or 3 cuts | | | |
| 1st cut | 145 | 120 | 95 |
| 2nd cut | 125 | 100 | 75 |
| 3rd cut or aftermath | 105 | 80 | 55 |
| For a 2nd grazing (not later than mid August) | 85 | 60 | 35 |
| Hay: 1 or 2 cuts | | | |
| 1st cut | 105 | 80 | 55 |
| 2nd cut or aftermath | 85 | 60 | 35 |
| For a 2nd grazing (not later than mid August) | 85 | 60 | 35 |

Source: Younie et al./Scottish Agricultural Colleges

*Chapter 12*

# Soil Fertility and Grass Production: Lime and Mineral Nutrients

---

Take not too much of a land, weare not out all the fatnesse, but leave it in some heart.

Pliny the Elder (AD 23–79) *Historae Naturalis*

It (grass) yields no fruit in earth or air and yet should its harvest fail for a single year, famine would depopulate the world.

Anon.

---

## LIME

The use of lime on acid soils confers a host of benefits and the farmer who wishes to keep his land 'in good heart' must have a sound liming policy. The benefit to crops from the application of calcareous materials to soils in Britain was referred to by Roman writers as early as the first century. Yet grassland surveys invariably indicate that around one field in four has a soil pH below optimum, whether the fields are in short-term, long-term or permanent grassland. The withdrawal in the mid 1970s of the lime subsidy, equal to about a quarter of the cost, was a major reason for the decline in liming in the UK, while the decreasing availability of basic slags, which have 50 to 70 per cent calcium carbonate, was also a contributory factor. The amount of lime currently applied nationally to British soils is below the amount needed to maintain the soils at a satisfactory pH for vigorous grass growth. The adverse effects of soil acidity become progressively evident on individual farms and fields. Most grasses, including perennial ryegrass, thrive in slightly acid conditions so complete neutrality (i.e. a soil pH of 7.0) is unnecessary. A satisfactory target on mineral soils is a pH of 6.0 and, on peats, as low as 5.5.

By countering excessive soil acidity, liming 'sweetens' the soil, thereby improving its physical structure and the chemical and biological conditions for the uptake of soil nutrients by plants. The mineralisation and subsequent availability of nutrients from soil organic matter is stimulated. Acidity can cause deficiency of certain trace elements, e.g. iron, boron, copper and molybdenum. Conversely, a major cause of injury to plant growth is the increased availability of aluminium and manganese as soil acidity increases (see Figure 12.1). The crumb structure and friability of heavy textured soils is improved

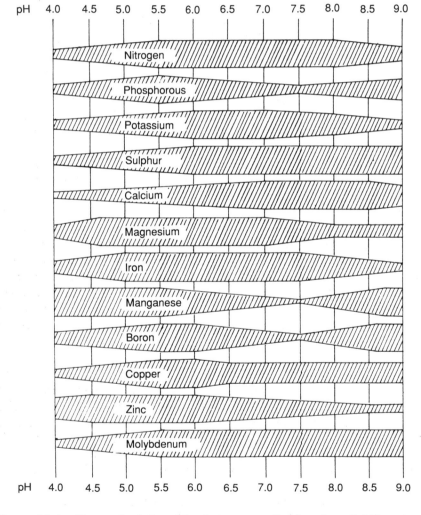

Figure 12.1  *General relationships between soil pH and availability of individual plant nutrients*
Source: Mengel and Kirkby

by liming, and this makes them easier to cultivate for grass production and utilisation. As well as its soil amelioration effects, lime supplies calcium (Ca) for the growth and multiplication of plant cells. Deficiency is most noticeable in young plants and leaves, which lack vigour and are stunted. Grasses contain 4 to 12 g/kg Ca in the DM. Forage legumes are particularly sensitive to calcium deficiency and will not thrive, and lack of calcium adversely affects their legume N-fixation process.

Heavy textured clay soils require more lime to reduce acidity than sandy soils, or to raise soil pH, since their higher clay and organic matter contents act as a buffer against change. Peaty soils are also highly buffered. Overliming by a single large application should be avoided since it can reduce the availability to plants of certain essential elements such as phosphorus, manganese and boron (see Figure 12.1). Apart from affecting plant growth adversely, the health of grazing animals may be put at risk because of trace element deficiency. However, overliming is much less common than underliming or lack of liming. Liming has little visual impact on sward growth so it may be tempting to neglect it in favour of the quick benefit, both visual and in terms of sward production, from fertiliser N; however, this benefit will decline as soil pH falls below 5.7 on mineral soils.

An application of lime should not exceed the equivalent of 7 to 8 t/ha of ground limestone (calcium carbonate). A programme of periodic liming spanning several years is recommended on very acid soils, some of which, with a pH of 4.0 or so, may indicate a requirement of 20 t/ha of ground limestone. On acid soils, the grass and weed species tolerant of sour soils attain dominance. Examples are Yorkshire fog, red fescue and bent grasses on lowland grassland or better hill land, and purple moor grass and sheep's fescue on poorer hill land. Weeds such as sorrel and spurrey are also indicators of acidity.

Many acid-tolerant grasses are not highly acceptable to grazing stock, nor do they have a high nutritive value, so herbage intake and animal output are adversely affected. The preferred better quality grasses such as perennial ryegrass are absent or present in limited amounts, and of stunted growth and low productivity. White clover, if present, will be sparse and dwarfed, neither contributing much to sward yield nor actively fixing nitrogen. Its root growth will also be restricted.

Where there is a history of soil acidity, earthworm activity and the rate of microbial breakdown of senescent leaves and roots is reduced and a typical mat of organic matter comprising dead and decaying plant material accumulates. This plant debris locks up plant nutrients, which under less acid conditions would be released and recycled by earthworms and microbial action. The mat itself inhibits light penetration to the base of the sward and hence the stimulation of grass tiller growing points is reduced. It is also an inhibitory environment for the germination and development of seeds introduced by oversowing techniques such as direct drilling. Clear-cut symptoms of soil acidity are easy to spot including mat formation or indicator plants which show that the sward would benefit from a dressing of lime as the first stage in its improvement. It is less easy in the absence of confirmatory soil analysis to diagnose 'subclinical' acidity since there may be a reasonable proportion of better grass species; however, they will not be performing at full potential.

As well as the parent material and its history of soil formation, the amount of rainfall is a major factor influencing soil acidity. Rain leaches calcium out of the soil into the drainage water by dissolved carbon dioxide, with losses greatest in soils with a high pH. Air pollution by sulphur dioxide and nitrogen oxides in heavily industrialised areas makes the rain weakly acidic and increases its leaching effects. Removal of silage or hay crops represents net loss of the calcium taken up by the plants, but the amounts are small in relation to other calcium depleting factors. Two cuts of silage amounting to 10 t DM/ha would remove 40 to 120 kg/ha Ca, for instance, equivalent to 100 to 300 kg/ha calcium carbonate.

Most modern nitrogenous and compound fertilisers with high concentrations of nutrients have little or no calcium in them to counteract their acidifying effects. Ammonium nitrate fertiliser with 33 to 35 per cent N and urea with 46 per cent N have strongly acidifying effects, especially in the surface horizons of the soil. Surface acidity is one of the causes of loose anchorage and stunted development of grass roots which can result in sod pulling during grazing and also in susceptibility to drought. The acidity progressively increases down the soil layers unless remedial action is taken. Once subsoils become acid their pH can be difficult to correct unless liming and deep cultivations are carried out, since liming materials applied to the sward surface are slow to percolate downwards. In the past commonly used ammonium nitrate—calcium carbonate mixtures containing 15 to 21 per cent N did supply calcium to offset the leaching effects of soil calcium lost as calcium nitrate. Approximately 200 to 300 kg/ha of ground limestone is required to neutralise the acidifying effect of 100 kg N/ha in the form of ammonium nitrate or urea fertilisers. Potash fertilisers also have an acidifying effect and 100 kg/ha of ground limestone is needed to offset the acidifying effect of 100 kg/ha $K_2O$. Cattle slurries rich in ammonium nitrogen and potash also contribute towards acidification.

The moral is that an *ad hoc* liming policy is not sufficient for modern day fertiliser use. In intensively used grassland, moderately to heavily fertilised with nitrogen, the amount and frequency of application of liming materials needed is greater than before, but precision is required as well. Input–output budgets can be used to construct a balance sheet on individual farms and fields. However, the foundation upon which planned, cost effective programmes of liming and fertilisation are built for grassland is routine soil analyses at least every three or four years.

Liming materials are sold on the basis of neutralising value (NV) and fineness of grinding. The NV is a measure of effectiveness relative to calcium oxide (CaO). The particles must be ground to a specific degree of fineness, with at least 40 per cent able to pass through a 100-mesh sieve (0.15 mm apertures). Ground limestone and ground chalk have an NV of about 50 per cent. Magnesium limestones, which have an NV of 50 to 55 per cent, are as effective as ground limestone in their action but are slightly slower acting. When other liming materials are purchased, their NV should be checked against that of ground limestone so that their rates of application can be adjusted.

## SOIL SAMPLING AND ANALYSIS

Laboratory tests for soil pH and the status of other soil nutrients are very accurate, but clearly the soil samples analysed must be representative of the fields sampled. If the previous agricultural history within parts of a field has differed, these parts must be separately sampled. Similarly, if different soil types occur within a field these should also be separately sampled. Sampling is best done by a skilled operator but it is possible on a 'do-it-yourself' basis if the principles of representative sampling are appreciated. Screw augers are not suitable for grassland; cylindrical corers or cheese-coring types should be used. The sampler should take 20 to 30 cores per area being sampled, using a 'W' or zigzag pattern. About half a litre of soil is required from mineral soil types and twice that from soils high in organic matter. Established grassland is sampled to a depth of 10 cm and this depth should include any layer of surface mat present. If the land is to be ploughed for reseeding it should be sampled to plough depth, 15 to 18 cm. If grassland is to be renovated by direct drilling, it is advisable to do an additional check on the soil pH of the top 2 to 3 cm, which may be very acidic if there has been a history of high fertiliser nitrogen usage.

Most soil sampling and soil analyses are carried out by advisory services. Provided they are given detailed information, especially on previous and future usage policies, and locations, so that soil type can be identified from soil classification maps, detailed forward liming and fertilisation programmes can be drawn up. Lime, phosphorus, potassium and magnesium status are normally provided from routine soil analyses. Additional nutrient or trace element analyses can be carried out if requested, or if required to solve a particular soil or sward problem. The SAC and ADAS classifications of soil analysis results are shown in Appendix 2.

# PHOSPHORUS

The value of applying phosphate fertilisers to grassland has long been known, yet surveys of fertiliser use on UK grassland indicate phosphate usage has remained low and relatively static. In recent years, the average amount applied annually to lowland grassland is reported to be 20 to 25 kg/ha $P_2O_5$. If the principle of balanced fertilisation is accepted as a means of optimising herbage production, then clearly phosphate usage is insufficient because analyses during soil fertility surveys have indicated that up to a third of grassland fields have soils deficient in phosphorous (P) status. Fields from which grass is cut and removed for conservation are the most prone to phosphorus deficiency. Hills and uplands rarely have satisfactory soil P levels.

Phosphorus is actively involved with a plant's metabolism and new growth, and it also plays a part in the plant's maturity process, although its content in the plant declines with maturity. Its role in aiding root and seedling development is widely known. When deficient, plant growth is weak during establishment, subsequent leaf size is reduced and hence grass production is lowered. Deficiency symptoms in grass are bluish-green leaves and purpling of young leaves. Concentrations of P in the herbage are normally in the 2 to 4 g/kg DM range. Thus P offtake is 2 to 4 kg/t DM. Two cuts of silage amounting to 10 t/ha DM will remove 20 to 40 kg/ha P (equivalent to 45 to 90 kg/ha $P_2O_5$). Table 12.1 shows the adverse effect on herbage production of limiting phosphate application.

Phosphorus is progressively fixed by soil minerals after application. Subsequently the crop relies on the slow release of phosphorus from the mineral compounds in the soil solution or the organic matter. Clay soils have a high rate of fixation on account of their high humus content and large surface area of soil particles. This means that clay soils may have large reserves of phosphorus yet its availability to plants may be limited. Conversely, availability can be high from limited supplies of phosphorus in sandy soils. Loam soils would have

**Table 12.1** **Relative herbage production from different rates of phosphate application averaged over equivalent potash rates: moderate soil P status at start of trial (production at highest $P_2O_5$ rate = 100)**

| Management | Harvest year | Applied $P_2O_5$ (kg/ha) | | | | |
|---|---|---|---|---|---|---|
| | | 0 | 48 | 96 | 192 | 384 |
| 3 cuts of silage annually using 125 kg/ha N per cut | 1 | 98 | 100 | 94 | 97 | 100 |
| | 2 | 95 | 99 | 98 | 93 | 100 |
| | 3 | 88 | 92 | 92 | 96 | 100 |

*Source:* Hunt

intermediate availability. Some leaching of phosphorus can occur from light sandy soils or from peats.

The cycling of nutrients, including phosphorus from excretal return on grazed swards, means that only a maintenance amount of phosphate needs to applied annually, e.g. 60 kg/ha $P_2O_5$ on a soil of moderate P status (ADAS index 2). Recommended phosphate rates in Scotland for different cutting systems are shown in Table 12.2.

## Phosphate and early grass

The response of grass swards to spring-applied fertiliser N is well known. It is less well known that on soils with very low soil P status,

**Table 12.2** **Fertiliser phosphate and potash recommendations for conservation**

| Management | Soil P status | | | Soil K status | | |
|---|---|---|---|---|---|---|
| | L | M | H | L | M | H |
| | $P_2O_5$ (kg/ha) | | | $K_2O$ (kg/ha) | | |
| *Silage: 2 or 3 cuts* | | | | | | |
| 1st cut | 100 | 60 | 20 | 130 | 90 | 50 |
| 2nd cut | 30 | 30 | 30 | 60 | 60 | 60 |
| 3rd cut or aftermath | 25 | 25 | 25 | 50 | 50 | 20 |
| For a 2nd grazing | 20 | 20 | 20 | 20 | 20 | 20 |
| *Hay: 1 or 2 cuts* | | | | | | |
| 1st cut | 80 | 40 | 0 | 100 | 60 | 20 |
| 2nd cut or aftermath | 20 | 20 | 20 | 40 | 40 | 40 |
| For a 2nd grazing | 20 | 20 | 20 | 20 | 20 | 20 |

*Source:* Younie *et al.*/Scottish Agricultural Colleges

the addition of phosphate can enhance the N effect further. In recent trials in Scotland spring herbage DM was increased by 29 per cent on soils very low in P status (ADAS index) relative to the amount produced with no added phosphate. A further 23 per cent increase was obtained when the phosphate and nitrogen were applied as an ammonium phosphate compound (26 per cent N:13 per cent $P_2O_5$) rather than separately as ammonium nitrate and triple superphosphate; the rates of application of nitrogen (90 kg/ha) and phosphate (45 kg/ha) were similar in both treatments. The uptake of P and its concentration in the herbage was greater too and this improved the mineral value of the herbage. Smaller but useful herbage production increases of 11 per cent on average were obtained from the same treatments on soils with a low P status. Implementing these findings on hill and upland soils, which are commonly deficient in phosphorus, would be particularly useful. However, many lowland grass fields are P deficient too and spring production could also be enhanced by phosphate application in early spring. Proprietary nitrogen:phosphate compound fertilisers are now available.

There are several reasons for the benefits. The release of P from the soil reserves in which it is fixed is slow at low temperatures, so the applied water-soluble phosphate enhances P availability to the plants. Active grass root extension to seek soil nutrients in general is encouraged by both the phosphate and the nitrogen. This is very important for the uptake of phosphorus since soil P is not a mobile nutrient like potassium, for example. The extra advantage from the nitrogen:phosphate compound is attributable to the closer proximity of both nutrients in the root zone than when they are applied separately and have different rates of movement from the soil surface.

## Types of phosphate fertiliser

Although most of the phosphate applied in the UK is by way of nitrogen:phosphate:potash compound fertilisers, a number of straight phosphate fertilisers are available. A simple classification is (a) water soluble and (b) water insoluble. The main straight water-soluble form now available is granular triple superphosphate (45 to 47 per cent $P_2O_5$) which largely superseded single superphosphate. Triple superphosphate also contains 16 to 21 per cent Ca. Water-soluble forms of phosphate contain phosphorus which is readily available to the growing crops immediately after application. The phosphate in compound fertilisers is also water soluble. Its source is ammonium phosphate, rich in nitrogen as well as phosphate. Because water-soluble phosphate can be fixed in the soil its residual effects are similar to those from water-insoluble types.

The main water-insoluble form is ground mineral phosphate (GMP). Its $P_2O_5$ content can vary considerably but will normally be 30 per cent or more. The best GMP comes from the soft mineral rock phosphates (mineral apatites) in North Africa. They are ground to a fine powder to increase phosphate availability to plants and then sold as a powder or in granular form. For effectiveness, the fineness of grinding should be such that 85 to 90 per cent can pass through a 100-mesh sieve with 0.15 mm apertures. GMP from hard rock phosphates is less effective. The best guide to quality is the stated total $P_2O_5$ content, the per cent of the total soluble in mineral acids and the per cent soluble in 2 per cent formic acid. High values for the latter test indicate a good degree of softness and also effectiveness for uptake by plant roots. GMP is a slower acting and longer term source than water-soluble types. Its effectiveness is greatest when used to maintain soil P status on grassland soils with a pH less than 6.0, and under high rainfall conditions—over 800 mm per annum—typical of hill and upland grazings GMP should not be used within six months of liming for reseeding, nor is it suitable for alkaline or chalk soils.

Many farmers remember basic slag with affection. It was a principal source of phosphate for grassland and additionally had both liming value and a significant content of some trace elements. However, with the demise of the older steel making processes, from which basic slag was a by-product, only limited imported supplies are now available. A good quality basic slag will contain 16 to 20 per cent $P_2O_5$, with 80 per cent or more being soluble in 2 per cent citric acid.

Other slow-acting phosphate fertilisers include aluminium calcium phosphate, 32 per cent $P_2O_5$, from West Africa, and mixtures of basic slag and GMP, so-called phosphatic slags. Mixtures of water-soluble and water-insoluble phosphates are also available; their value should be ascertained by checking the total phosphate content, the water-soluble fraction, and their fertiliser unit costs compared with other phosphate fertilisers.

## POTASSIUM

The use of potash on British grassland has remained fairly static in recent years, with an average annual usage on lowland grassland of 30 to 35 kg/ha $K_2O$. Potassium (K) is an essential element, needed to complement nitrogen if high grass production is to be achieved since the amount removed in grass growth is second only to nitrogen. Surveys of soil fertility have shown that about a third of grass fields, especially fields utilised for conservation, had inadequate soil K status.

Potassium is concentrated in the cell sap of the plant where it regulates cell water content and thus controls the loss of water by transpiration through the leaf pores (stomata). It is therefore important in maintaining cell turgor and strengthening the structure of the foliage. Potassium also plays a role in sugar formulation, assists root development and is closely involved in the transport of metabolites and nutrients within the plant. Its regulatory function has earned it the nickname of the 'traffic policeman'.

When the soil is potassium deficient even grass well fertilised with nitrogen will lack growth vigour, so production suffers. Older leaves turn light green as the potassium moves to younger growth and some scorching at the leaf tips and edges will be evident. Root development is restricted and this, together with less efficient control of leaf transpiration, results in wilting during drought. Potassium deficient grass also lacks winter hardiness. White clover is very sensitive to potassium deficiency; browning appears on its leaf margins, and the clover may eventually disappear from the sward. The concentration of potassium in grass DM can range from 25 to 40 g/kg. Thus potassium offtake is 15 to 40 kg/t DM, with 10 t/ha of silage DM thus removing 150 to 400 kg/ha K (equivalent to 180 to 480 kg/ha $K_2O$). Table 12.3 shows the serious adverse effect on herbage production from limiting potash application on intensively managed grass during a trial at Ayr.

Potassium is highly mobile in the soil solution and is readily taken up by plant roots. Heavy textured soils in which potassium is held on clay minerals normally have greater reserves and availability than loams and light sandy soils. The latter are therefore more prone to potassium deficiency and leaching. Calcareous soils are usually potas-

**Table 12.3**  Relative herbage production from different rates of potash application averaged over equivalent phosphate rates: moderate soil K status at start of trial (production at highest $K_2O$ rate = 100)

| Management | Harvest year | Applied $K_2O$ (kg/ha) | | | | |
|---|---|---|---|---|---|---|
| | | 0 | 48 | 96 | 192 | 384 |
| 3 cuts of silage annually using 125 kg/ha N per cut | 1 | 90 | 92 | 93 | 98 | 100 |
| | 2 | 66 | 82 | 91 | 97 | 100 |
| | 3 | 61 | 80 | 90 | 97 | 100 |

*Source:* Hunt

sium deficient. Peats can be high in total potassium content but availability to plant growth is low. When potassium in the soil is available in excess of grass requirements so-called 'luxury' uptake occurs. This may deplete the potassium available for subsequent grass growth if grass is removed in a silage cut. If there is surplus soil potash, luxury uptake will be greater with high than with low rates of N application.

As with phosphorus, potassium is returned to the grazed sward from animal excreta, so only maintenance dressings of potash are required annually, e.g. 60 kg/ha $K_2O$ on a soil of moderate soil P status (ADAS index 2). Recommended rates for cut swards under Scottish conditions are shown in Table 12.2.

Most of the potash applied to grassland is in the form of 'NPK' fertiliser compounds. Nitrogen:potash compounds are also available. The main potassium constituent of compounds is water-soluble potassium chloride, which is mined in several countries. It is also available as a straight fertiliser, muriate of potash, with 60 per cent $K_2O$. The latter can scorch grass if applied during dry weather. More expensive manufactured potassium sulphate, also water-soluble but with 50 per cent $K_2O$, is used on high value arable crops rather than grassland.

## SULPHUR

It is generally forgotten that grass growth requires sulphur (S) in similar amounts to phosphorus. Sulphur is part of the essential amino acids, particularly cysteine and methionine, needed to form plant protein. In the past, sulphur was present in several widely used fertilisers, although they were not applied specifically to correct sulphur deficiency. Ammonium sulphate contained 24 per cent S, for example, and single superphosphate 12 per cent S. These fertilisers have been replaced with forms such as ammonium nitrate and triple superphosphate which contain greater concentrations of nitrogen and phosphate respectively, but little or no sulphur.

The amounts of wet and dry sulphur deposited on the land from atmospheric pollution have declined due to government Clean Air Acts being increasingly implemented by industry; this trend is likely to continue. Nevertheless, land near industrial areas burning fossil fuels such as coal or oil, and with high rainfall, may still receive significant amounts by deposition (30 to 40 kg/ha S per year), although these are irregular in occurrence. Sulphur deposition can also come from distant industrial areas or even distant countries, depending upon the prevailing wind. Coastal areas receive limited amounts of sulphur from sea spray. On the other hand, low rainfall areas, land in the lee of prevailing winds and rural areas may receive only 5 to 10 kg/ha S per year.

Soil reserves of sulphur are present as available sulphate and as organic sulphur, some of which can be mineralised and made available slowly. Around 0.5 per cent of the total soil sulphur can be mineralised during the grass growing season. Apart from soil analysis of available sulphate-S as an indicator of soil sulphur status, the total N:total S ratio in grass can be used to predict potential yield response to applied sulphur. Response is likely at N:S ratios over 13:1. An average concentration of sulphur in grass ranges from 2 to 4 g/kg DM. Light textured sandy soils and chalk soils have the lowest sulphur reserves and heavy textured clay soils the highest, loams being intermediate. However, reserves may be exhausted by intensive silage cropping, even on clay soils, so it is becoming increasingly necessary to augment soil and atmospheric sulphur if satisfactory levels of herbage production are to be sustained.

Apart from reduced growth rate and production, S deficiency symptoms in grass include yellowing and stunting of the grass leaf blades. Nitrogen fixation is reduced in forage legumes. Multi-cut silage systems involving high fertiliser nitrogen inputs are at greatest risk from S deficiency since the sulphur offtake is 2 to 4 kg/t DM. Thus two cuts of silage amounting to 10 t/ha DM would remove 20 to 40 kg/ha S per year.

Not surprisingly, recent trials in various parts of the UK, where soils were classed as potentially sulphur deficient using the criteria above, have demonstrated herbage DM response to applied sulphur. An increase in herbage output of 10 to 30 per cent was obtained, mainly at second and/or third silage cuts, from S application compared with no applied S. Available sulphur in the soil, derived from atmospheric deposition and mineralisation from soil organic matter, was sufficient to meet the sulphur demand of first cuts. So no response to applied S was obtained then. The sulphur used in the trials was either in the form of gypsum (calcium sulphate) containing 18 per cent S in the sulphate form required by plant growth, or elemental sulphur marketed as a liquid or powder with 80 per cent or more S. Sulphur can be absorbed directly by grass leaves but the main uptake is by roots following oxidation of the sulphur to the sulphate form by soil bacteria. Some sulphate may be lost by leaching, especially on light sandy soils. The main atmospheric deposition is through sulphur dioxide gas, most of which is also converted to sulphate before plant uptake.

Following these trials, compound fertilisers enriched by declared

amounts of sulphur have been marketed. Standard compound fertilisers often contain some sulphur but the amounts are variable and
cannot be relied upon for sward sulphur needs. Other sulphate containing fertilisers, although not primarily used to supply sulphur, include
potassium sulphate and the magnesium sulphates, kieserite and Epsom
salt, containing 20, 23 and 13 per cent S respectively. Animal manures
are another significant though variable source of sulphur. Animal
slurry at 1:1 dilution with water contains 0.2 to 0.6 kg of sulphur per
cubic metre and farmyard manure 1 to 3 kg/t. Land which regularly
receives dressings of animal manure is unlikely to be sulphur deficient,
and neither is grazed grassland, because much of the sulphur in
ingested grass is returned as excreta.

In practice a single dressing of sulphur of 10 to 20 kg/ha in its
cheapest form, gypsum, in spring or after first-cut silage, will be
effective on soils of moderate S status and 20 to 30 kg/ha where there
is a more severe deficiency. A value of 8 mg/l available S or below
would be classed as very low status in the soil. If sulphur is applied to
swards intended for grazing, sufficient time should be allowed for the
sulphur to be washed off the leaves by rainfall or diluted by new leaf
growth since excessive sulphur ingestion can harm stock. Caution is
also necessary when applying sulphur to copper deficient soils since
increased S content of the herbage can adversely affect copper metabolism in the animals, with sheep especially affected.


## MAGNESIUM

Compound fertilisers containing magnesium (Mg) are not regularly
used on grassland in Britain, in contrast to some countries such as The
Netherlands. The role of magnesium is appreciated more in animal
nutrition than in plant growth, since hypomagnesaemia (grass tetany
or staggers) in stock is a high risk if the grass has a low Mg content.
The availability to stock of Mg in ingested grass is reduced if the grass
has high levels of nitrogen and potassium.

Magnesium, being an integral component of chlorophyll, is inti-
mately involved with photosynthesis in the plant. It also has a role in
various enzymes which act as catalysts in plant growth processes.
Magnesium is highly mobile and the leaf yellowing symptom typical
of deficiency appears on the older leaves first, in a way similar to
potassium deficiency. The concentrations of magnesium in grass DM
can range from 1.2 to 2.4 g/kg, giving offtakes of 1.2 to 2.4 kg/t DM
or 12 to 24 kg in 10 t/ha of silage DM (equivalent to 20 to 40 kg/ha
MgO).

High levels of available K in the soil, or from potash application, reduce magnesium uptake by the plants due to preferential uptake of potassium. This antagonism between potassium and magnesium is particularly important under grazing because of the risk to breeding stock of hypomagnesaemia. The risk is greatest at those times of the year, such as spring or autumn, when there may be a shortage of grass allied to weather or other stresses. Application of potash to grazed fields should be avoided at these times. Some progress has been made by plant breeders in the development of ryegrass varieties with an inherently higher magnesium content, but the problem remains of ensuring adequate levels of herbage intake during grazing.

Magnesium is held on the clay minerals in the soil, as are the other bases, calcium and potassium, so magnesium reserves and availability are greatest in heavy and medium textured soils and lowest on sandy soils. There is release of magnesium from the clay minerals to the soil solution by natural weathering, while some 5 to 10 kg per annum may be deposited by rainfall. Unlike potassium, but similar to calcium, magnesium is readily leached, especially on light soils. Magnesian limestone, which contains magnesium carbonate and calcium carbonate, is the most common and cheapest material used routinely to supply magnesium as well as to increase soil pH. Contents of magnesium are up to 12 per cent (20 per cent MgO) but should have a minimum of 9 per cent Mg (15 per cent MgO) to be regarded as a high quality magnesian limestone. Calcined magnesite with 50 to 55 per cent Mg is the most concentrated magnesium fertiliser. A typical dressing would be 75 to 100 kg/ha. Its application can speedily increase soil magnesium and plant uptake so it is used for quick effect and as a preventative against grass tetany on grazed swards. Water-soluble Epsom salt (10 per cent Mg) can also be used as a quick acting source; calcined magnesite or Epsom salt rather than magnesium limestone should be used on calcareous soils where soil pH is high.

## SODIUM

Sodium (Na) is not essential for grass growth but is nevertheless taken up by grass from the soil solution. In the plant it functions in a similar manner to potassium, in a regulatory role of cell turgidity and transpiration rate. Heavy textured soils have a greater supplying capacity than lighter soils. The sodium is held on the clay minerals but is easily leached from the soil. There is antagonism between potassium and sodium so sodium content in grass is depressed when soil potassium is at high levels. Rainfall can supply substantial sodium inputs, especially near the sea, where deposition can be 50 kg/ha Na or more.

The main sodium fertiliser is agricultural salt (sodium chloride), which has 37 per cent Na. Salt is relatively cheap and could replace part of the potassium in compound fertilisers. Sodium has long been recognised as essential to animal well-being. Salt is used in various ways, such as salt blocks or in water supply treatment, to provide dietary sodium directly to stock. Agricultural salt is sometimes applied to pastures, partly to increase herbage sodium and hence intake by animals, but also since it is believed to increase herbage acceptability. Salt application to swards is an inefficient way of increasing sodium in animal diets compared with direct supplementation.

Interestingly, grasses and legumes can be divided into sodium-loving (natrophile) and sodium-hating (natrophobe) classes. The former include perennial ryegrass, cocksfoot, Yorkshire fog and white clover; after absorption, sodium is translocated to their leaves. Natrophobes include timothy, tall fescue, red clover and lucerne. In these plants sodium is absorbed only slowly; most of it is not transported to the leaves but remains in the roots and basal parts of plants. Sodium concentrations in grass DM can range from 0.5 to 50 g/kg, there being severalfold differences in sodium content between sodium-loving and sodium-hating species. Thus sodium offtake can range from 0.5 to 5 kg/t DM and 10 t/ha of silage DM would remove 5 to 50 kg/ha Na.

## TRACE ELEMENTS

The trace elements essential to grass growth are iron (Fe), manganese (Mn), boron (B), copper (Cu), zinc (Zn), molybdenum (Mo) and chlorine (Cl). Cobalt (Co) is also needed by legumes for their symbiotic N-fixation. Most of these trace elements are also required for animal nutrition, as are iodine (I) and selenium (Se) which are absorbed passively by plants but are not needed for grass nutrition.

The trace element concentrations in grasses and legumes are influenced by many factors, including soil type, plant species, stage of growth, season, and the effects of other applied major nutrients. Trace element supply is augmented to a greater or less degree from rainfall, liming materials, fertilisers and organic manures. Soil contamination of grazed herbage and ingested soil provide some trace elements to stock when grazing is severe and sward height low.

Many soils in the UK have adequate contents of the trace elements required for grass and legume growth, since the amounts removed are small. However, leached sandy soils and soils with low reserves of organic matter have inherently low contents. Availability in soils is often lowered by liming and the consequent rise in soil pH, although molybdenum is an exception since its availability is increased by liming (see Figure 12.1). Knowledge of soil type and soil analyses are not always good predictors of trace element supply because of the complexity of interacting factors which influence it. Herbage analyses can usefully provide added information, especially when grown on soils with high contents of organic matter. The interpretation of soil levels of cobalt and their relationship to plant uptake and livestock nutrition is a good example. Account has to be taken of the soil pH, the soil drainage class (freely, imperfectly or poorly drained) and the organic matter content of the soil. Cobalt availability is lowered as soil drainage and soil pH are improved but rises with increasing soil organic matter.

Cobalt and copper are the trace elements most commonly applied to some types of grassland to boost soil and plant levels. It is sufficient to apply 2 to 3 kg/ha cobalt sulphate ($CoSO_4$) as a low volume spray or as a cobaltised fertiliser, and this is best applied in early spring before grass growth starts. Alternatively, the same quantity can be applied to a quarter of a field on a rotational basis. Provided lime has not been applied recently and the pH is below 6.0, either treatment lasts for three to four years. Copper sulphate $CuSO_4$) at 20 kg/ha can be applied to copper deficient soils and will be effective for four to eight years. Grazing stock should be kept off treated fields until the copper is completely washed off the leaves by rainfall, otherwise copper poisoning is possible; sheep are more susceptible than cattle.

When applying trace elements to overcome deficiencies it is important not to apply excess since toxity to plants or animals may occur. This excess can be harder to remedy than a deficiency. The specific element deficiency should be treated, so trace element 'cocktails' are unnecessary, and expensive. Treatment of a deficiency in animals may be more effectively remedied by direct animal mineral supplementation, e.g. copper needles. Application to the sward may be warranted, in addition to direct animal treatment, even though the percentage recovery of the applied element by the plants may be low. Care is needed to ensure uniform application because of the small quantities needed. It is best to seek specialist advice on any suspected trace element problems before taking corrective steps. Noteworthy from national surveys of fertiliser use is that only 1 to 2 per cent of grassland is given fertilisers other than lime and the 'big three', nitrogen, phosphate and potash. Considering the potential need for sulphur, magnesium or other elements in some situations, this current trend bodes ill for future grass productivity unless rectified.

## BLENDED SOLID FERTILISERS

Compound fertilisers can be blended from individual nutrients to form various nitrogen:phosphate:potash ratios which may not be on offer with conventional compound fertilisers. Blends can thus be tailormade to individual sward requirements. Fertiliser granules of the individual nutrients must be similar in size and density; otherwise there will be separation of the granules during spreading and the individual nutrients will not be distributed uniformly.

## LIQUID FERTILISERS

The choice between using solid or liquid fertilisers on grassland is a matter of personal preference and costs rather than technical efficiencies, since these are similar. Those based on urea, however, will be less effective in dry conditions, while aqueous ammonia (28 per cent N), which requires soil injection, is not suitable for stony soils. Liquid fertilisers, which comprise solid fertilisers in solution, are more easily applied but storage is more expensive and the concentration of nutrients is lower. While this may be regarded as a disadvantage, the uniform spreading of concentrated solid fertilisers can be a problem,especially when low rates are applied. Liquid fertilisers are surface applied to grassland, except for aqueous ammonia.

*Chapter 13*

# Using Organic Manures

---

Work returns to the husbandmen moving round in a circle, and the
year rolls itself round in its former track.

Virgil, *Georgics* 4, 212

---

Historically, in Europe, a high value and reliance were placed on
animal manures as plant nutrient sources. In western and northern
Europe increased densities of livestock and changes in winter housing
practices over the last thirty years have resulted in the accumulation of
large quantities of animal manures, stored mainly as slurry. Land
suitable for spreading slurry was not always close to where it was
produced, so it came to be seen as a waste product with a disposal
problem rather than as a useful resource by many farmers. This
view was strengthened by the availability and relative cheapness of
inorganic fertilisers, and their acceptance by farmers.

These attitudes towards animal slurry and farmyard manure have
changed in recent years, albeit tardily. There is more emphasis now on
lowering farm input costs and one way is using the nutrients in
slurry and farmyard manure more positively and rationally to replace
bought-in mineral nutrients. There are still economic constraints in
storing slurry, and problems of controlling the uniformity of nutrient
quality when spreading and of evenness of application.

The effects of animal manures, especially slurry, on the environment
need to be increasingly considered on both aesthetic and legislative
grounds. The main environmental problems are pollution of surface
and ground water by run-off or leaching of nutrients, nitrogen in
particular, ammonia volatilisation to the atmosphere and the release of
odours.

## ANIMAL SLURRY

Typical farm slurry is a mixture of livestock urine and faeces (excreta),
usually diluted with rainwater and also cleansing waters; fragments of
bedding material and waste feed may also be present. On some farms,

silage effluent is added to the slurry. As a farming topic, slurry handling and use generates more heated discussion than most, and slurry itself is regarded as a mixed blessing rather than an asset. It is not easy to attribute general nutrient values because of the variable dilution of slurry from farm to farm. The best way of estimating its value is from knowledge of the volume produced on a farm and the available nutrients known to be present in undiluted slurry.

## Volume produced

The volume of slurry produced by livestock depends on several factors including the type of animal, its size, age, breed, diet and liquid intake (see Table 13.1 for examples).

**Table 13.1   Volume of undiluted slurry from livestock**

|  | *Volume (m³) produced in:* | | |
| --- | --- | --- | --- |
| *No./type of livestock* | *1 day* | *4 weeks* | *26 weeks* |
| 100 dairy cows | 4.3 | 120 | 780 |
| 100 young cattle (250 kg live weight) | 1.0 | 28 | 180 |
| 100 fattening cattle (400 kg live weight) | 2.9 | 80 | 520 |
| 100 fattening pigs (70 kg live weight) | 0.4 | 12 | 80 |

Dilution of slurry should be minimised to that needed for efficient storage, handling and spreading onto the land. A 1:1 ratio of slurry to water is usually satisfactory. Excessive dilution simply means more storage capacity is required, the nutrient status per unit volume is reduced and more time and journeys are needed for disposal. The addition of water should thus be controlled as far as possible. Rainfall on the roofs and concrete areas of the buildings should be guided to ditches or soakaways, provided it remains clean. A winter rainfall of 500 mm on a 0.5 ha farmstead is equivalent to 2,500 m³ of water, which equals the volume of undiluted slurry produced by 30 dairy cows during winter. When calculating slurry storage capacity, a safety margin should always be built in; underestimating, with its cost-saving attraction, should be resisted.

## Nutrients in slurry

The plant nutrients in slurry depend upon the type of livestock, the rations fed and the storage conditions, but the amounts of nutrients can be substantial. Nitrogen, phosphorus and potassium are the main nutrients present but lesser amounts of calcium, magnesium, sulphur and trace elements have considerable value for plant growth too, a value which is often overlooked. Our farming forefathers knew the value of the urine fraction, which they used to collect and spread. Being rich in readily available nitrogen and potassium, it is worthwhile collecting it efficiently. The organic matter in slurry, mainly from the dung fraction, is also a useful addition to the soil, and this fraction is rich in phosphorus, calcium, magnesium, sulphur and trace elements, although microbial action is necessary before most of these nutrients become available to the plant. Table 13.2 gives a guide to the quantity of the main nutrients readily available in livestock slurry based on 50 per cent of the total nitrogen and total phosphate, and 80 per cent of the total potash being immediately available and assuming application in spring. Some of the remaining proportions go into soil reserves and become available after biological processing.

Table 13.2   Readily available nitrogen, phosphate and potash in *undiluted* livestock slurry

| | Nutrients (kg) produced in: | | | | | |
| | 4 weeks | | | 26 weeks | | |
| No./type of livestock | N | $P_2O_5$ | $K_2O$ | N | $P_2O_5$ | $K_2O$ |
|---|---|---|---|---|---|---|
| 100 dairy cows | 280 | 100 | 540 | 1,820 | 650 | 3,510 |
| 100 young cattle (250 kg live weight) | 80 | 20 | 140 | 520 | 130 | 910 |
| 100 fattening cattle (400 kg live weight) | 160 | 60 | 320 | 1,040 | 390 | 2,080 |
| 100 fattening pigs (70 kg live weight) | 50 | 30 | 30 | 325 | 195 | 195 |

Clearly, these quantities of plant nutrients can make a significant contribution to grass production on the farm, but only if slurry is regarded as a source of plant nutrients in its own right, and is used responsibly and not as a waste disposal problem. The monetary value of the main nutrients listed in Table 13.2 (26-week period) using notional fertiliser prices of 36 p/kg for nitrogen, 34 p/kg for phosphate and 19 p/kg for potash is £1,543 for 100 dairy cows, £404 for 100 young cattle, £902 for 100 fattening cattle and £220 for 100 fattening pigs (or £440 for the pigs over a year). Slurry can therefore be used to complement purchased fertilisers in a planned manuring programme, with a considerable saving of money. Knowledge is required of the soil nutrient status, of the slurry—by calculation using Tables 13.2

and 13.3—and of N needs for target levels of herbage production (see Chapter 11). When heavy dressings of pig manure are applied regularly, the soil levels of copper should be monitored; sheep should not be allowed to graze swards recently treated with pig slurry; and undesirable zinc build-up can occur in soils given heavy dressings of poultry manure.

## EFFECTIVE SLURRY APPLICATION

Slurry is usually applied in a diluted form. Table 13.3 shows the nutrient value of differing quantities of diluted dairy cattle slurry. Cash value can be visualised when looking at a purchased bag of fertiliser. It should be visualised when looking at a tanker load of slurry too.

Table 13.3   Available nutrients in various volumes of dairy cattle slurry *diluted in a 1:1 ratio* of slurry to water

| Volume | Nutrients (kg) | | |
| --- | --- | --- | --- |
| | N | $P_2O_5$ | $K_2O$ |
| 1 m$^3$ | 1.2 | 0.4 | 2.3 |
| Tanker, 4 m$^3$ | 5 | 1.6 | 9 |
| Tanker, 8 m$^3$ | 10 | 3 | 18 |
| 50 m$^3$ | 60 | 20 | 115 |

The 50 m$^3$ example in Table 13.3 is listed since this is the maximum volume recommended in official Codes of Practice for application to a hectare of land at one time. At least three weeks should elapse before a repeat dose of slurry in order to avoid sealing of the soil surface. Normally more than one application of 50 m$^3$/ha would be wasteful of the potash fraction since luxury consumption by grass would ensue, increasing the risk of hypomagnesaemia in cattle. Heavy application of slurry nutrients, in excess of those needed to build up soil fertility to a reasonable level or to meet specific grass requirements, must be avoided.

Recent Codes of Practice for water protection require that the maximum nitrogen loading per hectare in a year from animal wastes should not exceed 250 kg/ha of total nitrogen. There are also recommendations for the area of land needed to take the total nitrogen from various types of livestock, e.g. 0.16 ha for a dairy cow or 0.10 ha for a beef bullock when housed for six months.

Table 13.4 shows a typical example of how to adjust slurry and fertilisers when intensively manuring in March for first cut silage. In effect, the use of slurry approximately halves the cost of the potential fertiliser nutrient bill.

**Table 13.4   Example of dairy cattle slurry use in fertiliser programme**

|  | Nutrients (kg/ha) | | |
|---|---|---|---|
|  | N | $P_2O_5$ | $K_2O$ |
| Nutrients needed | 120 | 60 | 90 |
| 40 m³ of 1:1 slurry supplies | 48 | 16 | 92 |
| Additional fertilisers required | 72 | 44 | 0 |

When slurry is applied to grassland during winter some loss of manurial value occurs, particularly of nitrogen. The first potential source of loss is by surface run-off before the nutrients enter the soil. This occurs when the capacity of the soil to absorb the liquid slurry is exceeded, and is most likely in late autumn/early winter when poorly drained soils can be waterlogged. There may also be seepage into the drains. Should the run-off and seepage reach rivers and lochs, the nutrients pollute the water, lowering its quality and amenity value and endangering or killing wildlife. Run-off can also occur on sloping, frozen ground. Nitrogen may also be lost through volatilisation of ammonia gas, which disappears with the atmosphere. In addition, rainfall can leach slurry nutrients, especially nitrogen, into the ground water. The longer the slurry nutrients remain unabsorbed by plant roots, the more vulnerable they are to leaching and other processes, causing them to be wasted.

It has been estimated that about 75 per cent of the nitrogen can be lost following October/November application to grassland, 50 per cent following December/January spreading and 25 per cent after February/March application. For the similar application periods, losses of potash could be 20 per cent, 10 per cent and nil but less on loams and clay soils. Phosphate loss is minimal. Most farms have facilities to store slurry for at least part of the winter, possibly until January, so there is no need to incur the nutrient losses caused by earlier application. Future legislation may also prevent spreading slurry during the winter months, as is already the case in some European countries. Current Codes of Practice in the UK recommend that storage facilities must be big enough to hold four months' slurry.

## Winter versus spring application of slurry

Spreading slurry at intervals during the winter has certain attractions. It can be done with smaller machinery and less agitation equipment, while labour demand is spread over a longer period. There is less chance of undecayed solids being harvested with silage crops since the solids will have decomposed. Slurry contamination of grass on grazing fields and its potential rejection by stock will be reduced. Fresh slurry or slurry stored for short periods has a less offensive smell than slurry that has been stored for a long time; this can be important near urban areas.

Winter storage and application during early spring and summer permit large quantities of slurry to be spread at the time most effective for nutrient uptake by growing grass. However, it requires a greater labour demand over short periods, and larger equipment, unless contractors are used. There would be less risk of soil compaction and sward damage by equipment on heavy soils. There will be no stimulation by slurry nitrogen of winter grass growth, which is susceptible to winter kill, and the frequency of odours will be low. During storage some of the organic nitrogen will be converted to soluble available forms. The carry-over of pathogenic organisms will be reduced. Child safety should always be borne in mind when storing slurry and effective winter storage systems are usually safer than many of the *ad hoc* or temporary open lagoon types typical of systems involving winter slurry applications.

In spite of grants, the economic case for providing storage of all slurry over the winter is undoubtedly weak because the cost of providing adequate failsafe storage facilities is greater than the value of the nutrients saved by avoiding nutrient losses from winter application. Nevertheless, there is a considerable onus on farmers to reduce pollution risks to a minimum, since it is a legal offence to pollute watercourses.

This means maximising winter storage and making the most efficient use of slurry nutrients for plant uptake and production. It also requires adequate storage and efficient application procedures. The annual number of farm pollution incidents in England and Wales caused by slurry is higher than from any other source; in Scotland slurry pollution is second to silage effluent. The causes are mainly due to inadequate storage facilities, bursts and leaks, and faulty operations. Atmospheric pollution (unpleasant smells) is more and more socially unacceptable and complaints from the public often run at a high level.

## Alternative slurry handling

Technical solutions to the various problems are possible, although not always cost effective or necessarily suitable for all farm situations; slurry treatment before spreading can change the nutrient content and also the way the slurry is handled, for example. Liquid and solid fractions can be separated. The liquid fraction is more easily handled and contains a higher nutrient content per volume than the original slurry, but the solids have to be utilised separately. Anaerobic treatment of slurry in store decreases the solids, odours and, importantly, the biochemical oxygen demand (BOD); this is used to show the risk of causing pollution from organic wastes and is a measure (in mg/l) of the amount of oxygen needed by bacteria to break down the organic matter. Animal wastes have a high BOD, so if they pollute watercourses the oxygen is taken out of the water and life forms may be killed. In the soil, aeration is less adversely affected by anaerobically treated than untreated slurry, while grass scorching is also reduced. Aerobic treatment also reduces offensive odours and the BOD.

Injection of slurry into the sward rather than swash plate spreading reduces odour emission and ammonia volatilisation. Acidification of slurry is another means of reducing ammonia loss. However, acidification and injection both result in loss of slurry N by denitrification in the soil. This denitrification—the breakdown of nitrate formed from ammonium N in the slurry by soil bacteria—is restricted by the addition of a nitrification inhibitor, e.g. dicyandiamide, to the slurry. Its addition is particularly effective for autumn and winter slurry application.

Injection is not suitable for stony soils and, if carried out during very dry soil conditions, can induce sward damage near the injection slits. On heavy soils and dense swards, the slitting can cause tearing damage, while tractor wheelslip on sloping fields can cause soil smearing and rutting.

## Legislation

Both government and public are concerned with the environmental aspects of slurry handling throughout Europe, and more regulations are being enacted or prepared. Some European countries already have more stringent controls than in the past, aimed at reducing the risks to the environment from animal wastes. These include limitations to animal stocking rates on land, reducing the periods when slurry can be applied, especially in winter, preventing slurry use on steep or wet land, and compulsory injection of slurry into the sward.

## FARMYARD MANURE

Traditional farmyard manure is derived from straw or other forms of bedding which are used to absorb the dung and urine of housed animals. The nitrogen content of the manure depends largely upon whether or not the nitrogen-rich urine is efficiently trapped, together with the efficiency of storage in relation to avoidance of leaching by rainfall. Urine loss will also reduce the potassium content. Typical analyses of cattle farmyard manures in terms of available nutrients show 1.5 to 2.0 kg/t each for nitrogen and phosphate and 2.0 to 4.0 kg/t for potash. Manure is also a valuable source of organic matter for the soil.

When applied it should be done sufficiently in advance of grazing or cutting for silage so that the utilised grass is not contaminated by the manure. Sometimes it is applied to selected swards in winter to discourage wintering sheep from grazing them. Ploughing in a dressing of farmyard manure prior to reseeding will help to build up soil fertility and there is less loss of nitrogen by volatilisation than when surface applied to grassland. At any one time, dressings should not exceed 50 t/ha.

Various types of poultry manure are also applied to grassland, the main danger being in the use of excessive dressings. It is not always appreciated how rich poultry manure is in comparison with cattle farmyard manure, and its use can lead to scorching of the grass, while the excessive inputs of nitrogen can result in nitrogen loss by leaching as well as volatilisation. Broiler litter is richer in nitrogen than deep litter material, from which ammonia loss takes place. The available nutrient contents of deep litter are in the region of 10 to 12 kg/t of each. The maximum loading of 250 kg/ha total nitrogen in any one year applies to farmyard and poultry manures (and to sewage sludges) as per the Codes of Practice.

## SEWAGE SLUDGE

A large proportion of the sewage sludge produced in the UK is dumped at sea but some is applied to land, primarily grassland in western areas. It is available free from regional Water Authorities. The sludge supplied is untreated, or digested either anaerobically or by long-term storage. Anaerobic digestion increases the ammonia nitrogen content and lowers organic matter; it also reduces disease pathogens, such as those causing salmonellosis or brucellosis, and beef tapeworm (*Taenia saginata*) eggs, if present.

The main nutrient value of sewage sludge lies in the nitrogen and phosphate contents. In the year of application, the available nutrients supplied from 50 $m^3$/ha of liquid digested sludge would be approximately 40 kg/ha N, 20 kg/ha $P_2O_5$ and 5 kg/ha $K_2O$ part of these nutrients immediately available and part available later. Dewatered sludges have lower contents of available nitrogen than liquid sludges owing to loss of soluble nitrogen, but are higher in phosphate and organic matter. Sludges also contain calcium so their application does not lower soil pH. Potentially toxic elements (PTE) such as zinc, copper and nickel which can affect plant growth adversely, or cadmium, lead and molybdenum which can affect animals, are also present in sewage sludge, depending on the degree of contamination. Dried sludges should not be surface applied to grassland because of their high concentration of PTE but can be ploughed in before reseeding.

Rates of sludge application to land should be carefully controlled and monitored. Disposal authorities should provide analyses of sludges, and permanent records should be kept of applications made to individual fields. The levels of potentially toxic elements in the topsoil should be monitored every three to four years. Grass fields with a soil pH less than 5.5 are not suitable for sewage sludge since acidity increases the availability of PTE. The guidelines previously outlined for using animal slurries—concerning application rates, timing, hazards to soil and sward, and reducing risks to the environment—should be followed. However, there are limits to the concentrations of PTE in sludges which are acceptable, their rates of application, and the extent to which they are allowed to build up in the soil. Clearly, specialist advice is required if sewage sludge features strongly in a farm's manuring policy.

Topdressed swards should not be grazed for three to four weeks after application, because of the potential presence of disease organisms, or until at least 10 cm of new growth has taken place. There is less risk with the longer rest intervals typical of silage or hay making. The use of untreated sludges requires a safety policy of not grazing for periods of up to six months. In effect this means cutting for conservation over a season, unless the sludge is applied in winter. Largely on account of the restrictions in its use, its variable nutrient content and its occasionally irregular times of delivery, the use of sewage sludge is most suited to extensive grassland systems. In highly intensive situations, precision of nutrient application and predictability of production, both grass and animal, are major objectives.

## SILAGE EFFLUENT

Silage effluent can be regarded as a dilute manure. Its phosphate and potash content is similar to cattle slurry, but it has less nitrogen (Table 13.5). If applied to grassland in undiluted form its acidity can cause scorching of the grass and its BOD can deplete oxygen in the upper layers of the soil. Both effects depress grass growth. Dilution with water at a 1:1 ratio is advisable and a single application should not exceed 25 m³/ha.

Collected silage effluent is sometimes added to slurry stores for convenience and the mixture applied to grassland. During the addition of large quantities of effluent, and later agitation of the mixture prior to application, toxic hydrogen sulphide gas can be released. At high concentrations, this gas is not readily detected by smell and is lethal to man and beast, so areas close to where such mixing takes place should be well ventilated.

**Table 13.5    Available nutrients in silage effluent diluted in a 1:1 ratio of effluent to water**

| Volume | Nutrients (kg) | | |
|--------|------|------|------|
| | N | $P_2O_5$ | $K_2O$ |
| 1 m³ | 0.6 | 0.3 | 2.2 |
| 25 m³ | 15 | 7.5 | 55 |

# Chapter 14

# Feeding Value of Grass

---

I keep six honest serving-men
(They taught me all I knew);
Their names are What and Why and When
And How and Where and Who.

Rudyard Kipling (1865–1936), 'I keep six honest serving-men'

---

The chemical composition of grassland forage is affected by stage of growth, plant part, season of year, soil pH and soil nutrient status. As discussed in Chapter 12, lime and fertiliser can influence the uptake of nitrogen and major minerals, as can soil type and its inherent fertility, while trace element uptake is influenced by soil pH. Grazing and cutting management have significant effects on chemical composition by controlling stage of growth and proportion of leaf to stem, by altering botanical composition, especially in relation to forage legume content, and by recirculation of plant nutrients.

## CHEMICAL COMPOSITION

The water or moisture content of growing grass, which includes tissue water and external moisture from rain or dew, can vary from 850 g/kg in young leafy herbage to 500 g/kg in grass of advanced maturity; because of this variability, grass production is conventionally expressed in terms of dry matter rather than fresh matter. Most subsequent analyses of chemical composition are related to the dry matter baseline. In experimentation, precise determination of dry matter is achieved by drying samples of herbage in a forced draught oven overnight, the temperature depending on which chemical parameters are to be determined subsequently. This is done without delay after sampling swards in order to minimise respiration losses, particularly of sugars.

The chemical make-up of herbage is shown schematically in Figure 14.1. However, in analyses of chemical composition six main fractions are typically identified: water; ash, i.e. the mineral elements; crude

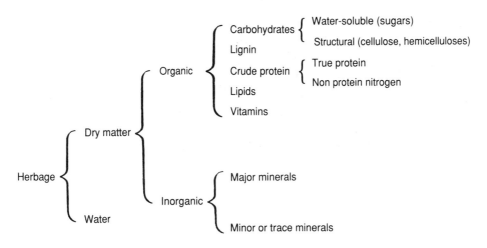

*Figure 14.1   Chemical make-up of herbage*

protein (CP) comprising the nitrogenous compounds; ether extract (EE) made up of lipids, organic acids and most of the vitamins; crude fibre (CF) comprising the least digestible cellulose, hemicellulose and lignin; and nitrogen free extractives (NFE) which is 1,000 minus the sum of the other components when all are expressed in g/kg. NFE comprises the more digestible cellulose and hemicellulose together with other carbohydrates such as sugars. In the past this type af analysis featured strongly when considering animal nutrition needs. While it still plays an important role the modern trend is to focus strongly on the analytical separation of the highly digestible plant cell contents and the less digestible cell wall material.

The cell contents principally comprise water-soluble carbohydrates (WSC) or sugar, protein, lipids, vitamins, organic acids and minerals. Cell walls consist mainly of cellulose and hemicellulose, which are the structural carbohydrates, and lignin; together these are often referred to as fibre or fibrous material. With advancing maturity, usually accompanied by a decrease in the leaf to stem ratio, the proportion of cell wall material increases in relation to cell contents. These changes have important repercussions on the nutritive value of herbage to ruminants, particularly in relation to herbage digestibility and availability of the nutrients, but also to their effect on voluntary intake of herbage by animals.

The dry matter is made up of an organic fraction of 90 to 95 per cent and a mineral fraction of 5 to 10 per cent. Carbohydrates make up the largest proportion of the organic fraction; their total content in the herbage DM can vary from 400 to 800 g/kg depending upon stage of maturity. They are the most valuable source of energy for ruminants. This energy is needed for activity, metabolic functions including maintenance and repair of tissues, provision of heat and eventually for production. Being highly digestible, the water-soluble carbohydrates are the most readily available for use by the animals. They are concentrated inside the plant cells and may range from 50 to 150 g/kg DM in young leafy grass, to 200 to 300 g/kg DM in more mature material. The WSC content in forage legumes, which unlike grasses also contain starch, are generally at lower levels. Sugar formation in grasses is reduced during dull wet weather (relative to sunny conditions) and by the heavy application of fertiliser nitrogen. The ryegrasses are superior in WSC content to other grass species, but especially in relation to cocksfoot.

The structural carbohydrates can range from 300 to 600 g/kg DM. They are digested to varying degrees by the enzyme action of the rumen micro-organisms, and as their proportions increase with plant maturity and stem formation, their digestibility diminishes because they become lignified or woody. The largely indigestible lignin only makes up a few per cent of the dry matter, although as much as 60 to 80 g/kg in mature herbage; however, it has a depressive effect on the digestibility of the structural carbohydrates because it becomes interlocked with them in the cell wall structure and this inhibits their breakdown by rumen enzymes.

Crude protein content is very sensitive to stage of maturity. It can range from 300 to 350 g/kg DM in young leafy grass to 50 to 100 g/kg in mature stemmy material. Leaf is superior to stem in protein content. Forage legumes are generally superior to grass in CP concentration but the levels in heavily N-fertilised grass can exceed those in legumes. The term 'crude protein' is derived from multiplying the N concentration in g/kg DM by 6.25. A high proportion, 70 to 90 per cent of the crude protein, is true protein made up of amino acids; these are metabolised from carbohydrate, nitrogen and sulphur. The true protein is mainly found in the cell contents. The remaining proportion of the crude protein, so-called non-protein nitrogen, is composed mainly of nitrate and ammonium salts.

Both true protein and non-protein nitrogen are metabolised into microbial protein during rumen fermentation by bacteria and protozoa. This becomes available to the animal in the digestive tract and is referred to as rumen degradable protein. A high proportion of herbage crude protein is of high degradability. True protein which is not broken down and utilised by the rumen is digested and absorbed further along the digestive tract; this protein is known as undegradable protein. Productive and rapidly growing animals often require supple-

mentary undegradable protein in order that their full protein requirements may be met. In essence, the animals require enough nitrogenous
substances to satisfy the needs of the micro-organisms and then
enough total protein, made up of microbial protein plus undegraded
protein, for their own needs.

Grass silage is an example when supplementary undegradable
protein is necessary for productive stock, even though the silage crude
protein content may seem adequate. This is because the highly
degradable protein releases ammonia so rapidly following microbial
action that much is absorbed and then excreted in the urine rather than
utilised by the rumen micro-organisms. Hay is a better source of
undegradable protein than silage but crude protein contents may be
inadequate for animal needs. A low ratio of carbohydrate energy to
crude protein can adversely affect protein digestion and utilisation;
conversely, a low ratio of protein to carbohydrate energy discourages
efficient carbohydrate digestion.

The lipids make up only a few per cent of herbage dry matter and
are highest in young leafy grass. They provide energy and also contain
some of the vitamins. Vitamins in the diet are necessary for the
well-being and productivity of animals. Vitamins A and C and the
complex of B vitamins are all synthesised by ruminants. Vitamins E
and K are present in adequate amounts in leafy green forage. Vitamin
A is synthesised from the precursor, carotene, which is present in
green herbage, leaf being richer than stem. Sun-bleached hay or hay
overheated in store loses most of its carotene. In winter, animals can
draw on carotene or vitamin A stored in their bodies during the
grazing period. They need supplementation eventually, if the diet does
not contain silage or good quality hay. If the diet is deficient in cobalt,
there is insufficient synthesis of vitamin $B_{12}$ and the animals lack thrift.
Vitamin D is made from skin sterols by the action of sunlight, the
sterols having come from precursors in the ingested herbage. Stock
not exposed to sunlight require supplementation.

Minerals are classed as major or trace according to the amounts
present in the herbage. A good number of the mineral elements are
essential for both plant and animal nutrition. However, some elements
such as selenium or iodine, absorbed passively by the grass, are
essential for animals but not for grass growth. Table 14.1 shows the
range of mineral contents in herbage. If stock require mineral supplements it is easier to provide them in enterprises such as dairy cattle
farming, where there is routine access and handling, than in extensive
enterprises such as hill sheep. Direct supplementation of minerals
may sometimes be a better way of overcoming deficiencies than
indirect treatment of the sward. Young leafy herbage is richer in most

149

**Table 14.1  Mineral composition of herbage**

| Major elements | g/kg DM | Minor elements | ppm |
|---|---|---|---|
| Phosphorus (P) | 2–3 | Iron (Fe) | 100–300 |
| Potassium (K) | 15–40 | Manganese (Mn) | 20–200 |
| Calcium (Ca) | 5–15 | Zinc (Zn) | 15–50 |
| Magnesium (Mg) | 1.2–2.8 | Copper (Cu) | 5–15 |
| Sodium (Na) | 0.5–5 | Cobalt (Co) | 0.1–0.2 |
| Sulphur (S) | 2–5 | Iodine (I) | 0.2–1.0 |
| Chlorine (Cl) | 4–20 | Selenium (Se) | 0.1–1.0 |
| | | Boron (B) | 1–10 |
| | | Molybdenum (Mo) | 0.2–3.0 |

Source: Whitehead

minerals than older stemmy material. Forage legumes and herbs are a better source of many minerals than grass; white clover, for example, is richer in calcium, magnesium, iron, manganese, copper, cobalt, molybdenum, boron and selenium.

## DIGESTIBILITY

The feeding value of herbage to animals is directly linked to the digestibility of its nutrients, that is, the proportion of nutrients consumed which is digested and absorbed by the animal as it passes through the alimentary tract. Following absorption into the bloodstream the digested nutrients can be utilised for maintenance and production purposes.

Digestibility can be measured *in vivo* by feeding animals over a period of two to three weeks and monitoring feed intake and faecal output. However, it can be estimated less laboriously by various *in vitro* or biochemical methods in the laboratory using small amounts of herbage. The accuracy of the various techniques has been checked against *in vivo* determinations. Table 14.2 shows the various ways digestibility can be expressed. An organic basis is preferred since it removes the effect of increased but variable mineral intake due to soil contamination either in grazed herbage or in the harvested silage crop. Excessive soil contamination reduces nutrient concentration, depresses the overall digestibility of the herbage ingested and reduces forage intake. As a proportion of the digestible organic material in the dry matter, D value is in effect an index of the concentration of digestible energy in the herbage.

**Table 14.2   Ways of expressing herbage digestibility**

*Herbage fed*

100 kg dry matter (DM) $\left\{\begin{array}{l} \text{90 kg organic matter (OM)} \\[1em] \text{10 kg minerals (ash)} \end{array}\right.$

*Dung*

25 kg dry matter (DM) $\left\{\begin{array}{l} \text{20 kg organic matter (OM)} \\[1em] \text{5 kg minerals (ash)} \end{array}\right.$

Per cent digestibility of DM (DDM) $= \dfrac{\text{Feed DM} - \text{Dung DM}}{\text{Feed DM}} \times 100$

$$= \frac{100 - 25}{100} \times 100 = 75$$

Per cent digestibility of OM (OMD) $= \dfrac{\text{Feed OM} - \text{Dung OM}}{\text{Feed OM}} \times 100$

$$= \frac{90 - 20}{90} \times 100 = 78$$

Per cent digestible OM in the DM $= \dfrac{\text{Feed OM} - \text{Dung OM}}{\text{Feed DM}} \times 100$
(DOMD or, popularly, D value)

$$= \frac{90 - 20}{100} \times 100 = 70$$

The main aim in grassland production should therefore be to achieve adequate yields of digestible nutrients and not merely dry matter. To obtain optimum yields of digestible nutrients and, in turn, optimum animal productivity, a compromise must be effected between low yields of highly digestible young herbage and high yields of mature stemmy herbage of low digestibility. This is illustrated in Figure 14.2. In most grass species a critical stage of growth occurs during the period around ear emergence. For some time before the shooting of the ears satisfactory yields of digestible nutrients can be procured because of the amount of growth made, the leafiness of the crop and the high digestibility of the young, fast growing stems and leaf sheaths. After ear emergence, when the plant matures rapidly with the formation of fibrous stem, flowerhead and eventually seedhead, the proportion of leaf decreases and the digestibility of the herbage falls steadily.

The effect of increasing maturity on the digestibility of individual

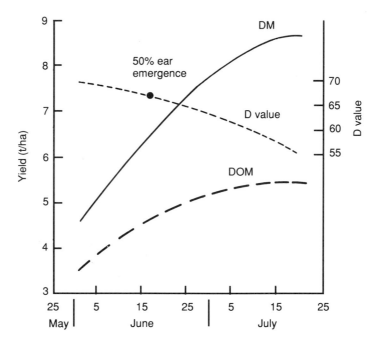

*Figure 14.2* Profile of DM production, D value and digestible organic matter (DOM) during primary growth of late perennial ryegrass
Source: Frame

plant parts during primary growth in spring is shown in Figure 14.3. Young flowering stem and leaf sheath are highly digestible but the digestibility declines rapidly with maturity, from 80 to 85 per cent to 60 to 70 per cent due to the build-up of structural carbohydrates and lignification. The leaf material is maintained at 80 to 85 per cent digestibility because of the higher proportion of cell contents to cell walls. However, among leaves of differing age, digestibility will range from around 50 per cent in dead or senescing leaf tissue to 90 per cent in young leaves. With increasing ratios of stem and sheath to leaf with maturity, the overall digestibility of the plant decreases.

Grass species and varieties within a species vary in the time of year at which ear emergence takes place. Figure 14.4 shows the characteristic relationships between stage of growth of representative species and D value in the west of Scotland. It can be seen that the varieties differ in D value at any given date because they are not at similar growth stages. Early heading varieties have reached a much lower digestibility by the beginning of June than later heading varieties, for example. The

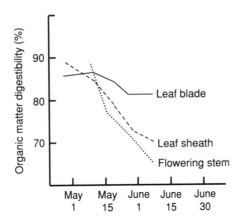

*Figure 14.3*   *Changes in digestibility of component parts of early perennial ryegrass with increasing maturity in spring*
*Source:* Terry and Tilley

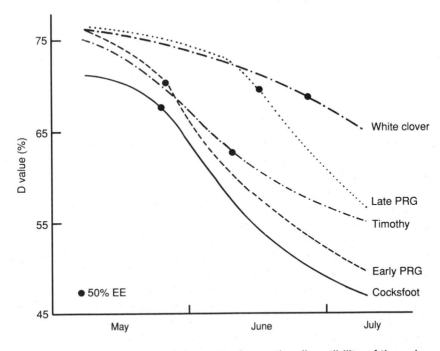

*Figure 14.4*   *Effect of advancing maturity on the digestibility of the primary growth of herbage species*
*Source:* Harkess

153

varieties also differ in D value even when they are at similar growth stages. In general it has been found that perennial, Italian and hybrid ryegrasses are 'high digestibility' species; cocksfoot and tall fescue 'low digestibility'; and timothy and meadow fescue 'intermediate'.

The perennial ryegrasses show a decline of about 0.2 units of digestibility per day up to 50 per cent ear emergence (EE) and thereafter a more rapid decline of about 0.5 units per day. In timothy the phase of rapid decline commences before 50 per cent EE. Italian ryegrass, which grows rapidly in early season before ear emergence, behaves similarly to timothy; on the other hand, the D value pattern of meadow fescue is similar to perennial ryegrass. Red clover and lucerne D values place these species in the 'intermediate digestibility' range. Their pattern of digestibility decline is more akin to timothy or Italian ryegrass than to white clover, on account of their stem formation with advancing maturity. White clover maintains a high digestibility even into July when grass D values have reached very low levels. This is due to clover's continuous growth of digestible petiole and leaf rather than stem.

Dates of grass ear emergence or heading are controlled mainly by the physiological response of the plants to temperature and day length. The dates vary from year to year, although not markedly so, depending upon early season growing conditions, especially temperature; low spring temperatures delay the initiation and development of the inflorescence growing points. The relationship between heading dates of different varieties is fairly consistent. Once the dates of 50 per cent EE of the early types are confirmed the corresponding dates for the later heading varieties can be predicted reasonably accurately. The spread of 50 per cent EE is greatest among the perennial ryegrass varieties which achieve this phase during a period of five to six weeks, while the other species achieve it during two to three weeks. Herbage digestibility prediction schemes are operated by the advisory services. Table 14.3 shows a less accurate but practical field guide for general purpose swards of mixed grass species.

Ear emergence is delayed in grasses by three to four days per 100 m altitude above sea level, depending upon field exposure. Latitude also has an effect, so specific stages of growth are attained two to three weeks later in Scotland than in southern England. Plant maturity and associated rate of decline in digestibility are speeded up in hot, sunny weather and delayed if cool moist conditions prevail. Thus the herbage production at a given D value is greater for the latter conditions. Silage and hay fields in upland areas are often grazed first in spring and this can delay ear emergence by up to seven to ten days. The dangers of spring grazing are reduced production and, if a field is grazed for

**Table 14.3  Prediction of D value (%) from grass growth stage and leafiness**

| | Percentage leaf | | | | |
|---|---|---|---|---|---|
| Developmental stage | 50 | 40 | 30 | 20 | 10 |
| Pre-emergent flower head extended up shoot 16–20 cm | 75 | 73 | – | – | – |
| Heads beginning to emerge (just visible) | 73 | 70 | 67 | – | – |
| Heads three-quarters emerged | – | 69 | 66 | 63 | – |
| Heads emerged and free from flag leaf | – | 67 | 64 | 61 | – |
| Heads fully emerged (flower stalks elongating) | – | – | 63 | 60 | 57 |
| Heads fully emerged (anthers visible) | – | – | 61 | 58 | 55 |

*Source:* Walters

too long before closing up, leaf growth suffers, flowering stem is dominant and herbage D value is consequently reduced.

For silage or hay conservation, advantage of the differing digestibility patterns of various species and varieties can be taken by using fairly simple mixtures of early flowering varieties in some fields, intermediate in some and late varieties in others. This enables a succession of swards which ear out over a period instead of simultaneously and permits the cutting of a succession of highly digestible herbages. Table 14.4 shows the main heading periods of the different grass species and their yields at a D value target of 67.

Appreciation of the effect of stage of growth on digestibility is particularly essential during the May–June period when the flushes of primary herbage growth, associated with stem formation and flowering, are being harvested for conservation. By cutting grass before or soon after ear emergence, silage with a feeding value of high potential for animal production can be secured. Similarly, by cutting grass in the early rather than late flowering or seed development stages, a high quality hay can be achieved.

Primary growths of grass in spring have a potential for high digestibility that cannot be achieved later in the season with regrowths following cutting. Growth rates are slower after the May–June period and consequent higher cell wall to cell content ratios reduce digestibility; while there can be an increase in protein, much of which is digestible, there is less of the highly digestible carbohydrates. Table 14.5 shows how difficult it is to maintain digestibility at high

**Table 14.4   Predicted dates of attaining 67 D value conservation cuts and the DM yields obtained (average data for England and Wales)**

| Species | Date | Heading stage | DM yield (t/ha) |
|---|---|---|---|
| Perennial ryegrass | | | |
|   Early | Late May | 50% EE | 6.9 |
|   Intermediate | Late May/early June | 50% EE | 7.2 |
|   Late | Early June | 50% EE | 6.8 |
| Italian and hybrid | | | |
|   ryegrass | Late May/early June | One week after 50% EE | 7.0 * |
| Timothy | | | |
|   Early/intermediate | Late May | Two weeks before EE | 5.1 |
|   Late | Late May/early June | Four weeks before EE | 4.3 |
| Cocksfoot | Mid May | 50% EE | 5.4 |

* Following an 'early bite' cut in April, otherwise would be a higher yield
*Source:* National Institute of Agricultural Botany

levels over the growing season, even with short rest intervals between cuts. The values shown are for freshly cut herbage, and would be a few units lower after the cuts were conserved for silage owing to losses during the silage fermentation process; the extent of the losses would hinge upon the efficiency with which the silage was made. The digestibility of regrowths falls more slowly in perennial ryegrass than in Italian ryegrass or cocksfoot, for example. If high quality silage is

**Table 14.5   D values (%) of N-fertilised perennial ryegrass cut frequently or infrequently**

| Cutting system | Cutting date | | | | | | | |
|---|---|---|---|---|---|---|---|---|
| | Apr 26 | May 21 | June 14 | July 9 | Aug 2 | Aug 27 | Sep 20 | Oct 15 |
| 8 cuts | 73 | 73 | 72 | 70 | 68 | 66 | 69 | 72 |
| 4 cuts | | 72 | | 66 | | 63 | | 69 |

*Source:* Frame *et al.*

the aim, cutting intervals should be four to five weeks for Italian ryegrass or cocksfoot in comparison with five to six weeks for perennial ryegrass.

Because conservation crops are cut at a later stage of growth than those for grazing it is easier to appreciate the implications of herbage digestibility profiles. Nevertheless, the same principles apply to herbage intended for grazing. Management has to ensure a succession of leafy digestible regrowths in rotational grazing systems and the maintenance of a leafy sward in continuous stocking systems. This involves measures such as controlling stocking rates and resultant grazing pressures, and removing surplus growth for conservation, all of which aim to prevent grass maturing and producing unwanted stem material and dead herbage.

Figure 14.5 shows the average digestibility levels of N-fertilised perennial ryegrass swards rotationally grazed by beef cattle and sheep in mixed grazing systems over three years at three stocking rates. D values of the herbage on offer, which was sampled close to ground level, declined with advancing season until the third grazing cycle and then levelled out. A similar pattern is evident for the residual herbage, but at lower digestibility levels. Two consequences of stocking rate were noteworthy. For the herbage on offer higher D values were associated with high rather than low stocking rate, especially in late season. This was because dilution of young grass regrowths by residual ageing herbage was minimal at high stocking rate; with low stocking, a build-up of rejected herbage and basal material of low

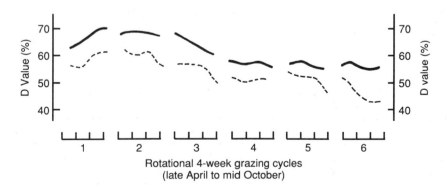

Figure 14.5   Digestibility of herbage on offer (—) and residual herbage (–––) in a rotationally grazed perennial ryegrass sward
Source: Frame and Dickson

digestibility occurred. Conversely, the digestibility of residual herbage was lower with high relative to low stocking rate because the sward was eaten down to a lower height, leaving mainly basal leaf sheath and stubbles.

In a trial in southern England perennial ryegrass swards were continuously stocked with ewes to achieve sward surface heights of 3, 5 or 7 cm—in effect corresponding to decreasing stocking rate intensities. Figure 14.6 shows that the digestibility of herbage eaten by the ewes decreased during summer as grazing intensity decreased (D values would be approximately equivalent to the cited *in vivo* OMD values × 0.9).

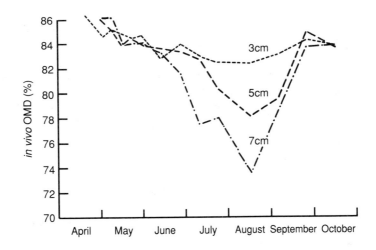

Figure 14.6   The effect of sward surface height on the in vivo *OMD* of herbage eaten by ewes grazing perennial ryegrass swards maintained at 3 cm (-------), 5 cm (– – – –) and 7 cm (–·–·–)
Source: Orr et al.

Table 14.6 gives details of a dairy cow grazing trial at Dumfries, and shows the adverse effect of advancing season on the digestibility of the herbage on offer in continuously stocked swards. It also shows the beneficial effect on D value of clover, when contributing 10 to 30 per cent to herbage DM over the season, in the comparison of a ryegrass/clover sward receiving no fertiliser nitrogen with a ryegrass sward given 350 kg/ha N per annum.

**Table 14.6** **Seasonal D values (%) of herbage on offer to dairy cows in N-fertilised grass or grass/clover swards sampled at 3-weekly intervals**

| Sward type | Apr 27 | May 18 | June 8 | June 28 | July 20 | Aug 10 | Aug 31 | Sep 20 | Oct 12 |
|---|---|---|---|---|---|---|---|---|---|
| Grass + N | 69 | 67 | 67 | 62 | 61 | 60 | 61 | 60 | 63 |
| Grass/white clover | 74 | 71 | 69 | 64 | 63 | 62 | 63 | 62 | 66 |

*Source:* Frame *et al.*

## METABOLISABLE ENERGY

Digestibility values are widely used as a means of describing herbage quality in advisory literature. OMD is favoured by researchers but D value has become the familiar term to farmers, especially when dealing with silage and hay quality. However, when compiling rations for livestock the energy value unit of measurement used to reconcile feeding value and animal needs is metabolisable energy (ME). It is measured in megajoules (MJ) and is closely related to digestibility; ME is approximately $0.16 \times D$ value.

Figure 14.7 illustrates the concept of ME.

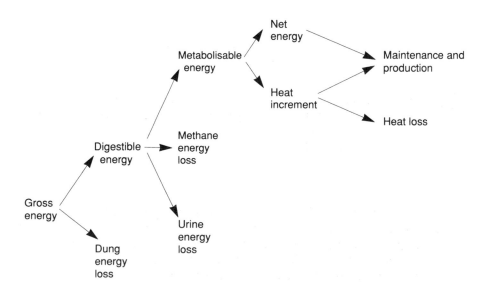

*Figure 14.7   Partition of dietary energy in ruminants*

Ingested grass is either digested or excreted in the dung. The digestible energy in the forage ranges between 60 and 80 per cent depending upon its stage of maturity. Some of the digestible energy is lost as methane gas in the rumen and some as urine, but the main proportion is available as ME. In turn this energy can be partitioned into the proportion known as net energy, which is available for maintenance and production, and the proportion lost, known as heat increment. Some of this heat is given off by the animal to the surroundings and is a real loss; some is also generated during the course of essential body functions such as the digestion process and is therefore not a real loss, but part of the nutrition process.

Table 14.7 shows the ME values for different qualities of grazing stage herbage, silage and hay.

**Table 14.7**  **Metabolisable energy values for grazing stage herbage, silage and hay**

| Herbage state/quality | ME (MJ/kg DM) |
|---|---|
| Grazing stage/ | |
| High | 11.5–12.5 |
| Medium | 10.5–11.4 |
| Low | 9.5–10.4 |
| Silage/ | |
| High | 10.5–11.5 |
| Medium | 9.5–10.4 |
| Low | 8.5–9.4 |
| Hay/ | |
| High | 9.5–10.5 |
| Medium | 8.5–9.4 |
| Low | 7.5–8.4 |

## Chapter 15

# Sward Growth and Development

---

In the world's audience hall, the simple blade of grass sits on the same carpet with the sunbeam, and the stars of midnight.

Walt Whitman (1819–1892) *Leaves of Grass*

I believe a leaf of grass is no less than the journey-work of the stars.

Sir Rabindranath Tagore (Hindu poet)

---

A typical perennial ryegrass sward is made up of a population of individual plants each consisting of tillers or shoots, most tillers usually bearing three leaves. The density of tillers per ha is greatest (30 to 40,000) in continuously sheep stocked swards and least (5 to 15,000) in swards regularly cut for silage or hay. Essentially, the sward composition is a function of the numbers of rooted plants, tillers per plant and leaves per tiller. It is possible to distinguish and count the number of individual plants per unit area or length of drill in the early weeks after establishment from seed but, subsequently, the density and intermingling of tillers and leaves make it only feasible to count the total number of tillers, such as when checking on the establishment success of sown species.

## GRASS TILLERING PROCESS

Immediately after germination, a sown seed develops an individual plant composed of roots, coleoptile or sheath, and emerging leaves. New daughter tillers with adventitious roots soon develop from axillary buds in the stem base at the rate of about one per week in summer. The tillers virtually lead a separate existence, though remaining connected to the parent plants, and metabolites may be transported to and from the parent plants, e.g. in times of stress after defoliation. Figure 15.1 illustrates a tillered perennial ryegrass plant.

The tillering process is strongly influenced by defoliation management. Decapitation of existing tillers by grazing encourages a new generation of tillers, stimulated by both defoliation and enhanced light

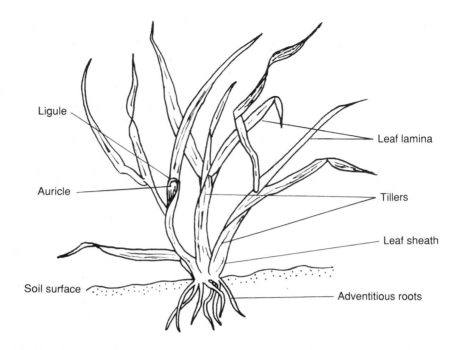

Ligule

Leaf lamina

Auricle

Tillers

Leaf sheath

Soil surface

Adventitious roots

*Figure 15.1   Tillered plant of established perennial ryegrass (see also*
*Appendix 3 for grass vegetative characteristics)*

penetration to the base of the sward. Continuous stocking management encourages a succession of tiller replacements and a dense, stable population of small, youthful tillers develops. Under rotational grazing or cutting the population is lower and older, but individual tiller size and growth rate is greater. Tiller numbers fluctuate in response to the rest periods of uninterrupted growth which suppress tillering, and to defoliation which encourages tiller initiation. Tiller numbers may be augmented by new seedlings from shed seed or from seed of unsown species in the soil.

There are spatial limits to the numbers of tillers which can co-exist within a defined area. Small tillers are crowded out by large ones and, if tiller growth is continuous, the sward base becomes shaded and tillering is reduced due to suppression of tiller development from axillary buds. The adverse effect is due to a lack of bud stimulation by light, and decreased photosynthetic activity in the lower leaves. There is strong competition for light, nutrients and water among sown plants and any invading weed grasses or broad-leaved weeds. Tiller survival ranges from a few weeks to several months and possibly even a year.

This general pattern of tillering is similar in 'creeping' grass species but, periodically, specialised tiller buds develop a sharp growing point and grow through the surrounding leaf sheaths close to the ground. These elongate to produce surface stolons, as in rough-stalked meadow grass and creeping bent, while below ground they produce rhizomes as in smooth-stalked meadow grass and couch grass. This creeping habit of growth is an advantage in colonising bare ground or spaces in swards.

The reproductive process in grasses during May/June is triggered by the previous sequence of environmental conditions, in particular the short days and low temperatures of winter. Initiation of leaf primordia on the stem apex is accompanied by the simultaneous appearance in their axils of reproductive tiller buds. This is evident in the form of a double ridge on the apex. The bud then develops to form an inflorescence which is carried upwards within its encircling leaf sheaths by elongation of the apex. Different grass species develop different types of inflorescence, e.g. ryegrass has a spike, fescue a panicle and timothy a spike-like panicle. Leaf primordia initiated before the change from vegetative to reproductive phase continue to develop at stem nodes until the final leaf—the flag leaf—appears at right angles to the stem. Its appearance signals that ear emergence will occur in the next few days followed by flowering or anthesis, i.e. extension of anthers and stigmata, and ultimately seed formation. This production of flowering stems in the spring coincides with the most rapid of growth in grasses.

When flowering stems are cut the next generation of tillers develop from basal buds which have remained dormant during flowerhead development although many buds will have died while the plant diverted photosynthates to the reproductive process. Regrowth from these basal buds is progressively retarded as the length of the flowering period is allowed to increase. A good example is the faster regrowth of tillers following silage cut at ear emergence, compared with the slow growth rate following late cut hay.

Continuous stocking which keeps the sward short curbs stem elongation, increases numbers of tillers per unit area and decreases numbers of reproductive tillers (see Figure 15.2). After the flowering period is over, the impetus of plant growth is again directed towards the production of vegetative tillers. When a stemmy sward is cut well above ground level, some buds on the stem nodes of the stubble may develop into aerial-rooted tillers, susceptible to total defoliation and thus adversely affecting sward persistence.

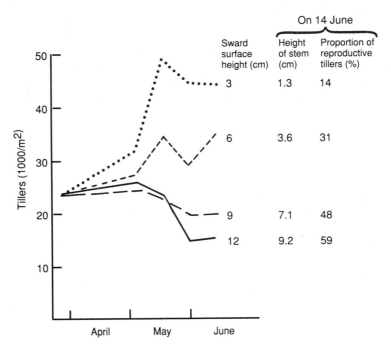

*Figure 15.2  The effect of sward surface height under continuous stocking
by sheep on reproductive development in a perennial
ryegrass sward*
Source: Penning *et al.*

# GRASS LEAF DEVELOPMENT

The vegetative part of a typical individual tiller is a series of con-
centric leaf sheaths with leaf blades or laminae on the outer visible
sheaths. From the outside inwards the sheaths contain rolled or folded
leaf blades at progressively younger stages of growth. At its inception,
a leaf is simply a single cell—the leaf primordium or initial—on the
growing point of stem apex. This small glistening apex, 1 to 2 mm in
size, is a zone of rapid cell division. The leaf primordia emerge from
ridges on the sides of the apex in opposite and alternate order in
grasses. Careful stripping of the visible tiller leaves reveals the inner-
most unemerged leaves and growing point (see Figure 15.3). A
horizontal cross-section of a tiller above the growing point only
reveals the tightly enclosed young leaves.

Emerged leaves comprise two parts—the lower tubular sheath and the upper leaf blade. At the junction there is a membranous collar—the ligule—which varies in size and shape according to grass species; it may even be hairy. In some species there may also be auricles, which are membranous projections at the base of the blade, e.g. in Italian ryegrass and tall fescue. Morphological features such as ligules, auricles, whether the leaf blade is folded or rolled before opening, its general shape, texture and hairiness are the main characteristics used to identify grass species in their vegetative state (see Appendix 3). Type of inflorescence is a major identification feature at later maturity.

When a young unemerged leaf is defoliated, both leaf base and sheath have the ability to resume regrowth by cell division. However, once a leaf blade has emerged from the sheath its capacity for cell division and regrowth ceases. The young tissue's characteristic for regrowth and the ground level position of the stem apex permit survival under grazing, as also does the restriction of stem elongation—highly vulnerable to defoliation—to a short reproductive phase in spring.

The rate of leaf appearance and development to maturity is largely determined by temperature, while an adequate nutrient supply, particularly nitrogen, increases leaf size. Nevertheless, leaf production is

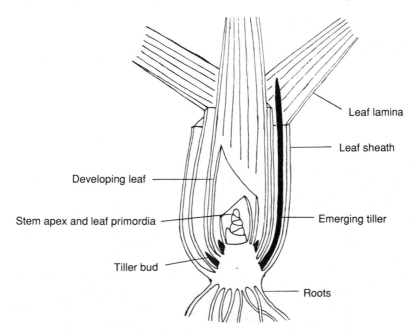

*Figure 15.3  Diagrammatic half section of a vegetative tiller*
Source: Jewiss

virtually continuous all the year round, except at temperatures near freezing. Successive leaves on a tiller appear at approximately 9 to 12 day intervals during warm summer conditions but at 25- to 30- day intervals or longer during winter. If ungrazed the leaf will survive for about 25 to 30 days in summer and for twice this period in winter. Normally each tiller bears three leaves—one recently emerged and actively growing, one fully developed, and one senescing or dying, senescence starting from the tip and progressing towards the leaf base.

There is therefore a continuous turnover of plant tissue within a sward and any leaves which are ungrazed senesce and become added to the dead plant litter at the sward base. Young leaves and tillers must penetrate the existing sward canopy to reach light, because of the position of their growing points. Extension growth must therefore be progressively greater the longer a sward is left ungrazed or uncut. Light intensity decreases from the top of the sward downwards and may be negligible near ground level. Photosynthetic activity is thus greatest in the upper layers of the sward, efficiency being highest in young leaves and declining with leaf age. Leaves which emerge in shade conditions subsequently photosynthesise less efficiently compared with those which emerge into light. There are three major consequences of this pattern of growth. First, the density of sward tissue decreases from the bottom to the top of the sward canopy; in effect the structure of plant material can be visualised as a triangle, being flatter in short swards under continuous grazing than under rotational grazing. Second, the bulk of dead decaying plant tissue accumulates at the base of the sward. Third, it is the actively photosynthesising leaves at the top of the canopy which are most susceptible to removal by grazing.

## WHITE CLOVER DEVELOPMENT

The morphology of white clover, the most common legume component of temperate seed mixtures, differs markedly from grasses, particularly in its stoloniferous growth habit and its canopy structure (see Figure 15.4). A clover seedling develops a short primary stem with an extensively branched tap root and trifoliate leaves. Ground-hugging, elongating branched stolons are initiated from buds in the leaf axils and petioles carrying leaves grow from a proportion of the nodes formed at intervals along the stolons. The nodes develop adventitious roots with numerous lateral branches. Early in the root's life rhizobial infection takes place and large numbers of nodules containing the N-fixing rhizobia develop, mainly on the finer

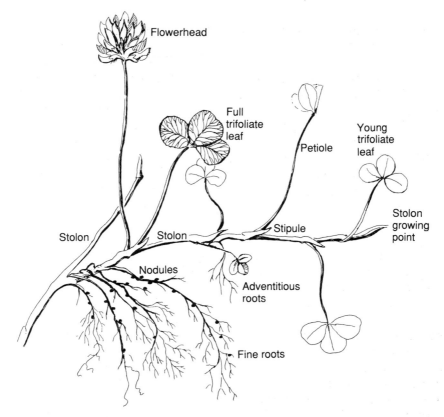

*Figure 15.4   White clover plant*

branches of the root system in the upper soil layers. Effective rhizobia and adequate photosynthetic leaf area are necessary for vigorous N-fixation. Further buds or growing points are initiated from the axils of the leaves and under good environmental conditions, e.g. a good light regime at the base of the sward, develop into daughter stolons or branches. Eventually the clover plant forms a rosette with a network of freely branched stolons along the soil surface. As the centre of the rosette dies, usually within one to two years, individual stolons become independent, supporting leaves and growing points and initiating further daughter stolons.

Thus the growth and survival of white clover in a sward are both strongly correlated with stolon development and replacement. Low temperatures, poor light conditions and, in summer, the initiation of inflorescences, all depress stolon production; also, stolon loss during winter can be considerable. Profuse flowering can reduce plant

167

persistence although subsequently shed seed may result in a new generation of clover plants. Small-leaved clover types are strongly stoloniferous and have higher growing point densities and better persistence than less stoloniferous, larger-leaved types, an advantage under severe grazing and in colonising bare spaces. Long rest intervals permit increases in length, diameter and weight of stolons, major reasons for the good performance of clover in grass/clover swards cut for silage. Fertiliser N application, which increases grass density, reduces the number of stolon growing points as a result of shading at ground level. Buried stolons resulting from stock trampling and earthworm activity, particularly in grazed swards, are a positive factor in clover persistence by their subsequent regeneration when conditions are suitable.

Leaf initiation and development are markedly affected by prevailing temperatures and light. The longevity of clover leaves is normally between four to five weeks from the time of emergence. In late autumn and winter the leaves on the older parts of the stolons generally die away leaving mainly undeveloped leaves near the stolon apices. Nevertheless there is a steady production and loss of leaf by senescence throughout winter followed by a steady increase in leaf growth and development when spring temperatures and radiation improve. White clover leaves in a grass/clover sward are not significantly shaded during summer since successive emerging petioles extend longer and keep pace with the height of the growing sward canopy. This ability is markedly reduced in spring and autumn since low temperatures inhibit petiole extension; clover is then at a disadvantage in relation to grass and therefore susceptible to adverse management or environmental conditions.

The horizontal arrangement of clover leaves and their position at the top of the sward canopy, advantageous for light interception, make them readily accessible for defoliation by grazing animals; however this susceptibility is counterbalanced by the soil surface position of the stolons which are not usually grazed and therefore constitute a source of energy reserves for the initiation of new stolon and leaf. Another aid to survival is clover's ability to become dwarf-like in form under intense grazing pressure and to assume normal form when the pressure is relaxed. Nevertheless the conflict between keeping sufficient leaf in the sward for photosynthesis and yet providing forage for grazing is apparent and a balance has to be maintained through the grazing management imposed.

Management of grass/white clover swards, including manipulation of the characteristics outlined above, which optimises white clover contribution, is discussed in Chapter 22.

## PHOTOSYNTHESIS AND TISSUE TURNOVER

The process of photosynthesis which controls plant growth is dependent on the interception of light energy by the leaves and subsequent elaboration of sugars from atmospheric carbon dioxide and from, or together with, water taken up by the roots. This reaction takes place in the green tissue of the plants, which contains chlorophyll, the green pigment necessary for the reaction. Sugars are assimilated into other carbohydrate compounds such as structural cellulose and also into protein. Sugar is lost by respiration during other growth processes, including root development, such respiration increasing as the herbage is allowed to accumulate. Surplus sugars can be stored in the plant.

In the past, the view was held that the leaf area index (LAI), which is the ratio of leaf area (one surface of the leaf) to unit area of land, should be maintained at a sufficiently high level to ensure maximum light interception and hence maximum herbage production. However, it has been increasingly appreciated that leaves have a limited life span and that a balance exists between photosynthesis, gross tissue production, respiration, tissue death and herbage utilised by grazing animals. This balance can be manipulated by maintaining various sward states, whether expressed as LAI, amount of herbage present or sward surface height.

Figure 15.5, derived from carbon balance studies of swards under continuous stocking with sheep, shows that rates of photosynthesis and gross tissue production are close to a maximum at a high LAI, but this is accompanied by a high rate of tissue loss by death; thus the amount of herbage harvested by grazing sheep is small. Conversely, in swards maintained at a low LAI, photosynthesis and gross tissue production are reduced, but there is less leaf and a greater proportion of leaf tissue is utilised. This increase in efficiency of utilisation compensates for the decrease in amount of herbage grown, and the amount harvested is increased. Neither LAI nor the amount of herbage present is easily determined on swards in practice but they are closely related to the more easily measurable sward surface height.

Parallel studies of the rate of growth and senescence on individual tillers (see Figure 15.6) arrived at similar conclusions to the above. It can be seen that the net rate of herbage production is at a maximum when the sward height is kept at 3 to 5 cm but the net production rate does not change much between 3 and 8 cm. The high tiller population typical of continuously stocked swards compensates for the lack of height and lower individual growth rate compared with fewer larger tillers in taller swards with a high LAI. A dense leafy sward is also

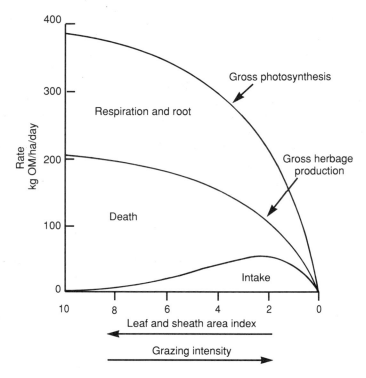

*Figure 15.5* *The effects of continuous stocking on the balance between photosynthesis, gross tissue production herbage intake and death*
Source: Parsons *et al.*

capable of efficient light interception even at relatively low LAI since many of the leaves are youthful and have high photosynthetic efficiency. Another consequence of enhanced sward density and leafiness, partly due to suppression of reproductive tillers, is a more uniform distribution of production over the season compared with the more marked peaks and troughs typical of rotational defoliation. In the latter, there is greater expression of the high growth rates of reproductive tillers in May to June, though not permitting ear emergence and flowering.

In managing swards, animal grazing behaviour and performance have to be considered as well as potential herbage production. Optimum sward surface height differs according to class of animal, due to differing grazing characteristics between classes (see Chapter 16), and their physiological stage. Herbage intake and performance by cattle is better on taller compared with shorter swards and vice versa for sheep. Research has shown that the optimum sward height for good dairy cow individual performance is 7 to 10 cm and for other cattle systems, including cow/calf or fattening beef systems, 7 to 9 cm. As the sward height is increased beyond these heights (see Figure 15.6) *net* herbage production decreases, so stocking rate potential is reduced. In sheep systems, lamb growth rates are best at a sward height of 5 to 6 cm and again potential stocking rate falls with increased sward height because senescence of herbage also increases. Maximum lamb output per ha will be achieved at a lower sward height of 3 to 4 cm, but at the expense of individual lamb performance. Sward heights are best used flexibly in response to specific performance objectives, whether per animal or per ha.

The bulk of research on sward height management has been carried out on continuously stocked swards but the same principles apply to rotational grazing with its variations in duration of regrowth interval

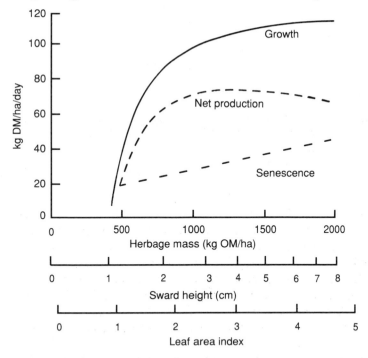

Figure 15.6   The influence of sward conditions (leaf area index, sward height, herbage mass) on rates of herbage growth, senescence and net production in perennial ryegrass/white clover swards continuously stocked with ewes and lambs
Source: Bircham and Hodgson

and intensity of grazing over the season. During rest intervals, rates of gross tissue production and senescence increase as progressively larger leaves become involved in the balance between photosynthesis and turnover of plant tissues. Under rotational grazing the guideline to follow is the residual sward surface height following grazing. In systems with three-to four-week rest intervals and sward heights at stock entry of 15 to 25 cm, target residual heights of grazed stubble are 4 to 6 cm for sheep and 7 to 10 cm for cattle. Grazing to lower residual heights increases the efficiency of utilisation but at the expense of individual animal performance since the lower sward layers are less digestible. Conversely, leaving higher residual heights reduces efficiency of utilisation because of lower stocking rates, but individual animal performance is improved.

With repeated grazings, tall rejected grass areas inevitably build up and these develop a high percentage of stem and dead material compared with well utilised areas. This has adverse repercussions on subsequent intake and animal performance, for example, in the second half of the season if there has been under-utilisation in the first half. The application of sward surface heights for efficient grazing in practice, the target heights for various livestock enterprises and the monitoring of sward state are dealt with in Chapter 17. Sward height studies have been conducted mainly on perennial ryegrass or ryegrass/white clover swards so further work is required on swards dominated by other grasses whose morphology and sward structure differ from ryegrass. While similar principles will apply, target sward surface heights will undoubtedly require modification from those found suitable for ryegrass swards.

## SEASONAL HERBAGE PRODUCTION

Figure 15.7 shows the seasonal distribution of herbage production from early perennial ryegrass swards of standard age, i.e. first full harvest years, cut over five consecutive seasons at Ayr (longitude 4° 33′ west; latitude 55° 28′ north). Figure 15.8 illustrates the variation in herbage production from the ryegrass swards due to season alone during the 5-year trial period. A rotational overlapping 4-weekly sequence of harvesting was employed and fertiliser nitrogen, phosphate and potash applied to ensure there were no soil nutrient limitations to sward growth. The asymmetric pattern of production obtained is consistent with that at many European centres, although in drier conditions than in west Scotland, a more pronounced dip in summer production is typical. Earliness of growth, duration of growth into

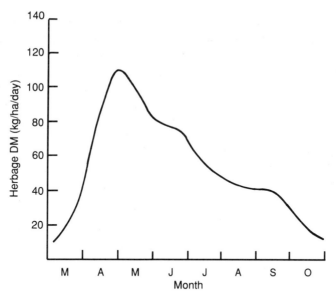

*Figure 15.7    Seasonal pattern of herbage production from early perennial
ryegrass (average of 5 years)*
Source: R. D. Harkess and A. Peeters, personal communications

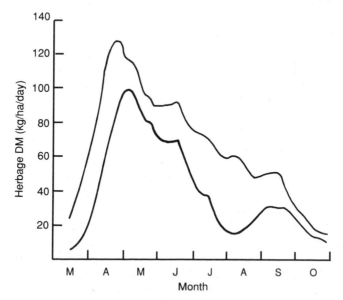

*Figure 15.8    Fluctuations in seasonal patterns of herbage production from
early perennial ryegrass during 5-year period (highest and
lowest growth rates taken within any single season)*
Source: R. D. Harkess and A. Peeters, personal communications

173

autumn and production peaks and troughs will vary with differing grass species, environmental conditions, the specific season and differing management practices (see Chapter 18).

In other work in south England on ryegrass swards grazed by ewes to maintain specific sward surface heights, the seasonal pattern of herbage production was more evenly distributed throughout the grazing season than under monthly cutting (see Figure 15.9), although the annual amounts harvested under cutting and continuous stocking managements were similar at 9.8 and 10.0 t/ha OM respectively.

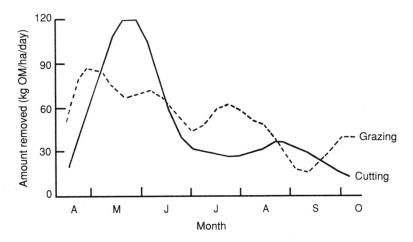

*Figure 15.9   Seasonal pattern of herbage removal under grazing and under cutting*
Source: Orr et al.

# The Grazing Process

Grasslands of various types account for over 3,000 million ha or about a quarter of the earth's land area. As differing grassland communities and associations evolved in the ancient past, so also did the diversity of herbivores grazing on them. Herbage species which could adapt to grazing survived and persisted while others disappeared. Types of adaptation included a good anchorage system of fibrous, adventitious roots, underground rhizomes or surface stolons, tillering from basal buds, the ability of leaves to elongate from their bases after defoliation, and the protection of growing points (stem apices) by leaf sheaths.

Grassland species used in Britain and many European countries evolved under relatively severe grazing pressure from domesticated cattle and sheep, but North American rangelands, for example, evolved under relatively light grazing by native herbivores. Rangeland species are thus less stable in the face of increased grazing pressures, and dramatic examples of their breakdown are evident in many parts of the world. For the foreseeable future, grazing by ruminants will continue to be the chief and most economical method by which grass is converted to animal products. Grazing in the UK is restricted to 5 to 9 months annually for dairy cattle and fattening cattle, whereas sheep in the hills and uplands graze all the year round. In the more favourable climate of New Zealand, 12-month grazing is the rule for all classes of stock.

The grazing process is characterised by the animal factors of selectivity of grazing, deposition of excreta and treading, with these factors affecting the growth and botanical composition of pastures both separately and interdependently. To obtain optimum nutrient intakes, which control potential levels of animal production, the herbage on offer to grazing animals must be sufficient to satisfy appetite, be acceptable and of high feeding value. A large proportion of the nutrients ingested by grazing stock is excreted in dung and urine. Excretion off the pasture, and the leaching, volatilisation and immobilisation of N from urine and dung patches are sources of loss from the soil fertility/grazed pasture nitrogen cycle. Urine deposition covers a large area of the pasture in contrast to dung, and contains higher contents of available nitrogen and potassium so is of considerable value in stimulating pasture growth. The adverse effects of treading (hoof trampling), which increase as grassland use is intensified, may be kept at acceptable levels by management that is designed to maintain dense, vigorous and long-term swards of tolerant pasture species.

## SELECTIVE GRAZING

The main factor influencing animal production from grazing is the animals' nutrient intake. This is determined by the amount of herbage available and its acceptability, the digestibility of the selected herbage and its overall nutritive value. Many plants and parts of plants may have a high nutritive value and digestibility, especially when young, but if intake levels are low because of poor acceptability they are obviously unsuitable for efficient systems of animal production.

Stock show preference by choosing some of the herbage on offer and rejecting other types and this is commonly referred to as herbage palatability. The term is not strictly correct, as palatability means agreeable to the taste, but grazing animals use other senses such as smell, touch or sight in making their selection. Conditioning from previous feeding habits, state of hunger and amount and range of forage material available also influence their choice.

For example, a lactating 600 kg dairy cow consumes 15 to 18 kg herbage dry matter (100 to 125 kg fresh grass) daily to meet appetite and nutrient requirements fully. This is equivalent to harvesting 100 to 120 $m^2$ of a sward with available herbage of 1,500 kg DM/ha. The cow may need 30,000 individual bites at 60 or more bites a minute to do this in grazing periods totalling 7 to 9 hours daily, and walking 3 km or more in the process. The importance of sufficient available and acceptable herbage may not always be readily appreciated by the grazier. The herbage DM intake of stock generally, sheep or cattle, is 2.5 to 3.0 per cent of body live weight. Stock appetite and intake are lowest in late pregnancy, because of physical limitations, and highest when animals are lactating heavily or growing rapidly.

## Herbage acceptability

Studies on the grazing of different herbage species or varieties grown separately or together, plus measurement and analysis of differing swards before and after grazing, have made it possible to describe the relative acceptability of herbage plants. Among commonly used species, white clover, meadow fescue and timothy are highly rated, Italian and perennial ryegrass slightly less highly, while cocksfoot, red fescue and tall fescue are least acceptable. Differences in acceptability between species or varieties of herbage plants are matched by differences between parts of the same plant, reflecting varying accessibility or presentation for grazing, physical or structural differences, or variable chemical composition.

Leafy, succulent young herbage is selected in preference to stemmy, dry, fibrous material. Parts most easily torn off the plant by tongue, teeth and mouth pad of the ruminant are selected. Animals graze in a vertical plane so, in a uniform sward, the upper layers are eaten first, the herbage being leafier and more digestible than older herbage in the lower canopy horizons. Close grazing is discouraged by the presence of accumulated dead or decaying herbage litter at the base of a sward. When mixed forages of equal attractiveness to stock are grown, easily accessible herbage is usually preferred to the less accessible material. For this reason, erect growing plants may be grazed in preference to prostrate growing plants. The intermingling of unacceptable with acceptable herbage depresses utilisation of the latter, especially in lower layers of a sward.

Coarseness, hairiness and spininess in plants lower their acceptability, while the presence of fungal disease, such as crown rust in ryegrass, renders infected herbage unattractive to stock. Varieties within a species may differ in acceptability because of differences in their susceptibility to fungal attack or physical structure. Some varieties of tall fescue and cocksfoot have coarse leaves, for example, while plant breeding has developed soft leaves in others. Grass species going through their cycle of development from leafiness to seedhead form pass from acceptable to less acceptable or rejected stages.

Herbage contaminated by dung or growing in the vicinity of dung patches is normally unattractive to stock, but the degree of rejection is lessened under high grazing pressures, and possibly because stock become more accustomed to tainted herbage. Slurry applications, particularly if not diluted with water, depress acceptability until the slurry is washed off the herbage by rain. Rainfall and a sharp frost also eliminate the smell of dung-fouled herbage and render it more acceptable. Cattle or sheep are not averse to eating herbage in the vicinity of

177

each other's dung, a finding put to advantage in mixed grazing of animal species. Herbage growing in urine patches may be preferred to other herbage because of its increased succulence and leafiness, caused by the growth boost from urinary nitrogen. Severe trampling and associated soil contamination reduce the attractiveness of herbage for grazing.

Various chemical aspects associated with the growth stage of sward components influence acceptability. Leafy herbage with a relatively high content of water-soluble carbohydrate (sugars) and low fibre content is highly acceptable. Conversely, herbage grown on poor soils and low in nutrients, such as nitrogen or phosphorus, is generally less acceptable. Relative to diploid ryegrasses, their tetraploid counterparts have a higher grazing attractiveness because of a combination of succulence, higher sugar content and enhanced digestibility.

The presence of bitter substances such as alkaloids or oxalates reduces acceptability. However, previously unacceptable herbage may be eaten after mowing and wilting, because of chemical changes in the material. Topping a sward which has developed a mosaic of well-grazed and poorly grazed areas will normally result in a fairly uniform leafy regrowth of improved acceptability (see Chapter 18). However, if there is a lot of cut material it is best removed by a forage harvester; otherwise it may rot and adversely affect both sward growth and herbage acceptability.

Certain poisonous weeds, e.g. ragwort, may be eaten after mowing or spraying with herbicide but stock should be excluded until the plants are completely desiccated. British pastures do not contain many weeds hazardous to stock, other than ragwort and bracken, and the greatest poisoning hazard is from browsing poisonous shrubs or trees on the boundaries of fields or on roadways. Toxic plants are sometimes bitter tasting, so are frequently avoided, and animals reared in areas infested with poisonous plant species often do not eat them. Stock new to the environment may eat them and be adversely affected, however.

## Grazing behaviour

Sheep, with their narrower and more pointed muzzles, compared with cattle, graze more fastidiously and readily select individual leaves and other plant parts. They graze a wide range of plant species given the opportunity, although choosing leafage almost exclusively. Cattle tend to tear off bunches of herbage material, gathering them with their prehensile tongue. The herbage may be of mixed composition, as regards both species and the proportions of leaf and stem. Cattle can

▲ Grass to butter and cheese via dairy cows. Part of the dairy herd at SAC, Auchincruive with cubicle shed/milking parlour in background and dairy technology building to the right *Chapter 17*

Fattening lambs utilising grass in rotational grazing system, south England
▼ *Chapter 17*

Assessing the stubble height of rotationally grazed pastures at Colonia,
▼ Uruguay *Chapter 17*

▲   Walled pastures for rotational grazing by dairy cows at Achada Valley, Terceira, Azores   *Chapter 17*

▲   Dairy cows strip grazing grass sward, south-west Scotland
*Chapter 17*

▲   Mixed dairy livestock grazing on common pastureland at Pico, Azores
*Chapter 17*

Buffer grazing area (left of picture) which can be grazed or cut for silage as
▼   required   *Chapter 17*

▲ Well fertilised grass crop being cut at the leafy stage of growth to ensure highly digestible silage with good intake characteristics (R. F. Gooding) *Chapter 19*

▲ Silage equipment showing application of additive, forage harvesting of wilted grass and efficient trailer filling (R. F. Gooding) *Chapter 19*

Outdoor clamp silage well sealed and consolidated to ensure air-free storage conditions
▼ *Chapter 19*

graze tall herbage more easily than sheep because of their physical size; this capability is of value in controlling tall, coarse growth in a pasture. Herds of cattle, preferably in an unproductive state, are sometimes used as defoliators to condition a sward for more productive stock. Differences in acceptability to animals can also form a basis for weed control, e.g. weeds maintained under cattle grazing such as ragwort may be grazed out or reduced by sheep, while goats will graze out thistles, rushes and gorse from infested pastures.

Social factors may play a role in grazing behaviour, especially under extensive grazing. Hill sheep form small social groups which graze only a part of the total area available, for example. Aggressive groups graze the most acceptable and often the most nutritious grassland communities while the less aggressive groups are forced onto poorer herbage. This pattern of use is maintained by the offspring of the various groups of ewes, who develop similar grazing patterns.

Within a group of apparently similar stock there may be marked individual differences in grazing behaviour, e.g. differences in time spent grazing, frequency of biting, herbage amount taken per bite and rumination times and speeds. Such individuality may be related to inherent differences in physical or physiological make-up. Physiological state also influences grazing behaviour, and animals in late pregnancy, in which the developing foetus restricts rumen physical capacity, have low appetite and short grazing times in relation to post-parturition.

Most supplementary feeding at pasture lowers herbage intake by shortening the time spent grazing. The decrease in herbage intake as a proportion of the amount of supplement is known as the substitution effect, although certain high energy supplements which enhance digestion of low quality herbage may not depress herbage intake. The substitution effect largely explains the poor milk yield response commonly found from feeding supplementary concentrates to grazing dairy cows.

Grazing animals divide their time between grazing, ruminating, lying down, loafing and walking. Most grazing takes place in daylight hours, especially in the early morning and early evening. If sufficient acceptable pasture of good nutritive value is available, stock may graze on average 7 to 9 hours a day. Rumination to break down ingested herbage and assist digestion follows each grazing period and occupies another 7 to 9 hours. The balance of the 24 hours is spent lying, loafing or standing. Stock seek shelter during storms and shade during hot sunny periods, resulting in reductions in grazing times and intakes.

If the amount of available or acceptable herbage is low, only a small amount of leafage is consumed at each bite; grazing times are therefore lengthened to appease appetite and rumination times may be correspondingly reduced, resulting in less efficient digestion. Conversely, if the total amount of herbage present is high but of low acceptability, grazing periods are reduced and rumination times prolonged. In practice, judgement by eye is widely used to assess whether or not herbage supply is matching demand since subjective measurement is not favoured. However, relationships have been established between sward heights and amounts of herbage present, and sward heights needed for optimal herbage intake and animal performance of different types and classes of stock (see Chapter 17). For example, Figure 16.1 illustrates that lactating ewes increased their intake up to sward heights of 5 to 6 cm on continuously stocked swards.

In the case of suckler cows, intake increased up to sward heights of 8 to 9 cm (see Figure 16.2).

For rotational grazing, relationships have been established between intake and residual herbage stubble; taking dairy cows as an example, intake is maximised when a stubble of 7 to 10 cm is left after grazing.

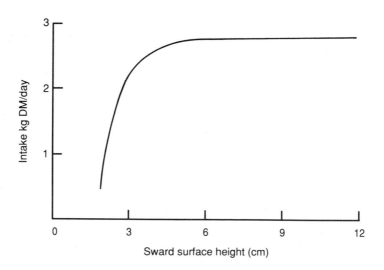

Figure 16.1    *The herbage intake of ewes suckling twins grazed on swards of different heights*
Source: Penning

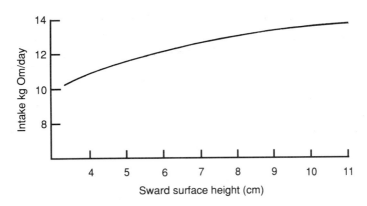

*Figure 16.2* *The herbage intake of spring calving suckler cows on swards of different heights*
Source: Wright

# EXCRETAL RETURN

Appreciation has recently grown of the role of the excreta of grazing animals in the soil fertility cycle, and also in environmental pollution.

## Nutrient circulation

The manurial value of excreta is largely accounted for by the nitrogen and mineral nutrients ingested from herbage, since a large proportion of these nutrients is excreted onto the sward after digestion. During a grazing season of 200 days between 3,000 and 3,500 kg DM is eaten by a 600 kg dairy cow. Nutrient contents in the herbage will be influenced by various factors, including the botanical composition of the herbage, especially if mineral rich white clover is abundant, the stage of sward maturity, inherent soil fertility, fertilisers applied, and season of year and the weather.

The relative proportions of nutrients retained and excreted vary with age, physiological stage and class of stock. Utilisation of nutrients is highest in productive dairy cows and lowest in store cattle or sheep. Of the ingested nutrients, dairy cows, for example, may excrete 70 to 80 per cent of the nitrogen, phosphorus and calcium and 80 to 90 per cent of the potassium, magnesium and other mineral constituents. These nutrients in circulation are not a net addition to the soil fertility cycle, since those excreted in readily available form can circulate several times in a season. Plant nutrients in grazed pasture are in effect

181

mobile, while nutrients locked up in ungrazed herbage only enter the cycle after a slow breakdown as plant material senesces and decays, although rainfall washes out some nutrients from living tissues into the soil. There can be considerable locking up of nutrients in extensively grazed rangelands. This is because stocking is geared to winter carrying capacity, which equates to under-utilisation of the pasture and resultant herbage accumulation during the growing season.

The pool of circulating nutrients in the soil/sward/animal complex can be augmented in several ways. Available soil nitrogen increases from the breakdown and mineralisation of soil organic matter. Weathering of soil parent material and soil itself slowly releases minerals, while the deeper roots of plants may bring minerals from the subsoil into circulation. The most important means of transfer of rhizobially-fixed clover nitrogen to grass is through animal excreta. Transfer of fertility takes place between swards grazed in rotation when they differ in nutrient content; the lower fertility swards may receive the excreta resulting from the ingestion and digestion of herbage from the higher fertility swards. Also, for example, more excreta is deposited on night-grazed than day-grazed swards so night sward fertility increases. Supplementary feed given to grazing stock is a source of additional nutrients through the animal. Applying fertilisers or organic manure is the major means of adding nutrients directly to the soil.

## Nutrients in dung and urine

There are major differences between dung and urine in nutrient content and in amounts and availability for plant growth. Dung mainly consists of undigested herbage cellulose and lignin residues, waste mineral matter and living or dead ruminant micro-organisms together with their metabolic products. The water content is around 85 per cent in cow dung and 65 per cent in sheep dung. Considerable quantities of silica may be present as a result of eating soil-contaminated herbage, although ingested soil also supplies some minerals. Urine is largely (90 per cent) water, plus nitrogenous compounds from the breakdown of protein, sugar substances and other end products of metabolism, with some mineral matter. The proportion of excreted N in the urine increases with increasing N in the diet. A typical analysis of the major elements in dung and urine is shown in Table 16.1.

Of the total nutrients excreted, dung contains 20 to 30 per cent of the nitrogen, almost 100 per cent of the phosphorus and calcium, 10 to 20 per cent of the potassium and 30 to 40 per cent of the magnesium and sulphur. Since dung and urine are deposited in small patches, there is a very high local concentration of nutrients in these patches.

Table 16.1  Major elements in dung and urine

|  | Dung (g/kg DM) | Urine (g/kg) |
|---|---|---|
| Nitrogen | 20 | 10 |
| Phosphorus | 10 | 0.3 |
| Potassium | 10 | 10 |
| Calcium | 10 | 0.6 |

Estimates for the three major nutrients place the localised rates at 700 to 800 kg N/ha, 250 to 500 kg $P_2O_5$/ha and 250 to 400 kg $K_2O$/ha for dung and 300 to 450 kg N/ha, 25 to 50 kg $P_2O_5$/ha and 700 to 800 kg $K_2O$/ha for urine. Value can also be ascribed to other nutrients, including trace elements.

The nitrogen and phosphorus in the dung are largely locked up in organic compounds and these require prolonged action by soil micro-organisms before becoming available for plant growth. Insects, beetles, worms and birds also influence the breakdown and eventual incorporation of dung into the soil. The smaller organisms are present in greater numbers, and are more active, in soils of high rather than low fertility. Hot weather retards the speed of decomposition while cool, moist weather accelerates it. Wet weather causes leaching of soluble constituents from the dung.

In urine, nitrogen and potassium are almost all in readily available form. Because of rapid hydrolysis of urea, which constitutes the major fraction of urinary nitrogen, and the high local pH engendered, a proportion of nitrogen is lost by volatilisation of ammonia. Weather is again important, since rainfall causes leaching of the urea, and of nitrites and nitrates from ammonia nitrification, while volatilisation is increased under hot, dry conditions. Nitrogen cycling was discussed more fully in Chapter 11.

## Pattern of excretal return

Typically a dairy cow excretes 5,500 to 6,500 kg dung (approximately 700 to 800 kg dry matter) during a 200-day grazing season. Daily output is 25 to 35 kg, made up of 10 to 12 defaecations, each 2.5 to 3.0 kg. Amounts vary with quantity of herbage ingested, its digestibility, and water intake. Slightly more than half the total weight is excreted in the evening and early morning. Dung is unevenly distributed over the grazing area but there is some concentration on night-lying areas, around feeding troughs or racks, water troughs, gateways and tracks. About 0.6 to 0.7 m² of pasture is covered daily by dung, or 100 to 130 m² in the grazing season. However, the area made less acceptable for grazing is several times greater, since there is rejection of herbage around the patch. Even after dung pats have decomposed the areas around them may remain unacceptable since the herbage is relatively mature. Local growth is also depressed by dung 'smother'.

Corresponding faecal output figures for sheep are 1 to 1.5 kg daily in 6 to 8 defaecations, each 0.1 to 0.2 kg. During the year, the total output deposited is 300 to 700 kg (approximately 200 to 400 kg dry matter). The area covered by the dung of one sheep is between 0.05 and 0.07 m². Deposition is also increased with sheep concentration on favoured areas such as where the animals are feeding, watering and lying.

The volume of urine excreted by cattle and sheep depends largely on their water intake. Estimates place the daily volume from a dairy cow at 20 to 25 l by way of 10 to 12 urinations. The area covered by a urination is around 0.3 to 0.5 m² depending upon volume, slope of the land and soil moisture-retaining status. About 1,000 m² may be covered per cow per season. Herbage growth rates in and around the patch are stimulated in accordance with the degree to which urine percolates to the sward root zone, but urine scorch damages herbage in hot, dry weather. The daily volume from a sheep is 1 to 2 l from 15 to 20 urinations. On pasture with 1,000 cow grazing days/ha during a season 40 to 50 per cent may have been covered with urine compared with 5 to 6 per cent for dung, although acceptability of up to 20 per cent of dung-affected pasture may be reduced.

Herbage species vary in their tolerance to or recovery from urine scorch or dung smother, the effects of which are less with sheep than cattle excreta. Clovers are more susceptible to urine scorch than grasses. White clover growing points on the stolons may be shaded out by the urine-stimulated grass, or by dung-fouled patches becoming tall and rank. In grass growing on urine patches, potassium content can build up and cause a lower content of other minerals such as magnesium and calcium. Mineral contents in the total herbage are affected by any botanical changes induced by excreta, since sward components differ in mineral composition.


## TRAMPLING

Sward trampling or treading by stock hooves is inseparable from the grazing process so, periodically, sward damage is inevitable. Stresses and strains on soil due to treading, and the frequency with which grazed swards can be completely trodden in a grazing season, are greater than is generally realised (see Table 16.2). The hoof stresses shown for static animals are increased considerably for walking or running actions and there is a soil shearing effect as well as soil compression. Loafing and lying about by stock for about two-thirds of each day is a further loading factor. Other permutations of stocking rate, length of grazing season and distance walked daily can be substituted in the table to estimate the extent of treading for various situations.

**Table 16.2   Estimates of extent of trampling on pasture**

|  | Intensive use | | Extensive use | |
| --- | --- | --- | --- | --- |
|  | Cattle | Sheep | Cattle | Sheep |
| Live weight (kg) | 500 | 75 | 500 | 75 |
| Total area of hooves (cm²) | 320 | 80 | 320 | 80 |
| Static hoof stress (kg/cm²) | 1.56 | 0.94 | 1.56 | 0.94 |
| Stocking rate/ha | 5 | 20 | 3 | 12 |
| Distance walked daily (km) | 3 | 3 | 4 | 5 |
| Step length (m) | 0.5 | 0.2 | 0.5 | 0.2 |
| Grazing season (d) | 200 | 200 | 160 | 160 |
| Average frequency of trampling (no./ha) | 10 | 24 | 6 | 19 |

*Source:* Frame

The damage to grass swards caused by trampling is realised, particularly in wet weather, when poaching damage to the soil/sward complex is obvious. Tracks, gateways, field margins, water or feeding trough areas and intensively stocked fields show poaching symptoms vividly. Subclinical wear and tear of everyday treading and its effects are less evident. Farmer surveys usually pinpoint poaching as a major problem of grazed swards, especially in wet areas.

Historically the grass types mainly sown to establish swards, or already present in permanent grassland, evolved under grazing and have adapted to various degrees of treading. The growing points are well-protected near ground level, for example, while leaf blades grow after partial grazing or damage and tillers recover from bruising or partial burial. However, if the tolerance level of grasses to treading is overrun, damage can be severe—in some cases irreparable—and herbage production is drastically reduced.

## Sward tolerance

Types of grass which tiller profusely, such as intermediate and late perennial ryegrasses, are the most tolerance of treading and have some ability to recover after poaching. The folded leaf structure of perennial ryegrass withstands the crushing pressure from hooves better than the rolled leaves of Italian ryegrass, meadow fescue or timothy. The creeping growth habit, by surface stolons, as with rough-stalked meadow grass, or by underground rhizomes, as with smooth-stalked meadow grass, confers an ability to regenerate from small pieces of stolon or rhizome. Plants with an exposed crown, such as red clover, are particularly prone to treading damage. Prostrate rosette-forming plants, e.g. ribwort plantain, and certain tap rooted perennials, e.g. docks, tolerate poaching and often colonise damaged areas. Plants such as annual meadow grass and chickweed are also early colonisers, mainly because of germination from a reservoir of seeds shed to the soil. Poaching may significantly alter the botanical composition of swards due to differential reaction of grass types to treading. The effects of trampling on soil structure and the various possible measures to reduce the adverse effects of poaching on the soil and sward were discussed in Chapter 9.

## Chapter 17

# Methods and Systems of Grazing

---

Be thou diligent to know the state of thy flocks, and look well to thy herds.

*Proverbs* 27: 23

The method of stocking is the governing factor in determining the botanical nature of a sward.

Martin G. Jones (1933)

---

The methods of grazing used in sward management for ruminants can be divided into two broad types: continuous stocking and rotational grazing. Each has its pros and cons. They vary in capital cost, labour to operate, simplicity of operations, degree of control of the stock and of the sward, and interactions between stock and sward.

## CONTINUOUS STOCKING

This describes free-range, uncontrolled grazing of stock on fields for a prolonged period, often the whole grazing season. Stocking rates may vary during this time; where they remain fixed for a particular period the term 'set stocking' is sometimes used. When grass is scarce, notably in the spring, the grazed area may be augmented by a 'buffer' grazing area. This is an insurance area which can be grazed, or, if not required, conserved. Buffer feed based on silage, hay or straw mixes can also be used to supplement grazing which is in short supply.

The term 'continuous stocking' is preferred to continuous grazing since individual tillers or leaves are not continuously grazed but are in effect rotationally defoliated within the sward. The frequency of defoliation depends on the stocking rate, since stock have free access to all the grazing area, but defoliation may be every 10 to 30 days. Continuous stocking encourages the formation of a dense, well-tillered sward, with up to 20,000 to 30,000 tillers per square metre, which makes for long-term stability, resistance to poaching damage, prevention of weed ingress and tolerance of drought.

Continuous stocking systems have fewer requirements for fencing, watering points, access tracks and gateways than rotational grazing systems. Less labour and capital are needed for installation and operations, and fewer day to day decisions. Field operations are easier, since there are bigger areas in which to manoeuvre machinery. However, there could be problems gathering large herds of dispersed dairy cows, and the system is not suited to farms with lots of small or scattered fields. Sward damage by poaching will be of lower intensity under wet soil conditions compared with that in rotationally grazed small paddocks. The seasonal distribution of grass growth is less variable and growth after summer better than equivalent swards rotationally grazed.

The main potential disadvantage of continuous stocking is the difficulty of achieving correct grazing pressures, i.e. matching stock numbers to the grass available, over the season and between seasons, given the seasonality of grass growth and year to year variation. However, this can be overcome and fluctuations in grass supply minimised, provided the condition of the sward is monitored, compared with stock requirements, and appropriate action taken. This includes adjusting stocking rate by reducing or increasing stock numbers or, alternatively, bringing in buffer grazing areas or buffer feeds.

The rate and pattern of fertiliser N application can also be modified. Under continuous stocking the nitrogen may be applied to a quarter of the total area in rotation weekly rather than to the whole area once a month. Continuous stocking is not suitable for certain sward types such as lucerne or red clover which lack persistence under grazing. On the other hand, perennial ryegrass, with its high tillering ability, is particularly suited to continuous stocking. In the UK the majority of farmers have decided that this system suits their circumstances best. The buffering techniques have given them confidence to stock at the high rate needed to match grazing pressure to the early season flush of grass production. For example, in moderately N-fertilised grassland, 12 to 15 autumn-born beef calves or about 2,500 kg total weight per ha may be needed to ensure efficient grass utilisation. Taking another example, 6 to 9 dairy cows per ha may be needed on heavily N-fertilised grassland. The stocking would reduce with the decreasing grass production as the season advances to approximately half the rate by autumn.

### The 1.2.3 system of continuous stocking

In the 1.2.3 system of continuous stocking, one-third of the total grassland area is grazed during the first part of the grazing season, while the remaining two-thirds is cut for silage. After regrowth of the silage area, stock are moved to it and the previously grazed area is cut for silage after a rest interval of 5 to 6 weeks. Thereafter the whole area is grazed for the rest of the season. The alternate grazing and cutting benefits the long-term persistency of the sward, since it avoids the repeated cutting and removal of heavy silage crops from the same area, which can weaken sward vigour. There is flexibility to alter the fractions cut or grazed. This system was developed for growing beef animals and young dairy stock but is capable of being adapted to suit other classes of animals.

Continuous stocking systems are widely practised in the UK now, in direct contrast to other countries such as New Zealand where rotational grazing is the most popular method. When continuous or set stocking was first widely used in the 1970s, the systems were equated with a low intensity of grass production and stocking rate and were therefore branded with the stigma of low animal performance. Intensification of grass management was generally accompanied by rotational grazing methods. Thus high animal output was wrongly associated with the system of grazing rather than the overall intensification, which included stocking rate. Eventually stocking rate was increased on continuous stocking systems, when it was shown that nitrogen could be applied while grazing. In most comparisons of grazing systems, where other factors have been equal, there has been very little difference in animal output. This is largely because, under grazing conditions, short, dense, well-tillered swards can be of similar productive capacity to taller, less densely tillered swards. However, trials conducted at the highest levels of intensity of grassland use have shown some advantage from well-controlled rotational grazing systems, occurring in periods of limiting sward growth.

## ROTATIONAL GRAZING

Under rotational grazing the area is divided into a series of fields or paddocks which are grazed in sequence, each use being followed by a rest period. The total length of the grazing plus rest period is called the rotational grazing cycle. A number of rotational methods are possible, varying from fairly rigid to extremely flexible, though all control access of stock to swards. In contrast to continuous stocking, the more erect swards which develop under rotational grazing are also more open, with tiller populations of 5,000 to 15,000 tillers per square metre. In rigid rotations, e.g. four paddocks each grazed for one week and rested for three weeks, decision-making is minimal. Paddocks can be subdivided for shorter periods than a week's grazing.

Nevertheless, in rigid systems, grazing pressure requires skilled manipulation to avoid understocking of paddocks in early season or overstocking in late season. Understocking may lead to good individual animal production but to poor animal output per ha; the sward, however, deteriorates due to depressed tillering and the appearance of rejected sward areas of rank and mature growth. Conversely, overstocking reduces individual animal performance though leading to increased animal output per ha. Overgrazing also weakens plant growth and sward deterioration results. In late season stock numbers must be reduced on the grazed area, or the grazing areas increased, and buffer supplementation may also be necessary.

In more flexible and effective systems, the number of paddocks grazed and the time spent grazing per paddock is varied. Normally, in early season, some of the paddocks are grazed for two to four days while the remainder are conserved. The latter paddocks are subsequently added to the previously grazed paddocks to increase the grazing area. This allows for the slower growth rate of grass later in the season. Silage fits into the system better than hay since it is cut at an earlier growth stage and since its regrowth is ready for grazing earlier.

An example is a ten-paddock system with six being grazed for three days each in rotation during early season while four are conserved. By late season all ten paddocks will be grazed for two to three days each. Grazing cycles longer than four weeks should be avoided since grass leaves not utilised within this period decay and die.

Grazing periods within the grazing cycle can be reduced to as low as half a day by the use of movable electric fencing within the paddocks. Because of the labour required, systems of ultra-short grazing periods, or strip grazing, are often used to ration grass closely, but only in the very early or very late season, when grass forms a less substantial part of the total feed intake. An additional complexity of strip grazing is the need to use a back fence to protect already grazed areas from regrazing. This is necessary where strip grazing continues in a paddock for more than four days.

Rotational grazing allows good forward budgeting of forage supply since the areas of growing grass in their various stages can be clearly seen and the amount of grass presented to the stock closely controlled. A high degree of management flexibility to match variability in grass growth is possible, including close integration with conservation; this entails regular monitoring of the sward and frequent decision-making to maintain herbage nutritive value at a high level.

## CREEP GRAZING

Both continuous and rotational systems of grazing can be adapted to include 'creep' grazing where young stock graze ahead of their dams to obtain first choice of the grass on offer. Creep grazing systems are most commonly used with ewes and lambs to allow the young lambs access to leafy nutritious herbage. In forward creep systems, the ewes are allowed into the creep areas after the lambs to clean up grass residues, but sideways creep systems prevent this. Some advantage to the lambs is gained from sideways creep grazing since the sward is less contaminated with parasitic worm larvae from the ewes, but 'clean' grass for lambs is not guaranteed since lambs and ewes do come together on the ewe-grazed areas. Creep grazing systems have not become popular because of the complexity of management.

The leader/follower system is a more acceptable form of creep grazing. Young dairy heifers or beef stock are grouped into two differing age classes, with the younger group preceding the older during the rotational grazing of a sequence of paddocks. The total time spent in an individual paddock should not exceed four to five days; otherwise young sward regrowth may be regrazed and recovery weakened. Since good liveweight gains are generally required in both groups of animals, it is essential to monitor performance of the follower group to check for satisfactory progress. Leader/follower systems have been tried for other classes of stock, e.g. dairy cows, through grouping them by milk yield, but the added complications of management has limited their uptake.

## 'CLEAN' GRAZING

Worm infection rather than potential grassland production is a limiting factor to the intensification of lowland sheep enterprises. Ewes and lambs are infected by the stomach worm (*Osteragia*) and the gut worms (*Trichostrongylus* and *Nematodirus*). If the infection by *Ostertagia* and *Trichostrongylus* is heavy, the resulting gastroenteritis causes serious losses in productivity and even death. Pasture contamination is heaviest in spring, from overwintered larvae, and in late summer, from worm eggs excreted earlier by ewes and lambs. Thus infection spreads rapidly during the grazing season. *Nematodirus*, a gut worm which causes scour and death in young lambs, is difficult to eradicate, because of the long period of dormancy on pasture of the eggs, which subsequently hatch giving rise to the infective larvae.

Effective worm control requires an integrated system of grazing management and strategic dosing with an effective anthelmintic. To achieve clean grazing for ewes and lambs, approximately equal areas of grassland must be altered annually between sheep, cattle and conservation (or arable cropping). The conservation or cropping is necessary in the rotation since pasture contamination can be continued from the cattle to the sheep. The critical feature is that ewes and lambs are not grazed in fields which carried lambs or hoggs the previous year. Different systems can be designed for individual farm circumstances. The long-term survival of its eggs means that *Nematodirus* persists on the pasture for many years, and it also affects calves, so it can be passed on to lambs in a clean grazing system. Nevertheless, clean grazing does reduce the infection and the need to dose. Over the years, clean grazing systems have demonstrated benefits of: higher ewe stocking rates; improved lamb performance and output; a reduction in anthelmintic use; and better grassland utilisation.

## MIXED GRAZING

Mixed grazing of animal species is an option on many types of farm, offering simplicity and adaptability. Experiments in grazing fattening beef cattle and sheep together show advantages in individual animal output and in output per ha over separate grazing by either species. Part of the benefit is attributable to different grazing habits: cattle graze the taller herbage and condition the sward for sheep, while the sheep prefer shorter herbage and are more selective, yet also graze closer to cattle and sheep dung. Further benefit accrues from dilution of parasite infestation, particularly helminths in the sheep. Growing beef cattle with their increasing herbage intakes over summer benefit from the removal of finished lambs.

On many dairy farms the grassland area for dairy young stock is under-utilised and better sward management is achievable by adding a ewe flock. Apart from mixing and grazing animal species on the basis of some fixed ratio there are situations where the temporary addition of a second species is advantageous. On upland farms where the sheep flock is the most important enterprise, suckler cows may be used to graze down sheep-stocked fields where sward height is increasing beyond the optimum for the sheep. In a more unusual example, it has been demonstrated that goats prefer eating rushes, thistles, gorse and mature grass in a sward, and this characteristic could be exploited in selected mixed grazing situations to improve the quality of the herbage on offer to sheep and cattle. In the present era, when diversification of animal enterprises is becoming a feature of European grassland farming, there is potentially more scope in the future for mixed grazing systems of different kinds.

## ZERO GRAZING

Zero grazing, sometimes referred to as 'forage feeding' or 'soiling', is the cutting and carting of grass to housed or partially housed stock, usually on a daily basis, thereby minimising the wastage inherent in the grazing process and giving a higher efficiency of grass utilisation. The system has the potential to give greater animal output per unit area than conventional grazing, but involves high capital costs of machinery. It is commonly used where large units of around 1,000 dairy cows exist, as in eastern Europe, and requires good slurry storage and handling facilities. Elsewhere it may be used on farms with scattered fields, distant from the farmstead, or for special purpose crops such as maize or kale. It may also be used on farms for short periods in early or late season to avoid the poaching of swards by grazing stock. The possible dangers of soil compaction and sward damage by heavy harvesting machinery should not be overlooked, however. The stock are totally dependent on the operator for the quantity and quality of forage supplied and a high degree of grassland management skill is needed to provide a continuous supply of leafy, highly digestible forage. Intervals between cuts normally range from three to four weeks in early season and four to five weeks later in the year. Fertiliser application to cut areas is made weekly. The system incorporates integrated conservation of herbage which has become too mature for feeding out fresh.

## STORAGE FEEDING

In storage feeding, the stock are housed and all the grass is fed in a conserved form, usually as silage. This system entails high capital costs for adequate housing and silage storage. It is not widespread in the UK and its success depends on efficient mechanisation, from field to feeding, and good silage-making procedures to minimise losses and ensure a high quality diet. Storage feeding is mainly suitable for large-scale units, including bull beef enterprises, and there can be major problems with storage and utilisation of the silage effluent produced and with the animal slurry output.

Partial storage feeding is the practice of grazing by day and offering silage *ad libitum* at night to the stock, which may be housed. It is sometimes used with dairy cows in spring or at other times of grass scarcity. The silage on offer ensures total feed intake is not limiting to animal output. For dairy herds, 4 to 5 t silage (about 1 t DM) per cow is required in addition to normal winter silage requirements of 8 to 10 t silage (about 2 t DM) per cow.

## CHOOSING A GRAZING SYSTEM

The choice of grazing system on a farm is a matter of individual preference as affected by such factors as existing layout of farm fields, field size, labour availability and ease of management, herd size, grass handling and storage facilities available, or any special constraints. It may be advantageous on many farms to adopt a combination of systems, to make best use of the grass resource. The more flexible the system the better it will cope with variable grass growth rates. All grassland management systems must be well planned in advance, including options to take in times of shortage. It must also be kept in mind that the system of grazing is only one link in the chain of overall grassland management which, however efficient, cannot compensate for other weaknesses such as a low level of grass production.

## MONITORING SWARD STATE

The primary aim of grazing is, and always has been, to match a variable seasonal grass production and quality with animal needs. It is important, therefore, to be able to assess whether or not the amount of herbage available to stock is sufficient both currently and in the near future. This assessment of the sward should preferably be made in an objective manner. Traditionally, walking the fields and judging by eye, utilising past experience, has been the method for day-to-day decision-making. Graziers can become quite expert in judging forage supply.

Judgment is based on sward height and density and is more accurate on short swards of uniform density than an open swards with variable density. It is least accurate on tall swards, or when the herbage has a high dry matter content during dry, sunny weather. In a grass/white clover sward, the amount of clover is often overestimated because of the plant's broad leaves. Since sward density is difficult to quantify, sward height has the greatest influence on eye appraisal methods. It is possible, though rarely done on farms, to check the judgment against cut and measured sample areas of sward.

Several versions of a simple weighted-disc instrument have been developed to estimate compressed sward height. The disc normally consists of a light aluminium or plastic plate and fits over a graduated stem held in a vertical position with its base at ground level. In some versions the plate is pushed up by the sward canopy, while in others the plate is allowed to fall. The height readings are influenced by a combination of sward height and density and therefore give lower readings than simple sward height measurements. Weighted-disc techniques are least accurate on stubble following rotational grazing. Trampled rejected herbage also makes accurate height measurement difficult. When relating the readings to herbage amounts cut from representative areas, it is necessary to use calibrations for different sward types and at different seasons of the year. Weighted-disc meters are used on farms in New Zealand and The Netherlands, but in Britain their uptake by farmers has been slow. Another method measures changes in the capacitance in a sward by electronic probes and these changes, in conjunction with previous calibrations against herbage yield estimates, give the current amount of herbage present. Electronic probes are commercially available in the UK but not yet widely used.

A major reason why farmers have not adopted objective techniques of herbage assessment is the lack of a standard method. However, modern research and experience, tying in the response of grass growth to grazing with animal requirements, output and behaviour, has come up with a simple practical answer—keep the sward surface at the right height above the ground. Relationships have been established between sward height and herbage intake per bite, rate of biting, grazing time and daily herbage intake per animal, together with their effects on sward and animal performance (see Chapter 15).

Sward heights can be measured in various ways—by ruler, graduated stick or even wellington boots with white-painted horizontal stripes. However, a highly objective and accurate method, using a specially designed 'sward stick', was developed at the former Hill Farming Research Organisation, now the Macaulay Land Use Research Institute.

Essentially, the sward stick measures leaf height by first touch of a small, transparent, plastic tongue fitted to a metal sleeve which moves up and down over an inner metal rod graduated in 0.5 cm divisions. The base of the rod is placed on the ground and the sleeve lowered until the tongue makes contact with any part of a green leaf; the height is then read off the scale. Seedhead or stem contacts are ignored and measurements are not taken at gateways, troughs or other areas unrepresentative of the sward. Heights must be taken on both heavily and lightly grazed areas. On continuously stocked swards, 30 to 50 readings are typically taken in the field, to the nearest half centimetre.

These are then totalled and averaged. Fifty readings are preferable on large fields, and where the sward has developed a mosaic of tall rejected herbage and closely grazed areas.

The standard method is to walk the field in a 'W' pattern and take readings every 15 to 20 paces, but transects in rectangular fields or diagonals in square fields can also be used, the important point being that readings should represent the whole field. Ideally swards should be measured twice weekly in the May–June period of rapid growth or in periods when heights have drifted from target. Otherwise, once a week will suffice, extending to once a month with winter grazing.

Optimal sward heights depend upon type and class of stock, including their physiological stage, whether at peak lactation or dry, and whether at store or fattening. Table 17.1 shows target heights for continuous stocking systems. The action needed when sward surface heights are drifting above or below target is shown in Table 17.2. It is simpler to remedy the situation when sward surface height is falling below target rather than rising above. Using a continuously stocked lowland sheep system as an example, the recommended sward height profile over the year is shown in Figure 17.1.

**Table 17.1   Target range of sward surface heights for continuous stocking systems**

| Stock | Sward surface height (cm) |
|---|---|
| Sheep (spring and summer) | |
| Dry ewes | 3–4 |
| Ewes and lambs | |
| (medium growth rate) | 4–5 |
| Ewes and lambs | |
| (high growth rate) | 5–6 |
| | |
| Sheep (autumn) | |
| Store lambs | 4–6 |
| Finishing lambs | 6–8 |
| Flushing ewes | 6–8 |
| | |
| Cattle | |
| Dry cows | 6–8 |
| Store cattle | 6–8 |
| Dairy replacements | 6–8 |
| Finishing cattle | 7–9 |
| Cows and calves | 7–9 |
| Dairy cows | 7–10 |

*Source:* Hodgson *et al.*

**Table 17.2 Approximate stocking rate changes (per cent of current stocking) needed when current sward height drifts from target**

|  | Change in height over previous week | | |
| --- | --- | --- | --- |
| Current height | Decrease | No change | Increase |
| High | 0 | +10 | +20 |
| On target | −10 | 0 | +10 |
| Low | −20 | −10 | 0 |

*Source:* Hodgson *et al.*

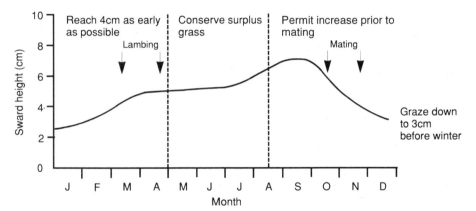

*Figure 17.1* Annual profile of sward height recommended for a lowland sheep system

*Source:* Maxwell and Treacher

In rotational grazing systems target sward heights are based upon the height of grazed stubble to be left (see Table 17.3), not the height of sward at entry which should usually be around 15 to 20 cm for dairy cows and cattle and 10 to 15 cm for sheep. These entry heights will vary with season of year and length of rest interval, but rest periods should not normally exceed 20 to 30 days, the shorter rest periods being typical of early season. Whichever grazing system is used, swards should be grazed down to a height of 3 to 4 cm in late autumn/early winter to minimise the risk of winter damage.

It should be emphasised that most of the trial work verifying the concepts of target sward heights has been done on perennial ryegrass-

**Table 17.3   Target grazed stubble heights for rotational grazing systems**

| Class of stock | Stubble height (cm) |
| --- | --- |
| Sheep | 4–6 |
| Store cattle | 6–8 |
| Dry cows | 6–8 |
| Finishing beef cattle | 7–10 |
| Lactating dairy cows | 7–10 |

*Source:* Hodgson *et al.*

dominant swards or ryegrass/white clover swards. Parallel studies are desirable for swards based on other grass species, e.g. timothy or cocksfoot or on permanent swards of more complex botanical composition.

First impressions by farmers learning about sward height rules are generally that the heights for continuous stocking seem very low; however, the compensatory factor is the large number of tillers per unit area, the general leafiness and sparsity of stems and seedhead, and the sustained vigour of growth, especially in the second part of the season (see Chapter 15). In contrast, stubble heights for rotational grazing seem high but, if lower heights are used, stock are forced to eat basal sheath and stem, including senescent or dead material. This is of low digestibility and feeding value, and animal productivity is reduced accordingly.

*Chapter 18*

# Seasonal Objectives and Management

---

The part of the holding of a farmer or landowner that pays best for cultivation is the small estate within the ring fence of his skull.

Charles Dickens (1812–70)

---

## EARLY BITE

Farmers have always been interested in the production and subsequent utilisation of 'early bite'. After a winter's indoor feeding of stock it is a relief to get them out to pasture to harvest their own food, and grazing is also cheaper than indoor feeding. Choice of field, grass type, fertiliser application and past grazing management are important factors influencing earliness of growth.

Fields with a sheltered, southerly aspect and light textured, free-draining soil which warms up with the rise in spring temperatures are best. Such fields can also carry grazing stock with a minimum of poaching damage. In contrast, north-facing fields with heavy textured, wet soils are slower to warm up and the rate of grass growth is slower. Even when enough grass is available for grazing on such soils it may not be possible to use it for fear of excessive poaching damage. On farms with heavy land, it is better to ensure that there is plenty of silage rather than relying on early bite.

Italian, hybrid and perennial ryegrasses are the most valuable and commonly used species for early bite. Within these species early growth is closely related to earliness of heading or ear emergence, although there are exceptions. The Recommended Variety Lists usually contain information on seasonal yield distribution so the earliest producing varieties can be identified and chosen. Italian ryegrass in its first full harvest year is the best early producer but its short-term longevity must be borne in mind. Timothy can produce valuable early growth in cold, wet upland conditions. Tall fescue is a species with good early growth characteristics but its other characteristics—slow establishment and a coarse forage—make it relatively unattractive for grazing stock. It is thus a minor species for UK conditions.

199

Fertiliser N application in late winter to early spring is necessary to stimulate early growth. Timing is important and can range from early to mid February in milder south-westerly parts to late March to mid April in upland northerly areas. Local experience is a valuable guide but, in general, N application tends to be seven to fourteen days later than optimum. Soil conditions often prevent application due to possible damage from spreading equipment, but specialised equipment with low ground pressure tyres is available. Application should only be made if weather forecasts for the immediate future are favourable.

When choosing the time of N application the objective is to be early enough so as not to lose growing days when the temperatures are sufficiently high for grass growth, yet not to be so early that temperatures are too low for grass to respond to N, or that the N may be leached or denitrified before being taken up by growing grass. Normally grazing is possible three to five weeks after N application. A reasonable aim for early bite for dairy cows is 1 ha per 10 to 12 cows. There should be silage or hay reserves in case of any hitches, and it is necessary to plan for the supply of grass to follow the early bite.

There are a number of forecasting methods based on temperature but sometimes predictions have to be overruled by local knowledge or unusual weather conditions. Grass grows when the soil temperature at 10 cm depth is 5 to 6°C, so a sustained period with temperatures at this level is suitable, provided it does not occur too early in the year. In recent years, the 'T-sum 200' system which originated in The Netherlands has been in use. The maximum and minimum temperatures in degrees centigrade are recorded daily from 1 January and the average worked out each day. The averages are then cumulatively summed, but ignoring any minus averages. Fertiliser N is applied when the total of the daily averages reaches 200°C, provided soil conditions for spreading and the future short-term weather forecast are satisfactory.

Temperatures are part of the information recorded at various official or grassland research meteorological stations, and advisory services can supply weekly updates. The information may also be available in the national farming press. However, such information is for fairly large areas and may be imprecise for individual farms where the thermometer can be installed facing north on a post or on a building, and the temperatures recorded at 9 a.m. each morning.

Trials have shown that the 200°C target need not be rigid and that there is a period two weeks before and after it when fertiliser N application will give 90 per cent of the optimum yield response. This gives flexibility to manoeuvre according to soil and weather conditions. Temperature monitoring does focus farmers' attention on grassland and alerts them to the need to prepare spreading machinery and have the fertiliser to hand. Traditionally, straight N fertiliser has mainly been used to stimulate early bite, but on soils with very low or low soil P status a yield advantage can be obtained from applying a compound nitrogen: phosphate fertiliser (see Chapter 12). The amount of N required on a soil of moderate N status will be in the range 60 to 80 kg/ha, depending upon the management objectives.

Autumn and winter management influences earliness of grass growth in spring. Ideally, surplus grass should be grazed off in late autumn and the sward then rested. This allows light to stimulate tillering at the base of the sward during early winter so that the resultant dense sward is able to respond well to spring-applied N. If winter grazing takes place, as with wintering hoggs for example, spring yield will be progressively reduced the longer the period of grazing. In this case, the fields designated for early bite production should be grazed first, preferably by December, and then shut off for the rest of the winter. Trials have shown the wisdom of this and, indeed, prolonged winter grazing can also depress first cut yields of silage. These effects are shown in Table 18.1, using a sward grazed by migratory geese in which areas were protected from grazing by setting out exclosure cages at monthly intervals during winter and also including areas from which geese were never excluded.

The response of grass to fertiliser N is less in early spring than later because of low temperatures, so the herbage is more expensive to produce. The grass should therefore be utilised efficiently, making use of the electric fence, for example, to ration its use and avoid wastage.

Table 18.1  Relative herbage DM yields and digestibility in spring following sward protection from winter grazing by migratory geese in west Scotland (early bite, 100 = 1.8 t/ha, silage, 100 = 7.5 t/ha)

| Months geese excluded from grazing | Early bite (26 April) | | Silage cut (1 June) | |
|---|---|---|---|---|
| | Relative yield | D value | Relative yield | D value |
| October | 100 | 74 | 100 | 65 |
| November | 73 | 74 | 84 | 66 |
| December | 71 | 73 | 84 | 64 |
| January | 49 | 75 | 85 | 66 |
| February | 36 | 73 | 84 | 67 |
| March | 20 | 75 | 76 | 68 |
| Geese never excluded | 6 | 72 | 57 | 68 |

Source: Patton and Frame

## MIDSUMMER GRAZING

Most grasses show a decline in their rate of growth during midsummer. This is linked mainly to the inherent seasonality of production of the grass plant species and their varieties, to the sward structure maintained and to growth producing factors, especially water supply and nitrogen. Certain animal production systems require particular management to ensure that adequate supplies of grass are available for grazing in midsummer; the summer calving dairy herd is a prime example. Various options are available to counter grass shortage in midseason but these have to be considered in relation to individual farm circumstances. Some influencing factors, summer rainfall for example, are not under the farmer's control.

Cocksfoot and tall fescue grass species exhibit better growth potential in dry summers than the ryegrasses, but have poorer intake and feeding value characteristics; they are therefore less suitable for highly productive stock but are nevertheless valued for dry situations in some countries. Among the large number of perennial ryegrass varieties available, some exhibit better midseason growth and digestibility than others and are therefore useful in specially formulated seed mixtures. White clover has good midsummer growth capability if grown with compatible grass companions and management favours the clover. A timothy/meadow fescue/white clover sward used to fulfil this purpose. A well-proven source of midsummer growth is the 'maiden reseed', that is, a sward established early in the spring and fertilised well which provides youthful growth vigour in the first summer. Italian, hybrid and early perennial ryegrasses are particularly suitable species for this purpose. However, the regular spring reseeding that is necessary obviously does not suit the many farms where management policy is largely based on long-term grassland.

Grass growth in midseason is influenced by the intensity of defoliation earlier in the year. Close early grazing will ensure the development of densely tillered swards capable of nutritious leafy growth into the summer season, since the ensuing tillers will be vegetative rather than reproductive. Conversely, understocking would lead to undesirable ageing and senescence of grass tissues and accumulation of rejected, stemmy growth, with an associated decline in energy, protein and mineral contents. Tillering capacity and generation of new leaf tissue are depressed in such swards. The soil surface and upper soil layers of the resultant open swards will be susceptible to drying out in hot summer temperatures and this will further reduce growth rate.

Flexible management in early season, using sward surface height guidelines, for example, is required in order to closely match grass supply and stocking rate. Topping can also be used as a means of sward control if the matching process develops shortcomings. Clean silage or hay aftermaths can also provide good midseason production, provided there is a period of regrowth before drought stress occurs in midseason. Aftermath growth is more rapid the earlier the stage of growth at which the conservation crop is taken. Tiller regeneration and sward growth can be very poor indeed following hay crops cut at a late stage of maturity, which is tempting for ease of hay making.

The amount and pattern of summer rainfall are major determinants of grass production in midseason. The availability of moisture to the plants will also be influenced by the moisture holding capability of the soil. Deep, heavy textured soils are best in this regard and shallow, light textured soils poorest. Where drought is a regular occurrence irrigation has a positive benefit and should be considered for at least part of the grassland, provided the economics of installing a system and having adequate water supplies are sustainable.

As long as there is adequate moisture in the soil, fertiliser N application will enhance sward production in midseason. The herbage production response per kg of applied N will not be as great as in late spring and early summer but, nevertheless, will normally be substantial and worthwhile. A typical dressing in soils of moderate fertility would be 50 to 60 kg N/ha. There is, however, no point in applying fertiliser N on drought-stricken swards since it will not be washed into the root zone for uptake by the plants. Aqueous ammonia injected into the soil in spring at a rate sufficient for the season will ensure that nitrogen supply in the soil is not limiting during dry conditions.

## TOPPING GRASSLAND

There are two schools of thought on topping. One says if topping is required, the sward must have been undergrazed. The other says it is a management tool for rejuvenating the sward. There are a number of techniques available to ensure that topping is not needed often. They include: grazing management aimed at matching stocking density to the amount of grass available; conserving surpluses; alternately grazing and conserving fields; mixed species grazing; the use of scavenger stock; or when a problem has arisen, 'mob stocking' with animals at an unproductive phase of their life cycle. The latter technique is really a form of animal rather than machine topping. Nevertheless, from time to time grazing management may not be fully in tune with growing conditions and mechanical topping will be desirable.

Topping established swards has several functions. It cuts down maturing tufts of grass which have been rejected by grazing stock. While this helps the aesthetic appearance of the sward—sometimes an objective—more importantly, it stimulates tillering and leaf production which make for sward density, production and nutritive value. Traditionally, topping is an occasional operation carried out in summer, and often later than is desirable because the problem has become acute. While topping should remain an infrequent rather than a routine operation, it should be used flexibly whenever there is a build-up of rejected tufts of grass—around dung pats, for example.

Small amounts of grass cuttings can be left on the sward surface to decay or be grazed. If there are large quantities, they should be removed by topping with a forage harvester to prevent them shading the sward and retarding growth. The height of topping can be adjusted to suit the sward height it is desired to maintain, for example, down to 6 to 8 cm on cattle-grazed swards and 4 to 6 cm on sheep-grazed swards. Topping can be undertaken to control weeds and is effective against tall growing annuals in establishing swards, where it can complement grazing. It is also useful against perennial weeds but care should be taken to remove these if there is a danger of seed spread from cut plants—flowering docks, for instance—or if the plants are harmful to grazing stock, e.g. ragwort, which is not normally grazed by cattle when it is growing but becomes acceptable after being cut down. Topping is thus a worthwhile operation when used flexibly and can help maintain a leafy productive sward.

## LATE BITE

Grass growth rate responds to autumn nitrogen, but if weather conditions are likely to be wet the problem may be how to utilise it without excessive poaching damage. As with early bite, fields with free-draining, light to medium textured soils are best. A fertiliser N dressing of 40 to 60 kg/ha can be applied up to mid September in southern areas, but in northern parts it is better to apply nitrogen no later than mid August. These modest amounts and cut-off times are necessary to avoid N leaching into drains and watercourses since swards regularly N-fertilised and grazed over the season will have residual N available. Also the herbage growth resulting from late autumn N application can become prone to winter damage, especially ryegrass swards if they are not well grazed off or if grazing is delayed. Old swards are at greater risk than youthful ones. The prolonged autumn cover inhibits tillering by keeping light from the shoot bases, and carbohydrate reserves are diverted from tiller bud formation to plant respiration. The removal of cover which exposes the remaining plant tissues to low temperatures is a physiological shock to plants. There are differences in autumnal growth vigour and winter hardiness among varieties.

Deferring the use of autumn grass—foggage or autumn-saved pasture—to provide winter grazing was more commonly practised in the past than now. Cocksfoot, meadow fescue and timothy proved the most suitable species for this since the accumulated herbage on ryegrass swards tended to rot earlier. Due to the senescence which takes place with time, the amount and feeding value of the foggage declines the longer it remains unutilised, especially if frost intervenes. With both autumn and winter grazing it is essential that grazing stock should have access to shelter in the event of inclement weather. Thus, if possible, fields with natural protection should be chosen. In the drive to prune animal production costs, extending the grazing season where feasible seems a logical aim to pursue because the expense of indoor handling and feeding operations is reduced.

## WINTER MANAGEMENT

Periodically, Mother Nature reminds us sharply that it does not pay to neglect winter management of grassland. Witness the damage to swards during severe winters in the past, and the spring scramble to sow alternative crops to fill the ensuing feed gap. Nor should the less dramatic, sometimes subclinical injuries which happen regularly be forgotten. These can be caused by a complex of weather factors, management factors and pest incidence, alone or in combination. Prudent grassland management is a year-round commitment and not only an exercise for the growing season. As the main growing season draws to a close, preventing, or at least minimising, potential winter damage to swards should be a major objective. This can be done by identifying causes and remedying them by management factors, some immediate and some longer term, as outlined below.

'Winter burn' is a desiccation of leaves and tillers; the tissues dry out, shrivel and become yellow-brown. The extent of damage can range from leaf tips only to virtually all the above-surface vegetation of the plant. Grass growing points, at or just below the soil surface, usually survive and spring production potential may not be markedly affected, although initial growth will be slower than from non-winter burnt material. Cocksfoot is a good example of a grass which often displays winter burn from spring winds or frosts.

At the other end of the damage spectrum, complete winter kill of sward shoots and roots can occur. In spring, winter killed areas may be evident as bare patches of soil or areas with dead grass, perhaps forming a mosaic with areas of live plants. Tillers on surviving plants will often be thinned out and plant growth vigour reduced. After a severe winter, whole fields can be bared. 'Winter proud' grass, around the previous season's dung pats, for example, and rejected by grazing stock, may typically become a decaying mass of dead herbage; beneath this, tiller thinning, plant damage or whole plant death is apparent.

Swards lacking density and growth vigour may be the result of poor establishment, overuse the previous season or winter kill. Nevertheless, it pays to check for infestation of leatherjackets. These larvae of the crane fly, hatched from eggs laid early in the previous autumn, feed on the grass shoots during winter, except when it is very cold. In addition to direct injury, tiller thinning and sometimes totally bare patches, the debilitated sward is rendered more vulnerable to frost damage. Severe leatherjacket invasion can only be controlled by insecticidal treatment.

Following prolonged snow cover the presence of 'brown patch' infection can indicate that snow mould fungus has been developing. Control by fungicides is uneconomic. The mould pathogen can also spread in the absence of snow when the sward's natural resistance is lowered by stress factors such as low temperatures, wet weather or a covering of slurry.

The caprices of weather are an overriding factor. Its effects interact with management factors, which are dealt with below. Winter temperatures have a dominant role, with freezing causing damage or death. Soil heave, due to alternate freezing and thawing, can shear roots and shoots, especially in young swards established the previous autumn. On badly drained/waterlogged areas, the sward may be suffocated under sheets of ice. In comparison with the steadier temperatures of continental climates, where a plant's winter dormancy is more definite (a winter damage avoidance mechanism), fluctuating temperatures are a feature of oceanic climates. Depending upon their region of origin and breeding, some grass varieties may be stimulated into growth during spells of mild weather. The plants are then in jeopardy when a frosty snap follows. Extended cloudy wet weather lowers the resistance of swards to low temperatures.

On bleak exposed sites, searing cold winds can dry out plant tissues and, if plant roots are frozen in the soil or broken, they may not be able to replenish plant moisture lost by evaporation through the leaves.

By constantly buffeting leaves or tillers together, wind can also cause physical damage to the tissues through chafing; water loss is accelerated, plant vigour weakened and cold stress resistance lessened. The wind factor is uncontrollable unless some practicable system of well-sited hedge or tree shelter belts fits the farm's situation.

The presence of high soil nitrogen levels in late autumn can cause major damage, because it can stimulate a flush of soft, high moisture material which, if not rapidly grazed or cut for silage, is prone to rotting in winter. Also, the individual plant is depleting its carbohydrate reserves when it should be conserving them, particularly the soluble sugars which lower the freezing point and increase frost resistance. Following late autumn or early winter defoliation, the low temperatures, low light intensities and short day lengths at this time of year provide the plant with little scope to build up its food reserves. On highly fertilised, intensively stocked farms a late season build-up of soil nitrogen is inevitable, so extra care is needed, avoiding autumn N application, for example, and grazing the sward short and uniformly.

The value of animal slurry as a source of plant nutrients, particularly potash and nitrogen, is widely recognised; however, slurry could be used more advantageously than it is at present. The nutrients are most effective when application is deferred to late winter/early spring. Applied earlier, slurry nitrogen can stimulate 'out of season' growth which is susceptible to winter damage and also the sward is put at risk when colder spells follow. If slurry is applied too thickly it can physically smother and weaken the sward, making it less able to recuperate or, indeed, survive. Wheeled traffic should be minimised in winter since it can injure or shatter frozen plant material, often causing serious harm, and can also damage soil structure with rutting, puddling or compaction effects. The associated sward will suffer varying degrees of short- or long-term damage.

Early spring growth rate and the production of grass are reduced by previous winter grazing, compared with no grazing. Depression is greater with an increased winter stocking rate, an effect attributed to poaching damage in addition to defoliation. Spring yield reduction is greater from January to March grazing than from previous October to December grazing since there is less time and less green photosynthetic tissue to replenish food reserves for growth initiation in spring. (Table 18.1 illustrates these features.) Rotational field grazing of wintering stock enables those fields designated for early bite to be protected against indiscriminate grazing and allows early application of fertiliser nitrogen. One difficulty in wintering hoggs from the hills on lowland farms is that fences are often not sheep proof.

The ryegrasses are the most prone to winter injury; Italian ryegrass is particularly vulnerable. However, in recent years, plant breeders have produced diploid and tetraploid perennial and Italian ryegrass varieties with enhanced winter hardiness. There is no excuse, therefore, for sowing susceptible varieties in seed mixtures, especially in upland areas or on exposed sites where rigorous winters are the norm. In Recommended Variety Lists, winter hardiness ratings are given for the ryegrasses. Timothy varieties are very winter hardy and usually winter green. Cocksfoot varieties show a range of tolerance. However, it should be realised that even the hardiest grass variety may not escape damage during really severe winters, particularly if some of the other causal factors outlined above are operating.

Table 18.2 summarises a number of winter management factors beneficial to sward productivity and persistence.

**Table 18.2   Key points for winter grassland management**

- Use winter hardy grass varieties in seed mixtures
- Do not apply fertiliser nitrogen after mid August
- Maintain balanced fertiliser application to ensure sward vigour
- Graze grass swards down to 3 to 4 cm before winter
- Avoid late autumn/early winter slurry dressings
- Control grazing by wintering stock to prevent soil/sward damage
- Minimise wheeled traffic on fields during winter
- Check for and control infestation by leatherjackets
- Maintain adequate drainage systems
- Plant hedge or tree shelter belts near exposed fields

*Chapter 19*

# Silage Making

That our garners may be full, affording all manner of store.

But good silage can be made easily from meadow-grass, rye-grass, trifolium, oats, vetches etc. The crops, however, ought to be cut just before they are ripe, as at that period the nutrients of the fodder will be more equally distributed throughout the whole structure of the plant. . . . The golden rule of silage making is—look to the consolidation of the sides, and the centre will look after itself.

Parry *et al.*, *Journal* of the Royal Agricultural Society of England (1886)

The making of silage has a long historical tradition, extending back into Greek–Roman antiquity. While it was first made in the UK in the late nineteenth century, a great expansion occurred from the 1960s and 1970s onwards. In the 1886 issue of the *Journal* of the Royal Agricultural Society of England, there is a reference to probably the first silage competition in Britain. It involved the judging of 27 silos and 9 silage stacks. Among other things, the judges referred to the benefits of cutting grass at an early growth stage, of chopping and consolidation, especially at the sides of the stack, and of sealing. In their summing up the judges concluded that: 'The chief advantage of silage making against hay making is its comparative independence of the weather; that the fodder is handled while green, without any risk of the tender and nutritious leaves being lost on the ground as in hay making; that the resultant silage is succulent and palatable; and that on purely grazing farms it is now possible to obtain a portion of the grass crop for winter use in such a state as to equal the effect of summer-fed grass for the purposes of the dairy.' This latter objective still remains an ideal a little out of reach, but not by very far, in the best silages.

The total amount of silage and hay dry matter conserved in the UK is currently between 14 and 15 million tonnes, with silage comprising over 70 per cent of this total. This is a reversal of the situation 15 years or so ago when silage made up only a third of a total of 10 million tonnes of dry matter being conserved. An expansion in silage making started in the early 1970s and has continued ever since, fuelled by advances in machinery and equipment, and in the technology of making and utilising silage efficiently. One could mention forage harvesters, precision chop harvesters, silage additives, plastic sheeting and big bales. Inherently, silage has advantages over hay since it is less weather dependent, is suitable for high nitrogen use to increase grass production, gives quicker regrowth after cutting and has potentially better feeding value. It is also more suited to an integrated grazing and conservation system than hay.

## SILAGE IN SYSTEMS

The period of most vigorous grass growth is May to June. According to the management system practised, 50 to 70 per cent of annual production can be obtained by the end of June. The 50 per cent figure would be typical of grazed fields and the 70 per cent for fields with a late first cut of silage. It therefore makes sense to maximise the area cut for silage during this period and to restrict grazing to the minimum area consistent with providing a sufficiency of leafy, high quality grass.

There is scope for improvement in this aspect since stocking densities during early season are often too low in practice; indeed, understocking is more prevalent than overstocking. Undergrazing results in large patches of mature grass developing with low acceptability and feeding value, and the sward's regrowth vigour is then lessened following rejuvenation of these patches, by topping, for example. Fear of running out of grazing is the main reason for understocking, but this fear can be removed by the use of buffer grazing areas in set-stocked grassland, or the flexible use of fields in rotational grazing systems accompanied by buffer feeding of silage, hay or a straw-based supplement at grazing.

Silage making need not be confined to the traditional early first half of the season. Rather, it should be seen as a sward management tool to conserve grass surplus at any time during the grazing season, in addition to its role of conserving feed for winter. Big bale silage makes up a significant proportion of the total silage now made and is ideal for 'opportunity' silage making to remove surplus grass at any time and aid good grassland management. In any case, the main aim is to make enough conserved feed to meet the stock's winter requirements, not to have a rigid policy based on a fixed conservation period or number of cuts. This total integration of grazing and conservation will make the

▲ Well made silage cut out by block cutter to minimise exposure to air
*Chapter 19*

▲ Poorly consolidated and sealed silage showing secondary aerobic
deterioration and wastage    *Chapter 19*

Well fermented silage, the product of a desirable lactic acid fermentation in anaerobic conditions (R. F. Gooding)
▼    Chapter 19

▲    Overheated silage as a result of aerobic conditions persisting in the silage, causing respiration and oxidation (R. F. Gooding)    Chapter 19

Mouldy silage because of air entry, often at the top, shoulders or sides of the clamp where sealing is faulty
▼    (R. F. Gooding)    Chapter 19

Butyric silage, the product of an undesirable butyric acid fermentation and protein degradation by clostridia
▼    bacteria    Chapter 19

▲   Wrapping big bale silage   *Chapter 19*

Tidy stack of wrapped big ▶
bales of silage, Lithuania
*Chapter 19*

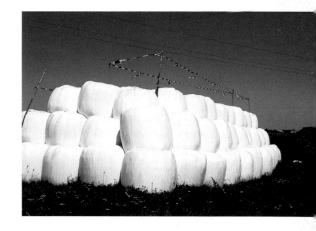

Well preserved, well presented silage of high D value ensuring adequate
▼   intake by highly productive dairy cows (SAC)   *Chapter 19*

▲    Stacked bales of hay awaiting transport from field to farm steading
*Chapter 20*

◄    Indoor clamp silage being critically inspected and the farmer being 'grilled' by the judge in a silage competition
(I. Stephenson)

best use of the grassland resource. There will be a minimal risk of shortage, either during the grazing season or in winter. It will also work wonders for overall confidence in the grass's potential.

It is not often realised how variable in length the winter feeding period is in different parts of the country. Low temperatures and wet weather can delay the date of turning out stock. Similar weather in autumn can advance the date of housing. The possibility of poaching damage to the sward is another influencing factor. Even within relatively small regions the recognised housing period for dairy cows can vary from 105 to 235 days. Local experience is the best guide to the likely winter feeding period and the amount of silage needed. The amounts needed per animal will also vary according to animal size, planned use of other feeds such as concentrates, the D value of the silage and its dry matter content. The moral is therefore to do the sums and build in a good margin for error. Within the objective of having sufficient silage for the winter feeding period, it is advantageous from both animal production and economic perspectives to aim for as high a D value as possible.

Depending upon the extent to which a buffer feeding system is adopted, an extra 3 to 4 tonnes of silage could be required per dairy cow or livestock equivalent. The daily amount of buffer silage needed per dairy cow may range from 5 to 10 kg during the first half of the season and from 15 to 25 kg during the second half—assuming silage of around 20 per cent DM; about three-quarters of these amounts would be needed for beef cattle. In estimating the amount of silage in the silo, the dry matter content of the silage is a key factor (see Table 19.1). Once feeding starts it pays to mark regularly on the silo the rate at which the silage is being utilised so that alterations in feeding policy can be made if required.

The principles for achieving good quality silage are well docu-

**Table 19.1   Average density of clamp silage**

| Dry matter (g/kg) | Silage density (kg/m³) |
|---|---|
| 180 | 760 |
| 200 | 725 |
| 220 | 695 |
| 240 | 670 |
| 260 | 650 |
| 280 | 630 |
| 300 | 615 |
| 320 | 600 |

mented and there are a number of rules which should be followed in practice. *The* golden rule is attention to detail at every stage of the system. The cost of failure lies in the losses which can be incurred at all stages of a silage system, from field to the animal's mouth. These losses can be considerable, as much as 30 per cent or more of the original amount of material cut—but many such losses can be avoided and unavoidable losses can be kept to less than 10 per cent. Farm surveys have shown a wide range of losses, averaging around 15 to 20 per cent for well-made silage but over 30 per cent for poor silage.

## FERMENTATION

Successful preservation of grass basically depends on a controlled bacterial fermentation in air-free (anaerobic) conditions. Lactic acid bacteria, notably *Lactobacilli*, *Streptococci* and *Pediococci*, which are mainly present on the leaf surfaces of the grass, thrive and multiply by utilising the sugars (water-soluble carbohydrates) in the grass as a substrate, quickly producing a desirable lactic acid fermentation; for a good fermentation there should be 3 to 3.5 per cent sugars in the grass being ensiled. This fermentation inhibits undesirable bacteria, notably clostridia and enterobacteria, especially when there is a rapid build-up of acidity. Once this has been achieved, namely, at a silage pH of 3.8 to 4.3, bacterial activity is halted and a stable, well-preserved silage results. The critical pH for inhibition of undesirable micro-organisms needs to be at the lower end of the acidity range for wet grass, 180 to 220 g/kg DM, but the higher end is satisfactory for drier ensiled material.

Grass with a high dry matter content, as a result of wilting or because it was cut at a mature stage of growth, also favours the activities of the lactic acid bacteria rather than the undesirable bacteria. A good fast fermentation utilises 3 to 5 per cent of the grass dry matter ensiled. To achieve this, exclusion of air and the maintenance of a relatively cool temperature—no more than 30°C—are necessary. A badly controlled fermentation can dissipate 10 to 15 per cent of the dry matter; in addition, the product will be less acceptable to stock, so intake and subsequent animal production will be reduced.

The chances of a good in-silo fermentation are greatest with grass harvested in sunny weather and in early summer, especially when the grass is at the ear emergence stage. The concentration of sugars in the grass is highest then—considerably higher than in young leafy material. As the plants mature further, structural carbohydrates such as cellulose and lignin increase and the content of soluble sugars decreases. Sugars are also at their highest level in the early afternoon of a typical day but the advantage is not sufficient to warrant organising a harvesting schedule around this time. While there is a clear correlation between good weather and high quality silage, the best silage makers also achieve good silage in bad weather seasons because they pay more attention to detail and to the principles of silage making than poor silage makers. Young leafy grass, fast growing grass heavily fertilised with nitrogen, and grass grown in dull wet weather with low light intensity are all low in sugars.

Among the commonly sown herbage species, Italian and perennial ryegrass are highest in sugars, timothy and meadow fescue intermediate and cocksfoot lowest. Tetraploid ryegrasses have slightly higher concentrations of sugars than diploid ryegrasses. Protein-rich legumes such as white clover or lucerne have low sugar contents. Herbage species also possess differing degrees of resistance to acidification during the ensilage process. This resistance, known as buffering capacity, springs from differences in chemical composition, including the content of organic acids present. In general, grasses have a lower buffering capacity than forage legumes. The high buffering capacity of legumes, allied to their low sugar and dry matter concentrations, is the main reason why legume rich material is difficult to ensile unless it is wilted and an effective additive used.

## RESPIRATION LOSSES

Respiration is another source of unavoidable loss in ensilage but the loss can be kept to a minimum by good silage making techniques. When grass is cut it continues to respire and plant enzymes break down sugars to water and carbon dioxide; there is also a breakdown of plant proteins. These processes occur in the field, and in the silo during the early stages until all the oxygen is used up and further air entry excluded. If respiration continues, indicated by rising temperatures in the silage, metabolite loss increases and overheating leads to a caramelised silage, highly acceptable to stock but of low feeding value. Following the breakdown of plant sugars there may be insufficient sugar substrate left to sustain a good lactic acid fermentation. This results in further losses from undesirable fermentation and an unstable silage. Losses of dry matter by respiration can range between 5 and 15 per cent.

In order to restrict respiration losses to a minimum field wilting should not be prolonged. Losses of dry matter can reach 1 to 2 per cent per day and a wilting period of no longer than 24 hours is desirable. To curtail respiration losses in the silo, it should be filled quickly with layers of about 2 m a day, consolidated, and sealed over with a plastic sheet every time filling is stopped for a few hours. This is an ideal which few farmers make sufficient time for but it can prevent loss, including a loss in feeding value of 2 to 3 units of D value. In large clamp silos, filling should be in a series of overlapping wedges, not thin layers over the whole silo area.

## WILTING

To wilt or not to wilt depends upon individual farm systems and objectives. Target grass DM values for ensiling are 200 to 250 g/kg for clamp silage, 300 to 400 g/kg for big bales and 350 to 450 g/kg for tower silage. Wilting is therefore most necessary for the latter two types of silage. It also reduces the amount of bulk and weight to be transported from the field. After the loss of moisture, the sugars in the grass are concentrated and this promotes a successful lactic acid fermentation in the silo. The silage becomes stable at a higher pH than for unwilted material. Another advantageous feature which is environmentally important is the reduction in effluent flow from wilted material. A model of dry matter losses under conditions of good management, derived from a series of European trials (see Figure 19.1), shows minimal losses when the herbage ensiled has a DM content of 250 to 300 g/kg.

Wilting increases field time and each day represents further losses of DM, especially in humid or wet weather. It should be omitted or curtailed if wet weather is persistent. If the swaths are conditioned or turned to speed up wilting time they are increasingly susceptible to nutrient loss, including fermentable sugars, by leaching from any subsequent rainfall. If consolidation in the clamp of a high dry matter silage is insufficient, the presence of air will encourage respiration and losses will occur until the oxygen is exhausted. Rolling the clamp does not always solve the problem since it may encourage air entry by a bellows effect. In Northern Ireland, wilting has been shown to increase DM intake by dairy cows but generally results in a decrease in milk yield and simple direct cut systems gave superior milk output per ha.

Short chopping grass to a 20 to 70 mm length at harvesting will restrict air movement in the silo, help consolidation and minimise rolling requirement. A short chop is particularly necessary for high dry matter grass. The chopping and laceration of the grass also speeds up the release of sugars from the grass sap and this aids the action of the lactic acid bacteria. Well-chopped grass also enables heavier loads to

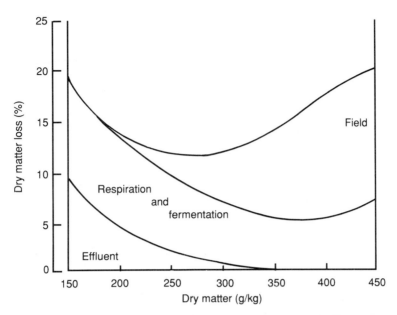

*Figure 19.1*   *Model of dry matter losses in well-managed silage systems*
   *Source:* Zimmer and Wilkins; Wilkinson

be transported and saves silo space, due to the high density of the silage.

## UNDESIRABLE CLOSTRIDIA

If wet or low sugar grass is ensiled, or if the plant sugars are decreased by respiration, the rate of lactic acid fermentation will be slow. Insufficient silage acidity will be developed and the scene is set for the multiplication of clostridia, the spoilage bacteria which also operate under anaerobic conditions. They utilise lactic acid and sugars to produce butyric acid. The amino acids in the protein are degraded, the digestibility of silage is reduced and ammonia gas is released. The net result of a butyric fermentation is a foul smelling, putrid silage unloved by man (or wife) or beast.

While some clostridia spores are present on grass foliage their numbers will be greatly increased if the sward surface is scalped and soil is picked up at grass harvesting. Other sources of grass contamination are animal slurries or farmyard manure previously applied to the

215

field at too short an interval before cutting time and picked up at harvest. Silage fields should be rolled in spring to press in stones and level wheel ruts, poached areas or molehills. The cutting height above ground level should be no more than 50 mm; however, a greater height should be used when cutting a previously grazed field, to avoid picking up dung. The cutting height should be monitored in wet weather since the wheels of the harvesting equipment may sink in and lead to scalping. There should be a clean concrete area at the silo for dumping the loads of grass from the fields and, as far as possible, mucky wheels should be kept off this area. The silo floor should be clean of old animal manure and rotting silage.

## ADDITIVE USE

Over a hundred silage additives are now marketed in the UK. As yet there is no EC approved scheme or official UK scheme for products—unlike Norway, say, where only a handful of additives are approved and can therefore be legally sold. However, the UK advisory services have launched a scheme whereby they will assess the efficacy of additives according to their performance in laboratory, field and animal production studies. Assessment will be based on properly designed studies carried out on a reasonable scale and capable of statistical analysis.

It must be emphasised that additive use will not compensate for poor silage making techniques or malpractice. Any response to an effective additive, whether in better silage or better animal production, will only occur when the rules of good silage making are followed. Additive use is not always necessary to ensure good silage, and it is not needed when the previously described factors and conditions conducive to a good fermentation are present. The use of an effective additive will be of greatest benefit in difficult wet conditions and on low sugar material.

Additives can be classified into groups according to their active ingredients and mode of action. The advisory services publish annual comprehensive lists of those available, together with their main ingredients, manufacturers' recommended rates and a guide to cost per tonne ensiled.

Fermentation inhibitors include additives based on organic acids (e.g. formic acid), inorganic acids (e.g. sulphuric acid) and various acid/formalin mixtures (e.g. formic acid/sulphuric acid/formalin). The acids and acid mixtures have a direct acidifying effect, lowering the pH of the silage, thus promoting the lactic acid bacteria and a fast fermentation, and lowering temperatures and the energy cost of the fermentation process. The organic acids also reduce the growth of undesirable micro-organisms, provided the acids are applied in sufficient strength. Ammonia/formic acid complexes release acid when mixed with grass; they were developed as a safer type of formic acid. The formalin constituent of mixtures, being a sterilant, inhibits micro-organisms in general. It also offers some protection against the breakdown of protein in the rumen of stock, enabling the protein to be used more effectively later in the digestion process. Salts such as sodium nitrite/calcium formate mixtures inhibit undesirable micro-organisms but need a good lactic acid fermentation, dependent upon adequate grass sugars, to act efficiently.

Fermentation stimulants act by encouraging the development of lactic acid bacteria. Sugar sources such as molasses supplement the sugars in the grass, thus giving a richer substrate for the acid producing bacteria. The sugar product must be added at the correct rate for the type of material being ensiled and the prevailing weather conditions. Bacterial cultures are available which supplement the naturally occurring lactic acid bacteria. Culture effectiveness depends upon satisfactory numbers of live bacteria of the right kind. *Lactobacillus plantarum* and *Pediococcus acidilactici* are particularly suitable and there is a good case for applying numbers as high as one million colony forming units per gramme of ensiled grass. Their effectiveness also depends on the ensiled material having an adequate level of sugars, at least 25 g/kg in the grass being ensiled.

Some trials have shown that silages treated with inoculants have improved animal performance relative to untreated or acid treated silages, although the fermentation as conventionally measured was not improved. The results are probably attributable to differences in the 'quality' of fermentation, but research has still to elucidate this. About half the silage additives now available in the UK are inoculants and their usage has increased mainly because of safety reasons, improved effectiveness and a growing interest in biological rather than chemical products in agriculture.

The development of enzyme additives is interesting since their use assists the breakdown of complex structural carbohydrates such as hemicellulose and cellulose. This increases the available fermentable sugars. A potential additional benefit from the breakdown of cell walls is an upgrading of silage digestibility but this has not always been demonstrated in trials. The ultimate additive may be a blend of sugar, effective lactic acid bacteria, and enzymes to continue the supply of sugar from the ensiled grass while at the same time upgrading its digestibility.

Several factors including cost, or better still, cost effectiveness, should govern the choice of additive. Unfortunately animal performance and economic data are not available for many of the additives on offer. Unproven products should be shunned. Clearly if safety is a prime consideration, sugar based and inoculant additives score highly under the right conditions of use. Present evidence would suggest that acid products are usually the most effective in difficult ensiling conditions.

No matter which additive is chosen it is essential that correct application rates according to manufacturers' label instructions should be used and the additive should be thoroughly incorporated. Silage additive pumps should be serviced before use so that they neither under- nor over-deliver.

In practice it is not always easy to decide when an additive is necessary and various guidelines have been proposed. For example, one advocates additive use when the sugar concentration in the grass being ensiled is less than 30 g/kg. The Liscombe 'star' system represents a practical step by step method of arriving at a decision (see Table 19.2). The number of stars accumulated by adding the star ratings for various influencing factors governs the decision (see the key in the table). The system advocates that, between three and five stars, a proven inoculant may be used although fermentation control will be less reliable than with acid-based products. While no additive is needed for fermentation control at five stars and above, some inoculants have produced improvements in animal performance on occasion.

## SEALING

In clamp silos the use of plastic sheeting as a final seal is all-important. Sound top sealing and using side sheets on the shoulders of the silage clamp is necessary, otherwise heating and moulding will occur. The surface and shoulders are particularly prone to aerobic deterioration. It is no use having an effective seal on top if air can penetrate the sides of the silo; these must also be of airtight construction. The plastic sheet must be weighted down evenly and firmly with tyres, straw bales or using other methods, to create a surface pressure of at least $100 \text{ kg/m}^2$. Excessive weight then and during consolidation, and storage in deep clamps, may exacerbate effluent flow in low dry matter silage. Adequate sealing also prevents the penetration of rainfall into silage in unroofed silos, for rain can cause considerable wastage.

If it is necessary to skim off a layer of wastage from the top of

**Table 19.2   The Liscombe 'star' system for deciding on silage additive use**

| | | |
|---|---|---|
| Grass species | Mixed swards, permanent pasture, clover dominant | * |
| | Perennial ryegrasses | ** |
| | Italian ryegrasses | *** |
| Growth stage | Young, leafy | 0 |
| | Stemmy mature | * |
| | Autumn regrowths | −* |
| Fertiliser nitrogen | Heavy (125 kg/ha +) | −* |
| | Average (40–125 kg/ha) | 0 |
| | Light (below 40 kg/ha) | * |
| Weather conditions (over several days) | Dull, wet | −* |
| | Dry, clear | 0 |
| | Brilliant, sunny | * |
| Wilting | None (150 g/kg DM) | −* |
| | Light (200 g/kg DM) | 0 |
| | Good (250 g/kg DM) | ** |
| | Heavy (300 g/kg DM) | *** |
| Chopping and/or bruising | Flail harvester or forage wagon | 0 |
| | Double chop | * |
| | Precision/fine chop | ** |

*Examples*

| | | | | |
|---|---|---|---|---|
| Perennial ryegrass | ** | Perennial ryegrass | ** |
| Leafy | 0 | Stemmy | * |
| Average N | 0 | Light N | * |
| Sunny | * | Dull, wet | −* |
| Wilted 250 g/kg DM | ** | No wilt | −* |
| Precision chop | ** | Double chop | * |
| Total | 7* | Total | 3* |
| Sufficient sugar | | Insufficient sugar | |

*Key:*

| | |
|---|---|
| 5 stars and more | − no additive needed |
| 4 and 3 stars | − use additive at recommended rate |
| 2 and 1 stars | − use acid additive or molasses at maximum recommended rate |
| 0 stars | − unsuitable conditions for making silage |

*Source:* ADAS Liscombe

the silage before feeding, this could represent a substantial loss of material. For example, 15 cm of unusable silage at the top and ends of a 1,000–tonne silo could amount to 100 tonnnes or 10 per cent. This figure could be in addition to existing losses of 15 per cent or more. To use an analogy with cereals, who would wish to throw away 25 bags of grain out of every 100 harvested?

## SILAGE EFFLUENT

Silage effluent or juice is a mixture of water-soluble plant constituents, organic acids, amino acids and minerals. It can also contain residues from silage additives if these are used. A typical rate of flow shortly after ensiling grass is 25 to 35 1/t per day, but peak rates can exceed these volumes, perhaps reaching 180 to 200 1/t per day if material with an exceptionally low dry matter content has been ensiled. These peak rates occur during the first few days of ensiling, but effluent may still be flowing over five weeks later (see Figure 19.2). Dry matter content has the greatest influence on the potential amount of effluent produced with negligible effluent flow at 260 to 280 g/kg (see Figure 19.3) and on DM loss in effluent (see Figure 19.4). Short chopping of

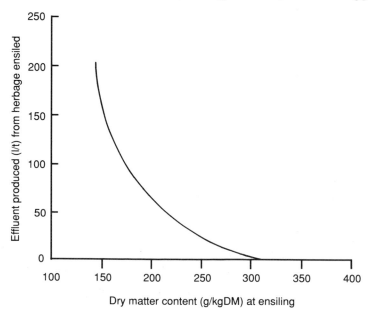

*Figure 19.2   Effect of dry matter content at ensiling on effluent production*
*Source:* Bastiman

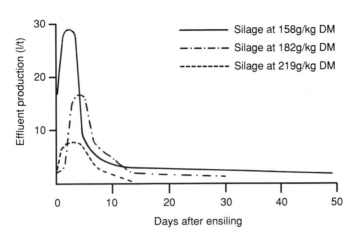

*Figure 19.3   Patterns of effluent production*
*Source:* Bastiman

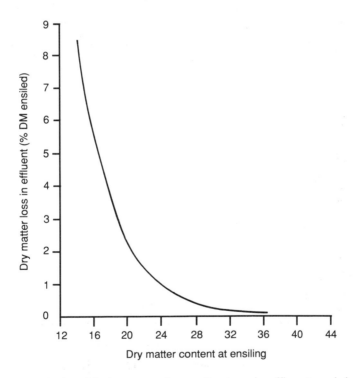

*Figure 19.4   Relationship between dry matter loss in effluent and dry
matter content of grass at ensiling*
*Source:* Bastiman and Altman

221

wet material, heavy consolidation during filling and deep silos all contribute to effluent flow. Acid or enzyme additives which break down plant cells also stimulate effluent release and increase its total volume.

Effluent is a powerful pollutant of watercourses and other sources of water supply if allowed to contaminate them. An example frequently cited is that silage effluent can have 200 times the polluting power of domestic sewage—due to its high biochemical oxygen demand. It is an offence liable to prosecution to permit effluent contamination of watercourses and supplies. On mixing with streams or rivers, resident bacteria are stimulated and use up all the oxygen with disastrous effects on river wildlife. All silos should have adequate effluent drainage and storage facilities. The general rule for storage tank size is at least three cubic metres capacity per 100-tonne silo capacity. However, the tanks should be checked regularly during silage making and emptied before they overflow. Clearly this is most needed in the early days of filling or after completion of filling, when peak flow rates occur.

Silage effluent can be used as a stock feed, 20 to 25 litres being equivalent in energy value to 1 kg barley. Intake should be limited to no more than 5 litres per 100 kg body weight in order to avoid feeding upsets from high nitrate or potassium intakes. Since effluent is a dilute source of feed nutrients, storage costs need to be kept low to warrant its use. A preservative such as formalin must be added when it is stored for later use. Silage effluent can also be used as a fertiliser, as outlined in Chapter 13, although its cost effectiveness is less than when used efficiently as a stock feed.

In the wet conditions typical of the western UK, wilting may not always be possible and direct harvesting of silage crops may be practised. This reduces field losses to a minimum. However, loss of dry matter and nutrients from effluent can be high (see Figure 19.1) and good effluent collection is environmentally vital. Alternatively, the use of suitable silage absorbents may be considered. Those absorbents able to soak up effluent most effectively may not be the best for improving the feeding value of the silage, however. Chopped straw in layers is an effective absorbent but it lowers the digestibility and feeding value of the overall silage. It also occupies a high volume of valuable silo space. Straw bales at the bottom of the silo are not recommended as they allow air entry and can actually increase effluent flow by acting as drainage channels. Cereal grains can be used but are not particularly effective absorbents. The incorporation of various forms of dried sugar beet with the grass during filling has proved successful for effective absorption and improved silage feeding value.

Rate of application is important and the amount needed per tonne of ensiled grass increases the lower the dry matter content of the grass being ensiled.

## AEROBIC DETERIORATION

When the silo is opened for feeding, the silage face is exposed to the air. Aerobic micro-organisms which have lain dormant in the silage may develop and cause deterioration and wastage. The signs are heating, caused by oxidation of sugars and organic acids, together with the presence of moulds and yeasts. The net effect is loss of dry matter and nutrients, and lower acceptability to stock. Avoidance measures include minimising exposure to air by using as narrow a clamp silo as practicable, by utilising 20 cm or more of face per day when self-feeding and by maintaining a smooth rather than a ragged face when feeding out (mechanically extracted for stock feeding). In this respect block cutters are better than front end loaders; the latter can disturb the inner layers of silage and allow air in. To avoid top wastage, top sheets should only be rolled back as far as the layer of silage which is about to be used. The key points or, as some would say, the commandments of good clamp silage making are presented in Table 19.3.

## BIG BALE SILAGE

The technology of big bale silage has gone forward by leaps and bounds since its inception in the late 1970s. Baled silage is obviously here to stay, estimates indicating that it now comprises 15 to 20 per cent of all silage made in Britain. Mechanisation of the process has been particularly innovative and effective, with cylindrical bales still dominating over other types. Big bales are best suited to farms without silage making facilities, as a supplement to existing silo storage and for the conservation of small quantities, say 250 to 500 tonnes. The bales can be made any time during the growing season, when there is grass surplus to grazing requirements. They can be fed out flexibly in situations which demand small quantities such as buffer feeding or the need to feed a limited number of stock, perhaps at a distance from the steading. Bales can also be sold, of course.

Where existing silage facilities are available, big bale silage is more expensive per tonne than clamp silage. Ancillary costs such as storage facilities, netting the stacks, vermin baiting and stockproof fencing have also to be borne in mind. However, the difference in cost is not as

**Table 19.3   Golden rules for clamp silage**

- Plan the system well ahead
- Preferably use ryegrass swards
- Implement a balanced fertiliser programme
- Roll silage fields in spring
- Aim for silage feeding quality to suit class of livestock
- Wilt for 12 to 24 hours if weather permits
- Avoid soil and/or slurry contamination
- Short chop to aid sugar release and consolidation
- Use an effective additive if needed
- Fill silo rapidly, spread material evenly and consolidate well
- Collect and utilise effluent efficiently
- Use effective absorbent if necessary
- Sheet over ensiled grass if filling ceases for 6 hours or more
- Use side sheets to reduce waste at shoulders
- Seal and weigh down plastic sheeting effectively
- Have silage analysed for feeding value
- Exploit silage feeding value to the full in livestock rations
- Avoid aerobic deterioration by good feeding practices at the silo face
- Avoid physical losses during feeding out

large as is commonly believed if the investment expenses of the clamp silo are brought into the equation. As further development and refinement of baling silage emerge, improved efficiency of making, handling and feeding should not incur ever increasing costs, otherwise alternative feeds may become more cost effective. It is as costly to make bad silage as it is to make good silage, so attention to detail at all operational stages is paramount. However, the system is basically a simple one and should preferably be kept so. Unless farmer-operated systems work to full capacity, use of contractors is a cheaper option.

The wrapping of bales has overhauled bagging and few exponents return to bagging once they have sampled the wrapping technique. Wrapping is faster, needs less labour input and transportation can be a separate operation. While low dry matter grass is not recommended for baling, the system can cope though bagging is better than wrapping for ensiled grass below 250 g/kg dry matter. Attaining physical stability of stacks is a problem with these silages and there can be considerable effluent loss when the bags are opened.

As with clamp silage a chain of factors controls the ultimate quality, starting with the early management of the sward. Grass for baling should preferably have a dry matter content of 300 to 400 g/kg. This can only be achieved from cut leafy herbage if it is well wilted, or from more mature, stemmier growth. Cutting at later stages of growth penalises silage quality. Salvaging a wet hay crop is not a recipe for good quality either. The quality target will depend upon the class of stock to be fed. Obviously productive animals require a higher quality feed than store stock. Big bales have proved a successful means of feeding silage to sheep, but high quality is particularly essential if intake is not to suffer. When batches of bales of differing feeding value are made, from differing seasonal cuts say, separate stacking of the batches will facilitate feeding management.

If prolonged wilting is needed to attain the desirable dry matter content in cut grass, it can lead to considerable field losses but little or no effluent loss afterwards. It is important to end up with conditioned swaths of the right width and shape for the baler. This ensures dense, well-shaped and uniform bales. In the baling and subsequent operations, the skill of the operator is a major factor in achieving success. With bales weighing 0.5 to 0.75 tonnes, safety precautions during all stages of handling must obviously be observed.

Bagging and sealing must be done promptly after baling with the objective of keeping air out in order to promote an anaerobic fermentation. The fermentation is more restricted than in clamp silage so acidity is less, pH levels of around 5 being common in well-preserved big bales. With a higher surface area to volume and lower density in comparison with clamp silage, bales are more susceptible to aerobic deterioration following the entry of air, so sound sealing is essential. Deterioration can be rapid if a bag is pierced, by whatever means. If bales are sealed by plastic film around a stack of them rather than individually, working out rate of usage in relation to minimum air entry is vital, otherwise the silage will deteriorate rapidly. Sheeting stacks containing a week's requirement of bales will overcome the problem.

In some cases a slow rather than fast ingress of air has led to the development of *Listeria monocytogenes*, the causal micro-organism of listeriosis. This can cause several diseases in animals, and particularly in sheep. The listeria, which are often present on grass and soil, multiply on mouldy silage at the outer surface of the bales. The removal of visibly spoiled silage before feeding out to stock will significantly reduce the problem.

Additive use on grass ensiled for baling has not been widely explored, but initial investigations indicated that some of the additives effective in improving clamp silage quality and resultant animal performance have shown promise. The use of an effective additive may be worthwhile when material with lower than optimum dry matter content is being baled.

The quality of big bales can be as good as clamp silage in terms of digestibility, metabolisable energy and protein content. These can be influenced in a major way by factors other than the system: stage of grass growth at cutting, for example. Generally, baled silage has a higher dry matter content, a higher pH and lower ammonia values. There is a range of suitable feeders for cattle, but physical wastage should be avoided. For sheep, cradle type feeders which allow the sheep to eat the bale from the underside as well as the sides enable better intake and less loss.

For bale storage the stacking height is controlled by the dry matter content of the bales. High dry matter bales may be stacked three or four high, but low dry matter bales only two high. A typical flattened pyramid shape is best, as square as possible to minimise the external area exposed. To prevent the plastic seal being torn, the stack should be netted against wind, and birds must be deterred. Baiting for vermin, mainly rodents, is a must. Regular inspection of the bales in important, so that any damage to the plastic cover can be repaired quickly. Air entry can cause total loss of bagged bales, but in wrapped bales the damage tends to be localised. A hard standing is essential for bale storage since regular removal of bales soon causes rutting of un-protected ground. The base can also prevent moles damaging the bottom layer of bales. While large quantities of silage effluent are unlikely, an effective draining, collection and utilisation system is still needed on the site. Storage areas should be well fenced to keep out any animals which may tear the protective plastic and expose the bales to the air.

The key points in big bale silage technology are shown in Table 19.4.

## SILAGE FEEDING QUALITY

Compositional analyses of silages of differing qualities are shown in Table 19.5. Most of the compositional factors influence silage intake by stock and consequent efficiency of animal production. For example, intake is enhanced by a low ash content, indicative of freedom from soil contamination; by a low content of ammonia, indicative of a good

**Table 19.4 Golden rules for big bale silage**

- Prepare swards as for conventional silage making
- Aim for silage feeding quality to suit class of livestock
- Wilt to a target 30 to 40 per cent dry matter content
- Form box-shaped swaths to match baler width
- Make dense, well-shaped and uniform bales
- Use an effective additive if needed
- Wrap or bag promptly after baling
- Store on a hard base at a suitable site
- Collect and utilise any effluent efficiently
- Stack bales closely together
- Net over the stack to prevent wind damage
- Control vermin, especially rodents
- Inspect stack regularly and repair any damage to bales
- Have silage analysed for feeding value
- Exploit silage feeding value to the full in livestock rations
- Discard mouldy silage which can cause listeriosis
- Avoid physical losses during feeding out

fermentation; but above all by high energy value, whether expressed as D value or metabolisable energy. Enhanced intake of good silage with high nutrient concentration is advantageous in animal response terms for the most productive classes of stock, e.g. high yielding dairy cows or rapidly growing stock. Feeding studies with dairy cows at the

**Table 19.5 Compositional analyses of silages**

| Parameter | Silage quality | | |
| | Good | Moderate | Poor |
|---|---|---|---|
| D value | 70 | 65 | 60 |
| ME (MJ/kg DM) | 11.5 | 10.5 | 9.5 |
| Crude protein (g/kg DM) | 160 | 140 | 120 |
| Ammonia N as proportion of total N (g/kg) | 50 | 100 | 150 |
| Ash in DM (g/kg DM) | 80 | 110 | 140 |
| pH | 3.8 | 4.3 | 4.8 |

Hannah Research Institute, Ayr showed an increased silage DM intake of 0.24 kg per one unit increase in D value (see Figure 19.5). High intake also leads to a reduction in the level of supplementary feed needed for a given level of animal output. An average reduction in supplementary feed of around 0.7 kg/day per unit increase in D value has been reported from several dairy cow experiments.

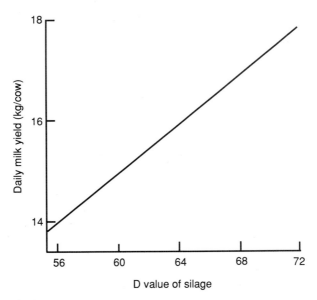

*Figure 19.5* *Relationship between the D value of grass silage and the average daily milk yield of cows eating silages of specific D value plus equal weights of barley concentrates*
*Source:* Castle

## HEALTH AND SAFETY

It is worthwhile stressing the health and safety aspects of silage making and utilisation. Silage making in particular is a highly mechanised job, so respect is necessary for the powerful and varied equipment involved. Sloping land imposes a particular hazard and has to be carefully worked; side-mounted equipment is safer than in-line, for example. All the equipment should be well serviced. Walled silos should be inspected regularly and the walls maintained in good condition. Many silos were built over twenty years ago, when filling and consolidating were carried out by lighter equipment than is now used. In some cases silo capacity has not kept pace with the amount being packed in, which is possibly exceeding the maximum safe loading capacity. Above-ground silos should always have guard rails or sight rails so that the filler knows his position relative to the edges. The weight limitations for filling and consolidating equipment should also be known to the operator. The equipment can have modifications fitted if necessary, such as extra front weights or extra tyres on tractors to prevent rearing backwards during wedge filling.

Given the widespread use of additives, many of which are based on corrosive acids, operators should be aware of any hazards, inflammable, corrosive or toxic, and must have the now statutory safety training. They should, of course, be familiar with the instructions on additive labels for storage, use and for health risks, including their avoidance and treatment if necessary. Protective clothing and appropriate first aid equipment should always be available.

## SILAGE COMPETITIONS

The aim of silage competitions is to raise standards of silage making and utilisation using marking systems designed to place emphasis on known factors which lead to efficiency and excellence. Competitions for the best silages are run annually by various bodies such as grassland societies and commercial companies. Competitions provide a focus and a stimulus towards better silage making in general, as the competitors vie with one another towards perfection. At award ceremonies, they learn from judges' appraisals and later from organised visits to the winners. Perhaps the best known major competition in recent years has been the National Silage Competition of the British Grassland Society, originated for the 1979 silage making season and run in association with various sponsors. Compositional analysis, visual assessment of quality, in-silo wastage, effluent control, and efficiency of silage production and utilisation are all taken into account by judges during winter visits. Tower and big bale silages can be entered as well as clamp silages. In the final stages, nine regional finalists emerge from approximately 1,200 initial entries at local grassland society level countrywide. Attention to detail at all stages in the silage making, handling and feeding operations shows through as the hallmark of each year's finalists. Table 19.6 shows the current appraisal criteria and marking used in the final judging; the marking system is periodically reviewed and modified in light of advances in silage technology.

Thirty per cent of the total marks are based on analyses for ammonia

**Table 19.6    Current scoring system used by the British Grassland Society for its National Silage Competition**

| | Marks |
|---|---|
| A. Silage analysis (maximum 30 marks) | |
|   1. Ammonia nitrogen (15%–5% in total nitrogen) | 0–15 |
|   2. Metabolisable energy (MJ/kg DM) | |
|     Dairy cows/beef animals (10.5–12.0) | |
|     Suckler cows and stores (10.0–11.5) | 0–11 |
|   3. Crude protein (12%–16% in dry matter) | 0–4 |

| | Marks out of |
|---|---|
| B. Silage inspection (maximum 22 marks) | |
|   1. Surface, shoulder and side waste | 11 |
|   2. Visual assessment of quality: | |
|     (i) Freedom from contamination and uniformity of preservation | 4 |
|     (ii) Absence of mould | 3 |
|     (iii) Temperature at face | 3 |
|     (iv) Freedom from evidence of overheating | 1 |
| | |
| C. Management of effluent and farm waste (maximum 11 marks) | 11 |
| | |
| D. Silage production and utilisation (maximum 32 marks) | |
|   1. Efficiency of labour and machinery use in silage making | 5 |
|   2. Efficiency of feeding procedure | 5 |
|   3. The contribution of silage to the diet and efficiency | |
|     of use by the animals: | |
|     (i) Animal production from silage | 12 |
|     (ii) Appropriate supplementation and productivity | |
|       of grassland for silage | 5 |
|   4. Sufficient silage | 5 |
| | |
| E. Overall impression (maximum 5 marks) | 5 |
| | |
| Total | 100 |

*Source:* British Grassland Society

N, ME and CP. Ammonia N as a percentage of total N is a good measure of fermentation quality and the maximum 15 marks are given if it is below 5 per cent (i.e. 50 g ammonia N per kg total N). Above 15 per cent gains no marks. High ammonia N indicates that an undesirable butyric acid fermentation has occurred with consequent protein degradation; preservation is unstable, further deterioration and losses of silage DM and CP can occur, and voluntary intake by stock is greatly depressed. Metabolisable energy, which is directly related to D value, gains maximum marks–11–for an ME of 12.0 (72 D value) in dairy cow enterprises and ME of 11.5 (69 D value) for suckler cows. Stage of grass growth at cutting is the most important factor governing ME, D value and CP. Silage CP of 16 per cent and above earns a maximum of 4 marks.

Another 22 marks are based on absence of visible wastage (11 marks) and visual assessment of factors influencing quality (11 marks). There are also 11 marks for the efficacy of silage effluent and farm waste control. These appraisals are made by the eagle-eyed judges. Surface, shoulder and side waste are all preventable by complete sealing, including side sheets on the shoulders to keep out air and rainfall, two main factors causing surface deterioration and shoulder waste.

The key to top marks for visual assessment of quality is prevention of air ingress at all times, maintaining as even a feeding face as possible and using silage at a rate which avoids secondary aerobic deterioration. Any signs of heating indicate that valuable feed is being destroyed. The judges also look for patches of poorly preserved silage caused by soil or slurry contamination. Table 19.7 shows an inspection guide to silage fermentation.

The adequacy of collection and storage facilities for silage effluent and farm waste and their subsequent disposal are assessed, particularly

**Table 19.7 Inspection guide to type of silage fermentation**

| Type of silage fermentation | Characteristics |
| --- | --- |
| Well fermented | Yellowish or brownish-green colour<br>Pleasant acidic smell<br>Firm texture apparent |
| Butyric | Olive green colour<br>Evil putrefactive smell<br>Soft slimy texture |
| Putrid/rotted | Greenish-black colour<br>Putrefactive smell<br>Wet slimy texture |
| Overheated | Dark brown to black colour<br>Burnt caramel to tobacco smell<br>Dry, disintegrating texture |
| Mouldy | Dark brown with white patches<br>Musty smell<br>Dryish, easily broken texture |

bearing in mind the need to avoid, or at least minimise, any form of environmental pollution.

The overall efficiency of silage handling from growing to cutting to feeding carries another third of the marks, including 17 for the efficiency of animal performance from silage. This is assessed by a series of searching questions from the judges. When good silage has been made it makes sense to feed it effectively, whether minimising physical losses by good feeding face practices or optimising its performance with matching nutritional supplements. The judges' calculators soon unearth any deficiencies or excesses in the ration being fed in relation to the silage quality analysis and planned or actual animal output. Lack of faith in the potential of excellent silage has been a common fault exposed in competitions. Contestants are also questioned on how they manipulated grassland management the previous growing season in order to achieve their production and quality targets. Obviously it is important that they know how much silage they planned for and have, and are able to calculate the rate of usage of existing supplies. The efficiency of use of labour and machinery during silage making is also evaluated.

A final 5 marks for the judges' overall impressions may not sound much, but they can be vital when competitors' marks for the standard criteria are running close. The impressions are based upon the quality of other silages on the farm, on stock fed on these silages and on general management. Factors such as difficulty of the environment for silage making, overall tidiness, making best use of limited resources or perhaps some new innovation also have an influence.

There is no doubt that entering a silage competition can make one think, plan ahead and act meticulously at all stages in silage production and utilisation.

# Hay Making

He who watches the wind will never sow, and he who keeps an eye on the clouds will never reap.

*Ecclesiastes* 11: 4

Good teddying is the chief poynte to make good hay.

John Fitzherbert *Boke of Husbandrye* (1523)

Hay has a long and ancient history of use as a conserved fodder for winter or other non-growing season. It was probably first used in hot dry countries where it was easily obtained, or even allowed to dry out as standing hay *in situ*. The use of hay in wetter climes hinged upon a plentiful labour supply to cope with the amount of manual handling needed to turn it into satisfactory feed. In the UK these methods involved gathering the partly dried hay into heaps in the field for further curing (drying) before transfer to the farm. The heaps were known by various local names such as pikes, cocks or rucks. On the farm the hay was stored in large stacks or hay sheds. A reserve of hay was often kept from year to year as an insurance against low yielding years, or for cash sale in emergencies. This way of making hay gradually changed, following the advent of balers in the 1930s. However, similar labour-intensive methods are still practised, for example, on small and mountain farms in European alpine regions and in Scandinavia, where a variety of structures including stakes, fences and pyramid-shaped hurdles are used on which to dry the cut grass and make hay of useful quality, despite weather difficulties.

While silage has now become more popular than hay, by a ratio of 2:1 in dry matter terms, the 4 million tonnes of hay DM made mean that it is still an important crop in the UK, making a substantial contribution to the feeding of ruminant livestock. There is also a thriving trade catering for the horse market. Hay has the virtues of ease of 'packaging', handling and transporting, and it can be a valuable cash crop.

233

## STAGE OF GROWTH AT CUTTING

Hay is customarily cut at a later stage of grass maturity than silage to facilitate curing, although there can be overlaps in the stages. Hay as well as silage could be made at or following ear emergence, for example. The livestock feeding policy on a farm or the equipment available for making hay may largely determine when the grass is cut. If some form of barn drying facility exists on the farm the crop may be cut for hay at an early growth stage, when the ratio of leaf to stem is still high.

Stage of growth at cutting is the major determinant of hay quality under the farmer's control, although wet weather may delay cutting at specific quality targets. There are unavoidable losses in the process of removing water from grass and, depending on the weather, these losses can total 25 to 50 per cent of the original dry matter and its nutrients. Cutting grass at late maturity simply ensures a poor quality end product, no matter how well the hay is made. The flexibility in feeding regimes which good quality hay gives should never be under-estimated.

Sward type is also controllable. Hay can be made from the whole range of seed mixtures sown, many of which are dominated by perennial ryegrass. The rate of water loss is slower in this species than in other sown species but it has better herbage quality features and this can result in higher quality hay. Timothy hay is less important now than hitherto; often cut late, it was therefore of rather low quality, but was prized for horses. By the use of seed mixtures which mature in succession, hay making can be spread over a period while still cutting the crops at a good quality stage. Long-term or permanent grass matures later than most sown mixtures so can be hayed last in the succession.

The time of cutting can also be manipulated by the extent to which spring grazing is practised before shutting up the fields for hay. Nitrogen dressings, including nitrogen from animal manures, rarely exceed 60 to 90 kg/ha since heavier rates lead to soft structured growth, which is high in moisture content and difficult to dry out.

## DRYING PROCESSES

Hay is usually discussed in terms of its moisture content rather than its dry matter content, whereas the reverse is true when referring to grass production generally. The reason is a traditional one which presumably arose because of the importance of water removal and a low target moisture content, equivalent to a high dry matter content, for satisfactory storage.

The principal aim in hay making is to reduce the moisture content of the cut grass from an initial 700 to 800 g/kg down to 150 to 200 g/kg. It should then be well cured and safe for storage. The evaporation of water is mainly a field process but storage in suitable open-sided Dutch barns or sheds allows further dehydration after stacking. Cold or warm air blowing in barns may also be used in the later stages of dehydration. Drying should be done as rapidly as possible in order to minimise the loss of herbage dry matter and nutrients. It is necessary to remove 2 to 3 tonnes of water per tonne of dried hay made.

Initially water loss is rapid, especially from the leaves, but as the crop becomes drier, the rate of drying decreases. Water release is rapid at the beginning because the leaf and stem pores (stomata) are open. These close as the herbage wilts and water has then to find pathways through the waxy surfaces of the leaves and stems. Stems, although lower in moisture content than leaves, retain their moisture more tenaciously in their plant tissue. Stem tissues are structured differently from those in the leaves; also, following the rapid leaf wilting, the natural transpiration flow of water from stems to leaves is interrupted.

The major factors influencing field drying are sunshine, wind and the relative humidity of the air. Warm sunny weather and moderate breezes provide ideal conditions for evaporation. Dull, humid, calm weather slows down the rate of drying, especially in the later, more difficult stages. The rate of moisture loss slows down with increasing humidity and an equilibrium may be reached when there will be little transfer of moisture to the atmosphere because it is already moisture laden. Air movement is needed to remove the moisture from the hay and ever lower relative humidities are required as the moisture content in the hay decreases (see Figure 20.1). This illustrates the difficulty of achieving the final drying out stage needed to attain 150 to 200 g/kg DM which is at equilibrium with 65 to 75 per cent air relative humidity; these humidity levels are often exceeded in the UK.

Rapid evaporation is essential in order to curtail respiration, which can result in substantial loss of dry matter, even in good hay making weather. Respiration is greatest in the earliest stages of drying when plant sugars are oxidised by respiratory enzymes to carbon dioxide and water. There is a release of heat, which can be noticeable in cut swaths that are left undisturbed. Protein is also degraded into amino acids which may subsequently be leached. While there is a range of machinery for rapid hay making in the field, this may not be available on every farm, so the area of hay cut at any one time should therefore

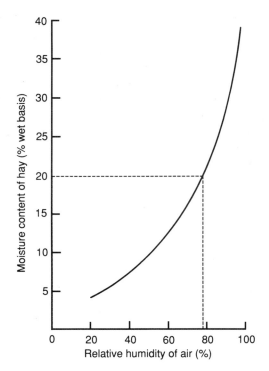

*Figure 20.1* *Approximate equilibrium moisture/relative humidity relationship for grass hay*
Source: Nash

be adjusted to individual handling capacity during intervals of good weather. This strategy requires fields of hay maturing at different times and good forward planning to ensure that the hay fields are prepared properly. Machinery should be well serviced and ready for action when the right weather appears. The mower should be set to avoid cutting too low, which may slow down the rate of the regrowth or even damage the sward by 'scalping', yet be low enough to ensure efficient utilisation of the standing crop. In practice, a cutting height of 4 to 6 cm is a reasonable compromise.

To speed up the drying process it is advantageous to start 'working' the cut swaths vigorously immediately after cutting, to assist crop moisture loss, beginning with the use of the mower conditioner. Various types of conditioning abrade, bruise or crush the cut material and wilting is speeded up. Thereafter the crop should be shaken (tedded) regularly, ideally twice or three times a day during good weather. It is better not to cut a wet crop since standing grass will dry more easily than grass in a cut swath. If cut swaths are left untedded for a period, they settle and cling to the ground so that drying is slower and uneven, especially with early-cut youthful material with a soft, moist structure. Later tedding cannot make up for the time lost and inclement weather may intervene before drying is completed. Figure 20.2 illustrates the drying benefits of early conditioning. Spreading and mixing the crop is beneficial since it encourages uniformity of drying throughout the material. These repeated operations allow good air circulation, which disperses released water vapour and ensures ventilation of the crop and even exposure to the sun.

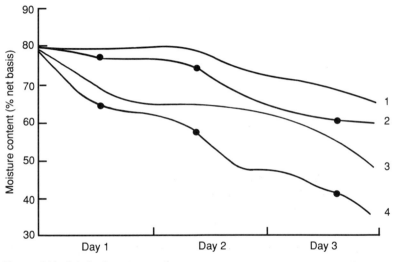

Key:    1 Undisturbed mower swath
        2 Mower swath, turned at times marked ●
        3 Undisturbed mower conditioner swath
        4 Mower conditioner swath, turned at times marked ●

*Figure 20.2    Field drying curves for various types of conditioning*
*Source:* ADAS Liscombe

Swaths set up for drying in good weather should be open and fluffy to allow air movement. Spread-out conditioned material should be windrowed into narrow swaths if there is a risk of it being wetted, whether by overnight dew or rainfall. Conditioned hay is more prone to loss of the soluble nutrients by leaching than unconditioned material, but even the latter is not immune during heavy rain. The frequency and severity of tedding at the later stages of field drying should be reduced to avoid loss of brittle leaves, which are the most nutritious part of the hay. This problem is exacerbated in legume-rich forage or with special purpose lucerne crops.

## INFLUENCE OF WEATHER

Hay makers must watch the weather and use local meteorological information and forecasts. No time should be wasted at the start of good weather spells during the main June–July period, because the chances of 5- to 6-day dry spells for hay making are less than the chances of 3- to 4-day spells. It will normally take 4 to 5 days of dry sunny weather to reach a safe baling stage. In some wetter northern and western areas there is only a 1 in 3 chance of dry 4-day spells in any week during the hay making season. The average is slightly better in southern and eastern areas. Since the weather generally gets wetter in the late summer and autumn, clearly hay making is an activity which is best confined, if possible, to the June and July months with their long daylengths and high temperatures.

The potential effects of different types of weather on the loss of dry matter in the field are illustrated in Figure 20.3; the losses would mainly be from respiration, leaching and leaf fragmentation. Extrapolation from the figure shows that when hay has reached baling stage, at around 30 per cent moisture content, the likely losses of dry matter are 8, 16 and 24 per cent for good, moderate and poor weather respectively. Some hays which have undergone excessive weathering from intermittent rain and sun over a long period may be of little or no nutritive value, including a shortage of minerals and vitamins.

The development of an additive which would allow hay to be dried and stored safely at higher moisture contents than currently required would be a major breakthrough. While there are promising additives the problem is the high rate of application, and hence cost. There are also difficulties in achieving uniform distribution of the additive when applying in the field.

## BALING

Baling is normally started when the hay has dried down to 25 to 30 per cent moisture content. As a rough guide, the hay rustles at this stage and is easily broken if twisted, and no moisture is extruded. If warm air blowing systems are available on the farm, the moisture content at baling can be 30 to 45 per cent; at this level there is no surface

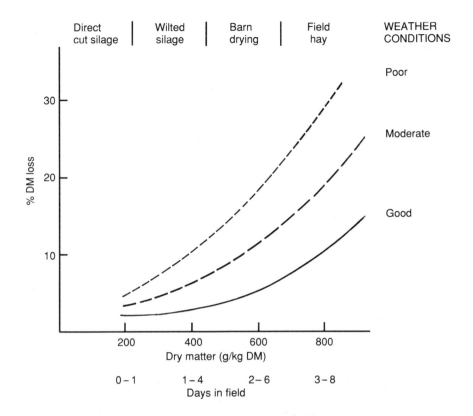

*Figure 20.3   Herbage dry matter losses during field drying*
Source: Honig

moisture and the leaves are dry, but moisture can be extruded from the stems when twisted. Baling at the higher moisture content reduces potential leaching and leaf shatter losses but will increase the risk of heating, unless the hay is subsequently dried properly.

Bales left to cure in the field should be made into small stacks and covered with plastic sheeting on top, to protect them from rain. It is easy to talk about actual percentage moisture content but it is not something that is easy to measure in the field. Experience is still the best guide, although moisture meters can be used. The problem with these is obtaining a representative sample because of the wide variability in the moisture content of the hay being made. It is necessary to take a large number of readings, 20 to 30, and average the results to give an accurate guide.

Bale density should be low enough so that heat from respiration can escape. Big bales have an advantage of lower density but the hay must be dried to a lower moisture content—20 per cent or so before baling—than for the smaller, standard sized bales, otherwise rapid heating and subsequent moulding is likely to occur. Big bales withstand adverse weather better than small bales in the field, but stacking for effective barn drying is difficult; their variability in density can also result in uneven drying out. Bale density is also important when barn drying systems are used since their effectiveness depends on efficient forced draught ventilation through the bales.

## BARN DRYING SYSTEMS

The advantage of barn drying lies in the shortening of field drying time and exposure to the vagaries of the weather. To be effective barn hay drying has to be seen as a positive measure, encouraging the making of good quality hay, and not simply as a form of insurance to salvage hay in bad weather. The costs involved in barn drying systems are not justified unless the end product is likely to be of high feeding value; this means cutting young, highly digestible grass.

A number of different types of drying systems have been devised. Electricity boards were particularly active in their development and installation. In most cold air systems, the bales are stored in open barns and then dried *in situ*. It may take two to three weeks of blowing to achieve the moisture content of 15 to 20 per cent for satisfactory storage. Blowing may need to be suspended at times when the relative humidity of the air is such that moisture is not being removed from the hay. The relative humidity of the air should be measured, using either a hygrometer or a dry bulb/wet bulb thermometer method.

When using warm air blowing each 1°C rise in the temperature effects a reduction of 5 per cent in the relative humidity of the air. In warm air systems, the bales are more commonly dried in batches in enclosed barns and then removed for long-term storage elsewhere. A degree of skill is needed to stack the bales correctly, so as to ensure uniform air flow, whether cold or warm. Also important is the correct matching of fan type and capacity for the volume to be dried. Completion of drying cannot be judged by external examination of the bales: they should be fan tested the day after it is thought ventilation has been sufficient. Apart from an absence of heating there should be no release of water vapour when the fan is started up.

Many trials and demonstrations have proved conclusively that superior acceptability and feeding value to stock compared with field drying can be guaranteed by using some form of forced draught ventilation at the final stages of drying. Despite this, possibly less than 10 per cent of the hay made in the UK is treated this way; in particular, there was little uptake of batch drying by warm air blowing. The most apparent reasons for the general lack of uptake of barn drying systems are: lack of a full appreciation of the benefits; the capital investment costs involved; electricity or fuel costs; the amount of handling needed; the fact that volumes being handled at any one time are of necessity limited; and faith that there will always be some weather suitable each year for field hay making.

## STORAGE LOSSES

The main losses during storage of hay are caused by moisture content being above the safe limit of 150 to 200 g/kg. Losses from respiration can range from 5 to 15 per cent of the dry matter. The main evident symptom is heat caused by the breakdown of sugars due to respiratory enzyme action. Moulding soon becomes visible. These effects can occur in the bulk of the hay or they may be confined to damp bales or patches resulting from uneven drying. Overheated hay is highly acceptable to stock but is of very low feeding value, while mouldiness is undesirable since it both reduces the hay's acceptability and lowers feeding value too.

Mouldy hay is a health hazard to man and beast on account of the presence of micro-organisms, including *Actinomycetes*, which can cause lung disease. The ailment in humans is referred to as farmer's lung. For this reason it is essential to wear respirators when dealing with hay which is mouldy and dusty. There may also be fungal mycotoxins in mouldy hay, which are harmful to stock after ingestion, causing either a lack of thrift or, in extreme cases, death. It is therefore worthwhile inspecting hay barns and stacks regularly for any overt signs of deterioration in the hay and taking appropriate remedial action if necessary.

## KEY POINTS

Table 20.1 summarises the rules of good hay making, while Table 20.2 gives the analyses of hays of different qualities.

**Table 20.1 Key points in hay making**

- Forward plan the hay making system
- Implement a well-balanced fertiliser programme
- Aim for a feeding quality to suit class of livestock
- Make best use of good weather spells
- Minimise field exposure time as far as possible
- Encourage release of moisture by frequent tedding
- Windrow if rainy weather threatens
- Bale at moisture content suited to method of final drying
- Make well-shaped bales of correct density
- Aim for stable storage moisture content of 150 to 200 g/kg DM
- Protect stored hay by well-constructed barns
- Inspect stored hay regularly and control any heating
- Analyse hay for feeding value
- Exploit hay feeding to the full in livestock rations
- Do not utilise mouldy hay which is a health hazard
- Avoid physical losses during feeding out

**Table 20.2 Compositional analyses of hays**

| Parameter | Hay quality | | |
| | Good | Moderate | Poor |
|---|---|---|---|
| Dry matter (g/kg) | 900 | 850 | 800 |
| Crude protein (g/kg DM) | 150 | 100 | 50 |
| Digestible crude protein (g/kg DM) | 100 | 60 | 30 |
| D value | 65 | 55 | 50 |
| ME (MJ/kg DM) | 10.5 | 9.0 | 8.0 |

*Chapter 21*

# Grassland Recording

---

If we could first know where we are and whether we are tending we could better judge what to do and how to do it.

Abraham Lincoln (1809–1865)

---

Appraising grassland by eye, guesswork or some other rule of thumb judgement is an instinctive pastime among farmers. Admittedly some grassland looks so bad that a glance is sufficient to show that it is not very productive, but the majority of fields cannot be judged that way. It is necessary to appraise sward botanical composition periodically over the season and thus a knowledge of grass species identification is necessary (see Appendix 4). The abundance of desired or preferred species is the most common criterion and perennial ryegrass often serves as a benchmark. A number of standard methods of measurement exist which involve taking observations over whole fields, but these are used by scientists more than by farmers.

The best way to assess the value of a sward for sustained animal production is by measurement. However, the central problem is that grassland cannot be judged on a single harvest alone since it is put to a medley of uses such as grazing, silage and hay. Further, grazed grass does not have a direct cash value but must be converted into animal products such as milk, meat or wool. It could be argued that lack of a simple yardstick of grassland output has often delayed uptake of new technology, because any advantages are difficult to quantify on the farm. It also explains why many farmers lack confidence in grassland and thus do not exploit it to its full potential for animal production.

Compare the rapid uptake of a new, high producing cereal variety, which is a cash crop, with the inertia that normally greets the introduction of a superior grass variety. Similarly, appreciation of the cash value of cereal grain means that it is usually stored and used efficiently on the farm for stock feeding, in marked contrast to the inefficiency with which grass is often grazed or stored and used as silage.

The weight of hay made per field or per farm is simply recorded by counting the total number of bales and weighing representative samples. The weights do not represent the amounts of grass produced and cut, however, due to field and curing losses, although the significance is that an estimate of hay production can be related to field management given.

Suitable weighing equipment on every farm will facilitate measurement of forage produced. A large-scale platform weigh cell is suitable for weighing sample trailer loads of grass for silage. Alternatively, when a silo is full the amount of settled silage can be estimated using density of material and volume occupied, for individual fields in proportion to the number of trailer loads per field. For both hay and silage the estimates made do not take into account physical losses which will occur during their feeding out. Other losses occur in the silo, such as secondary oxidation at the exposed face and heating losses in stored hay. Additionally the silage or hay will be utilised at different degrees of efficiency. For grassland production and utilisation under grazing it is necessary to use some form of grassland recording.

Whole-farm grassland recording resembles assessing the average milk yield of a dairy herd, whereas individual field recording is akin to measuring individual cow performance. A number of schemes have been devised but as each is made more scientifically acceptable, it also becomes more complicated and time-consuming. Most methods aim at measuring efficiency of use of grassland in physical rather than financial terms, but both can be linked in costing schemes for dairy, sheep or beef enterprises.

In the 1950s both farm and field grassland recording methods were available, based on assessing output in terms of energy as starch equivalent (SE). The energy value of all feeds other than grass and grass products was calculated and deducted from the total energy requirements of livestock and the residual utilised energy or USE ascribed to the grassland. SE requirements for different classes of stock and values for the various forms of feedstuffs were standardised.

## ANIMAL UNIT METHODS

In the 1960s the British Grassland Society devised and sponsored a simple system of measuring productivity from individual fields on dairy farms based on the 'cow day' unit of measurement. Field records were kept of numbers and class of stock, periods of grazing in each field, supplementary feeding and amounts of grass conserved. A cow day was judged to be equivalent to the amount of bulk feed that the average lactating cow in the recorded herd would eat in 24 hours. Conversion factors were allocated to other stock and also to winter-fed silage and hay (see Table 21.1) and field productivity assessed in terms

**Table 21.1** Conversion factors used to assess and compare productivity of fields in terms of the number of cow-day equivalents.

| Class of stock | Conversion factors to cow days |
|---|---|
| Milking cows | 1.0 |
| Dry cows | 1.0 |
| Suckler cows | 1.0 |
| Cattle over 30 months | 1.0 |
| Cattle 21–30 months | 0.75 |
| Cattle 11–20 months | 0.50 |
| Cattle under 11 months | 0.25 |
| * Sheep and lambs (kg) | |
| 18–36 | 0.07 |
| 36–54 | 0.10 |
| 54–72 | 0.13 |
| 72–90 | 0.17 |
| * Bulk feed (t) | |
| Hay | 68 |
| Silage | 21 |

* Conversion factors based on Friesian cows; factors adjusted upwards for smaller sized breeds
*Source:* Baker *et al.*

of number of cow-day equivalents. Although cow day and utilised energy systems gave similar results in general, the cow day system was recommended for individual field recording on the grounds of simplicity. Since the value of a cow day varied from farm to farm, being determined by size of animal, comparison between farms could only be made by converting cow days into energy terms. The major disadvantage of the cow day system was that it took no account of milk production; supplementary feed at grazing was also ignored, except when it formed a substantial part of the daily ration, and adjustments were then necessary. Few farmers took up either the cow day or SE system due to the amount of record keeping necessary, except when assisted by an organisation with facilities for processing the records.

Other grassland recording systems based on livestock units (LSU) or livestock unit grazing days (LUGD) have been devised. A mature productive animal such as a dairy cow is taken as the standard unit, and other classes of stock are given factors based on live weight and associated feed energy requirements for conversion to the standard. The factors can be refined further from knowledge of stock live weights, performance and nutritional requirements. Organisations do not use the same sets of conversion factors, so comparisons of the results of surveys by different organisations require careful interpretation.

245

## UTILISED METABOLISABLE ENERGY

Metabolisable energy (ME) is the modern unit of energy in megajoules (MJ), in which stock needs and feed values are measured. As with SE, the ME values of home-grown and bought feed are subtracted from the ME requirements of livestock, the balance—utilised metabolisable energy (UME)—representing the energy output from grassland. Output is usually expressed in Gigajoules (GJ), 1 GJ equalling 1,000 MJ. UME is more easily and accurately calculated for dairy cow enterprises than for other stock enterprises because of the daily contact and control of herd movements, availability of daily milk production records and the easily measured supplementary feed given. UME has advantages over other simpler measures such as stocking rate or milk output per hectare since it is a guide to overall effectiveness of grassland use. It has been found to be closely related to gross margins and to potential profitability of milk production.

The calculations necessary to estimate UME on a dairy cow basis are given below and a worked example is shown in Table 21.2.

1. Calculate the total annual ME required per cow. Published animal nutrition tables are available which provide daily maintenance ME requirements for cows of different live weights, including allowances for growth and pregnancy, and ME requirements for milks of different composition. Maintenance and production requirements are added to obtain the total ME requirement.
2. Calculate the total ME supplied by all feed inputs other than farm grass, e.g. concentrates, brewers' grains, using published tables of ME values for commonly used feeding stuffs.
3. Subtract non-grass feed ME from the total ME requirement of the cow. The balance is UME from farm grass.
4. The UME value divided by the total ME required per cow, and expressed as a percentage, gives the extent to which grass is used in the cow's diet.
5. UME per cow multiplied by the average annual stocking rate of dairy cows on farm grassland (grazing plus conservation) gives UME per hectare of grassland.

UME on a dairy herd basis is calculated in a similar manner (see Table 21.3). Records of the average annual number of dairy cows and annual amounts of bought-in feed and home-grown fodder other than grass are needed together with the grazing and conservation areas used to feed the dairy cows. Where a grassland area has been utilised by both dairy cows and other stock it must be apportioned suitably.

**Table 21.2   UME calculations on a dairy cow basis**

|  | GJ/year |  |
|---|---|---|
| Maintenance (including growth and pregnancy; average live weight 600 kg): 68 MJ per day × 365 days | 24.8 |  |
| Production (milk composition 3.9% fat, 8.7% SNF): 6,000 litres × 5.3 MJ/litre | 31.8 |  |
| Total cow energy requirement |  | 56.6 |
| Non-grass feed energy input: Concentrates (1,500 kg × 11.5 MJ/kg) | 17.3 |  |
| Brewers' grains (500 kg × 2.9 MJ/kg) | 1.5 |  |
| Total |  | 18.8 |
| Balance (UME/cow) |  | 37.8 |
| Reliance on grass (37.8    56.6 × 100): 67% |  |  |
| Average annual stocking rate: 2.5 cows/ha |  |  |
| UME/ha (37.8 × 2.5) |  | 95 |

**Table 21.3   UME calculations on a dairy herd basis (100 cows)**

|  | GJ/year |  |
|---|---|---|
| Maintenance (including growth and pregnancy; average live weight 600 kg): 100 cows × 68 MJ per day × 365 days | 2,482 |  |
| Production (milk composition 3.9% fat, 8.7% SNF): 100 cows × 6,000 litres × 5.3 MJ/litre | 3,180 |  |
| Total herd energy requirements |  | 5,662 |
| Non-grass feed energy input: 100 cows × 1.5 t concentrates × 11.5 MJ/kg | 1,725 |  |
| 100 cows × 0.5 t brewers' grains × 2.9 MJ/kg | 145 |  |
| Total |  | 1,870 |
| Balance (UME/herd) |  | 3,792 |
| Reliance on grass (3,792    5,662 × 100): 67% |  |  |
| Grassland area (grazing plus conservation): 40 ha |  |  |
| UME/ha (3,792    40) |  | 95 |

Dairy farm costings have revealed a wide range of UME values achieved, from 40 to 120 GJ/ha. The average value lies between 60 and 80 GJ/ha, although occasionally values in excess of 130 GJ/ha are achieved. In contrast, the results of a survey of UME output on hill farms in Northern Ireland indicated an average of 15 GJ/ha and a range of 4 to 41 GJ/ha, with lower outputs being associated with a high ratio of hill swards to inbye swards on a farm.

The average UME output of whole-farm grassland can be estimated from annual values of the following, measured from one spring to another:

A. Average ruminant livestock numbers, including their production
B. Quantities of home-grown farm fodder crops sold
C. Quantities of fodder crops in hand at end of year
D. Quantities of purchased ruminant feedstuffs
E. Yields of home-grown fodder crops produced
F. Quantities of fodder crops in hand at beginning of year

These items are converted into ME equivalents using annual factors for the different classes of stock, calculating the ME value of their production and growth, and using standard ME values for the feedstuffs. The data are transferred to a balance sheet as shown in Table 21.4. The average UME per hectare of farm grassland is derived from the formula A + B + C minus D + E + F divided by the number of hectares of grassland.

The use of computers has facilitated schemes by official and commercial advisory services for monitoring and improving the financial performance of all types of animal production systems. Assessment of

**Table 21.4   UME balance sheet for farm grassland**

| | GJ/year | | GJ/year |
|---|---|---|---|
| A. Livestock | 12,550 | D. Bought feed | 1,900 |
| B. Crop sales | 650 | E. Crops grown | 1,650 |
| C. Crops in hand | | F. Crops in hand | |
| (end of year) | 550 | (beginning of year) | 95 |
| | | Balance | 9,250 |
| | 13,750 | | 13,750 |

$$\text{UME/ha} = \frac{9,250}{\text{Ha of grassland}} = \frac{9,250}{120} = 77 \text{ GJ}$$

*Source:* Frame

UME output of grassland is usually done as part of the calculations, mainly for major cash-generating enterprises such as dairying or fattening beef stock. Monitoring can also be extended to include replacement stock for a particular enterprise or, indeed, all the ruminant stock on the farm, but appropriate records must be kept of the grassland areas or fields used, stock numbers and type, supplementary non-grass feed and animal performance, e.g. as liveweight gain. Output from individual fields or from the whole-farm grassland can be estimated as required.

## INTERPRETATION

The ME of feeds other than grass is allocated full value since it is assumed these feeds are efficiently used; any inefficiency in their use reduces the residual UME. However, in practice, costly non-grass feedstuffs such as concentrates are normally fed effectively. A more accurate residual UME will be obtained at higher levels of grass in the diet since the potential source of error in allocating full ME value to non-grass feeds will be reduced.

Grass production and utilisation are the main factors influencing ME. Low grass production efficiently used on a farm can result in a similar or better UME than in high grass production where the grass is inefficiently utilised. The inefficiency might result from too low a stocking rate, for instance. A high rate of fertiliser nitrogen application may not necessarily result in high UME; factors may exist, such as soil acidity or a poor soil potassium status, which limit the production response of the sward to fertiliser nitrogen, or the extra grass produced may be inefficiently used.

UME gives a basis for diagnosing reasons for low output and pointers to necessary remedies. Small fields often give higher UME outputs than large fields because a farm's fields are seldom of the same size but they may carry the same stock numbers for similar periods of time. Output from grassland decreases with increasing altitude because of the reduced length of the growing season and the effects of poorer soil and climate. Comparisons between fields sown to different seed mixtures have given conflicting results when the management of a field has been affected by the manager's bias. Similarly, general comparisons of long-term or permanent grass versus leys require careful interpretation because these sward types vary widely or are managed differently. The amount of variation found in utilised output from field to field on the same farm and within farms in the same area indicates the scope for improvement which exists.

This is also evident when comparing grassland output from different farms. Comparison between farms requires caution in view of different management skills and different stock potential. Dairy cow genetic potential, calving index and the pattern or onset of disease problems can affect milk production, for example, and the UME per hectare of dairy farms' grassland is strongly associated with milk production per hectare. Changes in UME output between years on a farm can shed light on the effect of new management practices. Model calculations can be used to assess the potential effects on UME of making changes in stocking rate, supplementary feed levels or potentially different milk production levels.

*Chapter 22*

# Management of Forage Legumes

The corner-stone in land improvement in all parts of the world is the leguminous plant. Let it once be possible to establish a swarding and edible leguminous herb and it becomes easy to create a tolerably good sward—a sward with a definite stock carrying capacity.

R. G. Stapledon, *The Land Now and Tomorrow*

Legumes make possible an ecologically sound nonexploitive and yet productive agricultural system, with which a hopefully stabilised (human) population can live in permanent balance or better.

J. S. Gladstones (1975)

White clover in grassland is increasingly being utilised in practical farming systems in Britain, but uptake has been slow and lags well behind research findings. Red clover is undervalued and almost relegated to museum status, while lucerne is still used in the southeast. Birdsfoot trefoil and sainfoin may have special purpose uses in certain areas but their potential has not been fulfilled. Yet most of the agronomic limitations to the successful use of forage legumes have been overcome and the main problem now lies in how best to integrate legume rich swards into simple but effective economic systems of animal production. Nor should the potential of legume-based swards be overlooked in systems producing 'organic' animal products, for which demand is increasing. There is also scope to use legumes in multiple cropping systems or as green manures prior to cropping.

This chapter gives an overview of the agronomy and management requirements of the main forage legumes, concentrating on the positive approach needed to overcome limitations and achieve reliability. Now that present extensification objectives in EC countries have caused renewed interest in grass/legume and legume swards, the time is ripe to exploit the valuable characteristics which forage legumes possess, not only in British farming but in those other countries in the world with a temperate climate where forage legumes have latterly been neglected in favour of manufactured fertiliser N.

# WHITE CLOVER

## Varieties

Smaller-leaved varieties are best for frequent and severe sheep grazing, while the larger-leaved clovers perform best under infrequent and lax cattle grazing or cutting. A simple blend of recommended clovers is normally best in seed mixtures for grassland: 1.5 kg/ha each of a small-leaved variety (e.g. Aberystwyth S 184) and medium-leaved variety (e.g. Menna) for sheep grazing and 1.5 kg/ha each of a medium-leaved and large-leaved variety (e.g. Alice) for cattle grazing and/or cutting for silage.

Most of the seed currently sown in the UK is Grasslands Huia, a variety from New Zealand that has a majority share of the total European consumption of clover seed. There is no shortage of UK or European varieties, many superior to Grasslands Huia, but supplies of seed have been restricted. However, seed of British-bred varieties is increasingly being produced under contract in more favourable climates abroad, including the United States of America and New Zealand. Attempts are being made to find suitable seed producing regions in Europe to add to Denmark, still a major producer of white clover seed.

## Compatibility with grasses

Grass species differ in their suitability as companion grasses for white clover. Meadow fescue is highly compatible, for example, but cocksfoot much less so. Table 22.1 shows the differing compatibilities of some grasses. In general, grass species and varieties which create densely tillered, close-knit swards are the least conducive to clover

**Table 22.1** **Effect of grass companion on annual DM production from grass/white clover swards**

| Grass species | Total herbage (t/ha) | White clover (t/ha) | White clover (%) |
|---|---|---|---|
| Crested dogstail | 8.3 | 4.4 | 53 |
| Smooth-stalked meadow grass | 8.3 | 4.0 | 48 |
| Red fescue | 9.1 | 3.6 | 40 |
| Creeping bent | 5.3 | 1.9 | 37 |
| Yorkshire fog | 7.9 | 2.5 | 32 |
| Perennial ryegrass (control) | 8.5 | 2.8 | 33 |

*Source:* Frame

stolon proliferation and plant development due to inhibition of the growth of clover growing points by shading at ground level. However, the adaptation of a grass species to a region or for a specific purpose currently takes precedence over its compatibility with clover when sowing mixtures. Work has shown that, for perennial ryegrass, the main grass species sown in the UK, early or late ear emergence types of varying sward density did not greatly affect clover growth. However, tetraploid ryegrasses permitted a better clover contribution than diploids, under grazing and cutting, and 50 to 75 per cent by weight of the grass component as tetraploid has been successfully used in seed mixtures.

To promote optimum production from both grass and clover components of a mixture, a co-adaptation concept is being developed by plant breeders at the Institute of Grassland and Environmental Research, Aberystwyth. In effect, both components are selected from co-existing genotypes found in a long-established sward. Initial work has shown good prospects, at least in the early life of the sward (see Table 22.2), though adequate seed supplies of selections are not yet available.

**Table 22.2  Effect of co-adaptation on annual DM production from grass/white clover swards**

| Perennial ryegrass companion grass | Total herbage (t/ha) | White clover (t/ha) | White clover (%) |
|---|---|---|---|
| Aberystwyth S23 | 7.4 | 5.1 | 69 |
| Co-adapted | 10.3 | 7.1 | 69 |

*Source:* Evans *et al.*

There is considerable room for flexibility in the grass:clover seed rates in seed mixtures, but the higher cost of clover seed can limit clover seed rate. A high clover seed rate or low grass seed rate initially improves the establishment and presence of clover, but a management which favours proliferation of the clover stolons can eventually compensate for a low clover seed rate. Nevertheless, it is prudent to aim for a good clover presence early in the sward's life by sowing an adequate clover seed rate. A typical seed mixture for drilling would be 18 to 24 kg/ha of perennial ryegrass and 2 to 3 kg/ha of white clover.

### Establishment needs

Shallow sowing of white clover is essential—at about 10 mm—so broadcasting the seed mixture on a firm ring-rolled seedbed, or drilling the grass component separately and then broadcasting the clover, is suitable. An establishment target should be 150 white clover plants/m² three months after sowing, developing to a 30 per cent ground cover twelve months later. Early rather than late season sowing and direct seeding rather than undersowing are desirable to give the clover plants a chance to establish satisfactorily before the onset of winter. The fate of clovers and of clover stolons over winter has been insufficiently researched, but where clover plants or stolons are killed by a hard winter there will be a time lag in clover contribution the following season. Management favourable to clover will then be essential to regenerate a vigorous clover content. A stolon network on the soil surface of at least 20 m/m² of sward in early spring is desirable to be sure of a satisfactory baseline presence.

Successful establishment of white clover has been achieved by various forms of direct drilling into an existing grass sward, or a grass/clover sward where the clover has become depleted. Guidelines required for success are particularly: adequate bare space for the clover seed to germinate and flourish; a satisfactory soil pH of 5.8 to 6.0; application of water-soluble phosphate fertiliser; anti-slug pellets; good soil–seed contact at drilling; and controlled post-sowing grazing after drilling to limit competition from the existing grass plants (see also Chapter 7). Open sward conditions can be created by a pre-sowing application of a low dose of paraquat, band spraying of glyphosate or by taking two heavy cuts of silage before drilling.

## Nitrogen fixation

Fixation of atmospheric nitrogen by clover takes place in nodules which develop on the root system. These nodules are formed following infection by rhizobial bacteria which live in the soil. Nodules formed by effective N-fixing strains of rhizobia are plump and, when cut open, are pinkish or reddish in colour due to haemoglobin. Small, hard nodules, which are pale white or yellowish when cut open, are a sign of ineffective rhizobial strains. Adequate numbers of effective rhizobia are usually present in lowland and upland mineral soils so that rhizobial inoculation of clover seed before sowing is not necessary, although it is recommended on all hill soils. Trials showed that inoculation was particularly essential on deep peats and wet, peaty podzols where effective strains of rhizobia were either absent or else present in low numbers.

Rhizobial inoculant can be obtained direct through seed merchants and are host-specific, that is, only particular strains are effective for each species of forage legume. Research work is now aimed at isolating separate strains for individual varieties of legumes. The commonest medium of inoculation is a powdered peat suspension of bacteria. This is mixed thoroughly with the clover seed, moistened with water and a sticking agent a few hours before sowing. It is essential that inoculant is stored in cool, dark conditions, and preferably in a fridge. Furthermore, the inoculated seed must not be exposed to sunlight prior to sowing. An alternative is to spray the inoculant in water onto the emerging clover seedlings during early sward establishment. Clover seed pre-pelleted with rhizobial inoculant plus a filler of lime and phosphate can also be sown, a technique often used in New Zealand.

The amount of N fixed has been found to vary considerably, according to growing conditions. While it can reach 280 kg/ha, an average amount for a sward in which white clover is well established would be about 150 kg/ha. Good summer temperatures and sunlight are favourable factors. Any factor which adversely affects clover growth, such as low temperatures, availability of mineral N from fertiliser input or excretal return, will depress nitrogen fixation. A good target clover content in practice would be 25 to 35 per cent of the annual total herbage, which would translate into 10 to 20 per cent clover content in the sward in early season and at least double this in midseason and later. Visual assessment of clover usually overestimates the actual amount because of the horizontal nature of clover leaf growth and its concentration in the upper layers of the sward.

Urinary nitrogen following ingestion of N-rich, clovery swards is the most rapid route by which the rhizobially fixed nitrogen is transferred to the sward. Some loss to the system takes place as leached nitrogen through the soil or as volatilised ammonia from the urine. There is also transfer of nitrogen to the soil through the death and decay of plant parts: nodules, roots and from surface leaf or stolon litter. The underground transfer is accelerated when the clover plants are put under stress by defoliation or drought. Underground transfer is the main route by which soil nitrogen is built up in the long term in grass/clover swards.

## Soil fertility requirements

White clover can tolerate a wide range of soil types, fertility and climatic conditions but it does not thrive at pH levels less than 5.8 or in poorly drained soils. In grass/clover pastures, clover is the more 'sensitive' component to deficiencies of plant nutrients such as phosphorus, potassium or trace elements; it soon declines and becomes replaced by grasses and weeds when nutrient deficiencies are prolonged. This is partly due to the lower root mass of clover compared with grasses and a different root morphology. Clover is also more susceptible to severe drought than grass and does best on moisture-retentive, albeit free-draining, soils. Nevertheless, it can tolerate and grow under moderate drought as well as, if not better than, grasses. It is highly responsive to irrigation.

Application rates of lime, phosphate and potash should be determined by routine soil analyses before establishment and every three to four years thereafter. At a moderate soil fertility (ADAS index 2), about 75 kg/ha each of phosphate and potash are needed at establishment. Peaty soils are often deficient in available potassium. Nitrogen should be omitted unless establishing clover after a run of cereal crops, when 40 kg/ha N will act as a starter dressing for the seedlings; a small starter dressing will also be beneficial on hill soils of low fertility. Grazed swards only require 30 kg/ha each of phosphate and potash in midseason as annual maintenance dressings. On swards cut for silage, additional phosphate and potash (about 40 and 80 kg/ha respectively per cut) are required to replenish soil reserves.

## Effect of fertiliser N usage

Repetitive application of fertiliser nitrogen to a grass/clover sward during the season increases total herbage production, but at the expense of reduced white clover content and clover production (see Table 22.3 and Figure 22.1). The mean herbage response from many trials is 8 to 9 kg DM per kg N, about half that from an all-grass sward. Negative responses can occur at low to moderate nitrogen rates when the loss in clover production is not sufficiently compensated by the gain in grass production. The magnitude of the total herbage response to applied nitrogen increases as white clover content in the

Table 22.3  Effect of fertiliser nitrogen rates on annual DM production from grass/white clover swards

| Fertiliser N (kg/ha/year) | Total herbage (t/ha) | White clover (t/ha) | White clover (%) |
|---|---|---|---|
| 0 | 7.8 | 4.1 | 53 |
| 120 | 8.7 | 2.4 | 28 |
| 240 | 10.0 | 1.1 | 11 |
| 360 | 11.7 | 0.5 | 4 |

Source: Frame and Boyd

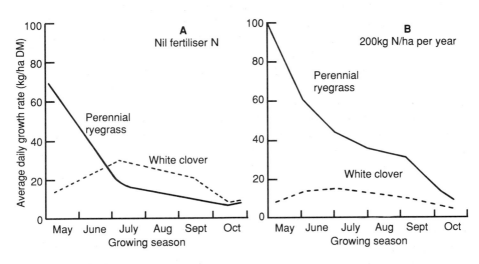

*Figure 22.1   Average daily growth rates of grass and white clover at two annual fertiliser N rates: (A) nil N, (B) 200 kg N/ha/year*

sward decreases; however, the amount of nitrogen fixed by clover decreases. It is generally reckoned that a grass/clover sward given no fertiliser nitrogen yields as much (7 to 9 t/ha DM) as a grass sward given 180 to 200 kg N/ha/year; however, this requires a satisfactory clover performance during the year with a 25 to 35 per cent clover contribution to the annual herbage production.

With low spring and autumn temperatures, clover leaf growth is less than grass. Thus the concept of tactical nitrogen application in spring and/or autumn has emerged on the premise that increased grass growth at these times will be less competitive to clover than at other times of the growing season. Spring-applied nitrogen does improve total herbage production in early season, but white clover performance in midseason is adversely affected. Clover can recover provided moderate N rates (about 50 kg/ha) are used in spring, but only if there is an adequate initial presence of clover plants and stolon. Fertiliser nitrogen application in late summer/early autumn also improves grass production and decreases clover production, but to a lesser extent than spring nitrogen; with spring plus autumn nitrogen combinations, clover is most depressed when a high proportion of the nitrogen is applied in spring (see Table 22.4). Slurry can be used on grass/clover swards, but it should be used in a planned way, mainly for supplying potash, and not excessively, since it is also rich in nitrogen; also heavy applications may physically smother clover.

**Table 22.4   Effect of tactical N application in spring, autumn or spring plus autumn on clover content (%) in grass/white clover swards**

|  |  | Autumn N (kg/ha) | | | |
|  |  | 0 | 25 | 50 | 75 |
| --- | --- | --- | --- | --- | --- |
| *Spring N* | *0* | 46 | 43 | 38 | 41 |
| *(kg/ha)* | 25 | 39 | 35 | 35 | 33 |
|  | 50 | 34 | 31 | 30 | 29 |
|  | 75 | 20 | 26 | 27 | 27 |

*Source:* Frame and Boyd

## White clover dynamics

White clover content is likely to be much lower on commercial farms than in experimental plots, especially under simulated grazing. These differences also occur in New Zealand, where conditions are better suited to clover than the UK. Under experimental conditions, there is usually a strong emphasis on the factors favouring clover's needs, for example, adequate pH, soil P and K status, which may not be the case on farms. Intensive nitrogen use on farms, whether repetitively or tactically, can readily depress clover to an extent from which it is unlikely to recover.

It should be appreciated that, in addition to applied fertiliser nitrogen, excretal return from grazing livestock boosts the soil nitrogen pool, the net effect of which helps to tip the 'balance' in the grass/clover sward towards grass dominance. Clover recovery can then only occur if soil nitrogen is depleted, for example by a silage cut, assuming there is a nucleus of surviving clover plants or a reservoir of buried stolons or seed. Where this is not so, the best option is to reintroduce clover by reseeding with a grass/clover sward or to direct drill clover into the existing grass-dominant sward.

## Weed, disease and pest control

Most herbicides effective against broad-leaved weeds will damage or kill white clover and there has been interest in developing clover-safe herbicides, such as those based on benazolin, bentazone, cyanazine or linuron. Timing of application is critical since the young seedlings must be adequately developed. In the established sward, complete clover safety is rarely possible, particularly when perennial weeds have to be controlled with herbicides such as mecoprop, 2,4-D or related mixtures. Problem perennial weeds such as docks or thistles are best controlled by glyphosate before grass/clover swards are established.

Fungicide control of clover diseases is not feasible or economic and the use of resistant varieties is advocated. Clover pests include slugs and leatherjackets. Slug damage is sometimes significant and control by the use of chemicals such as methiocarb in anti-slug pellets is necessary when direct drilling clover into an existing grass sward. Leatherjackets can damage white clover but effective prediction models of infestation are available and some pesticides such as chlorpyrifos are effective. Recent research has focused on the detrimental effects of stem eelworm and clover cyst nematode but there is no economic method of chemical control.

## Cutting for silage

It was long believed that infrequent defoliation systems reduced clover in a grass/clover sward because clover would be shaded out by the taller growing grass; in fact, clover leaves are very efficient at competing for light and, although grass output increases with increasing rest interval between defoliations, white clover is not necessarily reduced. However, if fertiliser N is applied, the proliferation of growing points from the stolon branches is inhibited due to increased shading at ground level; the N-stimulated grass also becomes more competitive for soil water and nutrients. A rest interval within an intensive sheep grazing system and a subsequent silage cut has also been shown to benefit white clover persistence and productivity. Beneficial factors are reduction of grass tiller density, resulting in less competition and more light reaching the base of the sward, and depletion of soil N with cutting and removal of a silage crop. Much of the benefit to clover is in increased stolon length and weight during the rest interval and the stimulation of growing points subsequent to harvesting. Recent research results have shown that grass/clover swards could be cut for silage in successive years without detriment to the clover and giving DM production of about 70 to 80 per cent of that from grass swards receiving 300 to 350 kg N/ha/year, i.e. 9 to 10 t/ha versus 12 to 14 t/ha. The grass/clover herbage ensiled was usually higher in digestibility, protein and mineral content than the grass herbage (Table 22.5).

The tolerance of grass/white clover swards to flexible management is an important advantage for its practical exploitation. Traditionally, white clover has been regarded as a plant primarily suited to grazing but, as indicated above, it is more suitable for conservation than was

259

**Table 22.5**  Composition of grass/clover and grass herbage at ensiling (g/kg DM unless otherwise stated)

| Parameter | Grass/clover (nil N) | Grass (340 kg/ha N) |
|---|---|---|
| Dry matter (g/kg) | 156 | 160 |
| Crude protein | 161 | 149 |
| D value | 71.5 | 68.8 |
| ME (MJ/kg DM) | 11.4 | 11.0 |
| P | 3.9 | 3.5 |
| K | 25.7 | 19.4 |
| Ca | 10.1 | 6.0 |
| Mg | 3.9 | 2.5 |

*Source:* Roberts *et al.*

previously thought. It can be more persistent than red clover or lucerne, the usual legumes grown for conservation, and maintains its digestibility better. Also, its shorter growing season relative to grass is of less significance for conservation than for grazing. Integration of grazing and silage in a cattle grazing system could be best on an alternate year basis, since the chances of dung contamination in the silage are high when a cut is taken following a rest interval after grazing.

## Grazing swards

Continuous heavy grazing by sheep may reduce white clover persistence and production compared with rotational grazing, but clover can persist in well-managed continuous stocking systems, which include the use of suitable clover varieties with compatible grasses (see Table 22.6). Cattle grazing has less severe effects on clover. The more uniform excretal N return from sheep compared with cattle and the selective grazing of sheep are major clover depressing factors. Some

**Table 22.6**  Seasonal variation in clover content (per cent on a DM basis) in total herbage on continuously stocked sheep swards

| Year | May | June | July | August |
|---|---|---|---|---|
| 1 | 4 | 9 | 20 | 26 |
| 2 | 17 | 26 | 32 | 36 |
| 3 | 14 | 17 | 29 | 32 |

*Source:* Vipond and Swift

New Zealand research suggests that the ideal system may be continuous stocking in spring and rotational grazing in summer.

Table 22.7 shows examples of the high quality of grass/clover swards, continuously stocked by dairy cows and sampled every three weeks, compared with grass swards receiving 350 kg/ha N per annum. Stocking rate on the clover swards ranged from 3 to 4 cows/ha in the first half of the season to 2 to 3 cows/ha by late season. Herbage from the grass/clover swards was usually higher in phosphorus, potassium and calcium content. The grass herbage was often superior in nitrogen and magnesium content but consistently higher in sodium; nitrogen content was higher in grass/clover herbage in late season when a high proportion of clover was present.

Table 22.7   **Annual quality of herbage on offer in two grass/clover (GC) and N-fertilised grass swards (GN) grazed by dairy cows (N and minerals as g/kg DM)**

| Sward | D value | N | P | K | Ca | Mg | Na |
|-------|---------|------|-----|------|-----|-----|-----|
| GC*   | 64      | 33.9 | 4.0 | 24.9 | 8.6 | 2.8 | 1.5 |
| GN    | 61      | 27.3 | 3.0 | 18.1 | 6.3 | 3.3 | 2.8 |
| GC**  | 66      | 27.1 | 4.0 | 24.3 | 6.5 | 2.8 | 1.0 |
| GN    | 64      | 28.0 | 3.7 | 22.8 | 5.0 | 2.8 | 2.0 |

* Annual clover content 44%   ** Annual clover content 19%

*Source:* Frame *et al.*

There is evidence that late autumn/early winter grazing of a grass/clover sward is beneficial to subsequent clover growth because clover stolon growing points benefit from the better light regime. However, overgrazing in late winter/early spring is detrimental to clover vigour, especially if the stolons are exposed through lack of grass cover and winter kill occurs. Clover production and persistence have been successfully maintained in recent grazing research, using the guidelines and management outlined in Tables 17.1 to 17.3, Chapter 17. Optimal heights for grass/white clover swards are 4 to 6 cm for continuously stocked sheep and 6 to 9 cm for cattle, with some adjustments upwards for productive stock. There is less information on rotational grazing but residual heights should be approximately 4 to 5 cm for sheep and 7 to 9 cm for cattle. French rotational grazing methods for cattle favour 4-week rest periods in early season and 6-week rests in late season.

## Management guidelines

The key points for successful management are summarised in Table 22.8.

**Table 22.8   Management guidelines for grass/white clover swards**

| | |
|---|---|
| Seed: | Choose compatible companion grasses<br>Sow 3 to 4 kg clover seed/ha in seed mixtures<br>Use a blend of white clover leaf types |
| Establishment: | Sow shallowly (10 to 12 mm)<br>Attain soil pH of 5.8 to 6.0 and adequate P and K status<br>Preferably sow direct in spring |
| Production: | Make minimal tactical use of fertiliser N<br>Maintain adequate soil pH, P and K status<br>Use clover-safe herbicides for weed control |
| Utilisation: | Avoid heavy continuous stocking with sheep<br>Provide clover recovery periods, e.g. use of rotational grazing, insertion of a silage cut in a sheep grazing system<br>Utilise efficiently in autumn or, alternatively, graze moderately in early winter |

## Systems output

Milk production systems based on grass/white clover swards, using lower inputs and costs compared with heavily N-fertilised grass swards, have proved economically viable. Annual output data from a farm-scale study of dairy systems at Dumfries are shown in Table 22.9. In work elsewhere, the output of lamb or beef from grass/clover swards has matched the output from moderately N-fertilised grass swards. A key factor is improved individual animal performance, which compensates for lower herbage production from the grass/clover swards. Increased land area—which suits land extensification policies—or supplementary feed can be used to offset any forage production shortfall. In practice, many sheep and beef farms use low rates of fertiliser N and both herbage production and animal output could be improved by the introduction of grass/white clover swards. The results of a grazing trial at Edinburgh (see Table 22.10) are a good example of the successful use of grass/white clover swards for lamb production. The need for a compatible

262

**Table 22.9** Dairy cow performance from systems based on continuously stocked grass/clover (nil N) and N-fertilised grass swards (350 kg/ha)

| Year | Grass + clover | | | Grass + N | | |
|---|---|---|---|---|---|---|
| | 1 | 2 | 3 | 1 | 2 | 3 |
| Cow numbers | 71 | 70 | 70 | 70 | 72 | 69 |
| Grassland (ha) | 36 | 36 | 46 | 36 | 36 | 36 |
| Stocking rate (LU/ha)* | 2.4 | 2.4 | 1.9 | 2.4 | 2.4 | 2.4 |
| Silage DM (t/ha) | 9.9 | 9.3 | 11.4 | 12.0 | 11.5 | 11.9 |
| Milk sales (litres/cow) | 5658 | 5605 | 5719 | 5764 | 5532 | 5724 |
| Concentrates (kg/cow) | 1709 | 1554 | 1096 | 1412 | 1185 | 1101 |
| Concentrates (kg/litres) | 0.30 | 0.28 | 0.19 | 0.24 | 0.21 | 0.19 |
| Milk fat (g/kg) | 40.0 | 39.9 | 40.5 | 39.5 | 40.7 | 40.4 |
| Milk protein (g/kg) | 31.9 | 31.9 | 32.4 | 31.8 | 31.3 | 31.8 |

* Including young dairy stock

*Source:* Bax and Thomas


**Table 22.10** Ewe and lamb performance on grass/clover (nil N) and N-fertilised grass swards (160 kg/ha) continuously stocked to maintain 4 to 6 cm sward grazing height during April to August

| Parameter | First year | | Third year | |
|---|---|---|---|---|
| | Grass + clover | Grass + N | Grass + clover | Grass + N |
| Daily liveweight gain (g/d): | | | | |
| Lambs | 262 | 209 | 272 | 223 |
| Ewes | +51 | −49 | +24 | −45 |
| Per cent lambs finished | 65 | 7 | 80 | 29 |
| Ewes with twins grazed/ha | 10.4 | 17.4 | 14.8 | 16.8 |
| Lamb gain/ha (kg) | 804 | 1,051 | 1,129 | 1,111 |
| Per cent clover in herbage DM | | | | |
| April–June | 6 | – | 16 | – |
| July–August | 23 | – | 29 | – |

*Source:* Vipond and Swift


tetraploid perennial ryegrass and a small-leaved clover persistent under sheep grazing was emphasised. Animal output improved markedly on the grass/clover sward after the first full grazing season when the clover became better established.

# RED CLOVER

Apart from its traditional but declining use as a minor component (3 to 4 kg/ha) in seed mixtures, red clover has shown considerable potential as the dominant component in swards utilised for silage cropping, as outlined below.

## Varieties

There has been no red clover breeding work in the UK for many years, and only two of the seven varieties currently recommended in the NIAB Recommended List were bred in the UK, the remainder being European-bred. When the first tetraploid varieties appeared in the 1970s they were more productive and persistent than most diploids, but tetraploidy *per se* is not now a guarantee of superiority in all situations because of improved diploids, and individual varieties should always be chosen for the purpose required.

## Compatibility with grasses

Compared with a pure-sown red clover sward, a red clover/grass mixture gives higher total herbage DM production and digestibility but lower clover content and lower CP concentration. Intermediate perennial ryegrass is the most suitable and favoured companion grass, but timothy or meadow fescue have also proved satisfactory. The effects of companion grass on total herbage production and quality are greatest at the first harvest, usually in June, when the grass component is more dominant than at later harvests.

## Establishment needs

When sowing red clover in monoculture a seed rate of 12 to 15 kg/ha is used, while in a red clover/grass mixture, a suitable blend is 12 to 15 kg/ha red clover plus 4 to 5 kg/ha grass seed. Sowing between April and July has given significantly higher DM production and red clover content in the following harvest year than sowing later; the adverse effects of late sowing are not fully offset by increasing red clover seed rates. In late season sowings, clover seedlings do not fully develop before winter and may then be subject to winter kill. At 6 to 8 t/ha, DM production in the establishment year of spring sown red clover swards is about 60 per cent of that obtained in the first full harvest year.

Direct seeding rather than sowing under a cover crop is preferable since a late harvested or lodged cereal grain crop depresses the undersown clover, though arable silage is a suitable cover crop. Regardless of time or method of seeding, sowing shallowly, not more than 15 mm, is recommended. A density of 200 plants/m$^2$ in October of the sowing year is a suitable establishment target. Rhizobial inoculation of seed is rarely advocated in the UK since there is no evidence of benefit in the soil types and situations where red clover is likely to be sown.

Direct drilling can be used to rejuvenate an existing red clover sward which has thinned out and become less productive and where decline in red clover population and vigour has been due to ageing and not to disease or pests. Spring or late summer are suitable drilling times and the rules outlined in Chapter 7 should be followed.

### Soil fertility requirements

Red clover can be grown under a wider range of soil fertility conditions than lucerne, but neither species will thrive on acid or waterlogged soils. Development is best at a soil pH of 6 and above, since clover is particularly sensitive to manganese toxicity in acid soils. Adequate phosphate and potash in the seedbed are necessary, as indicated by soil analysis. If soil N is likely to be low, after a succession of cereal crops, for example, a starter N dressing of 40 kg/ha will encourage initial clover seedling development.

About 80 to 90 per cent of the annual production is obtained from two cuts taken in June and late July/early August, and annual phosphate and potash application should be split over these cuts. Typical annual requirements, based upon soil of moderate fertility (ADAS index 2) and a removal of 12 t/ha DM, would be 100 to 150 kg/ha phosphate and 250 to 300 kg/ha potash. An application of 50 m$^3$/ha of cattle slurry (1:1 dilution with water) would supply available phosphate and potash of 20 and 115 kg/ha respectively.

Red clover or clover/grass mixtures do not require fertiliser nitrogen if the clover is abundant and growing vigorously. While the nitrogen will increase the total production of the mixture, it will be at the expense of reduced clover in the sward (see Table 22.11). However, nitrogen will sustain total herbage production into a third or fourth harvest year when the clover has thinned out.

### Weed, disease and pest control

Weeds are more likely to be a problem in pure stands of red clover

**Table 22.11    Effect of fertiliser nitrogen application on herbage production from a red clover/grass sward**

| Harvest year | N rate (kg/ha) | DM (t/ha) | | Red clover (%) |
|---|---|---|---|---|
| | | Total herbage | Red clover | |
| 1 | 0 | 12.7 | 8.9 | 70 |
| | 180 | 13.1 | 6.8 | 52 |
| 2 | 0 | 10.8 | 6.4 | 59 |
| | 180 | 12.6 | 2.7 | 21 |
| 3 | 0 | 7.9 | 1.5 | 19 |
| | 180 | 12.7 | 0.4 | 3 |

*Source:* Frame

than in clover/grass mixtures. A weed-free seedbed is a first essential and perennial weeds in the previous crop should be eradicated prior to sowing. Most herbicides effective against broad-leaved weeds will damage or kill red clover. Clover-safe herbicides based on 2,4-DB or MCPB are suitable for post-emergence control of broad-leaved weed seedlings but timing of application is critical. Sufficient development between first and third trifoliate leaf stages of the clover seedlings is necessary.

Clover rot has been a limiting factor to red clover production in southern and eastern Britain and its incidence might increase if red clover were more widely grown. However, varieties are now available with high resistance, which represents an advance from the early 1970s when enhanced resistance to clover rot in tetraploid varieties was first shown.

Stem eelworm can be a serious problem in susceptible varieties if contaminated seed is used. Seed should have been fumigated, e.g. with methyl bromide, by the seed merchant. Initially, eelworm affects patches of clover which die out, later enlarging to the whole stand after a couple of years. Varieties with resistance to eelworm, clover rot and to red clover necrotic mosaic virus (RCNMV) are available and should preferably be used.

Damage by slugs during establishment can be reduced by the use of methiocarb pellets, which should always be applied when red clover is sown after grass. Several chemicals, e.g. chlorpyrifos, applied as a spray, are effective against leatherjackets. Other insect pest damage should be identified and treated; weevil attack can be controlled by fenitrothian sprays, for example.

266

## Utilisation

Standard management is to take a silage cut in early to mid June, at the clover's early flowering stage, with a second cut seven to eight weeks later. The autumn regrowth can be grazed in October, or possibly cut for a third time, although this harvest may only contribute 10 to 15 per cent of annual production. Typical annual DM in harvest years one to three from a range of UK experiments was 9 to 18, 9 to 15 and 4 to 14 t/ha. Data are sparse for later harvest years since, except under exceptional circumstances, red clover does not persist or produce significantly after the third full harvest year. On some occasions DM production ranged between 6 and 12 t/ha for the third to sixth years. Cutting more frequently than three times a year results in lower DM production and, although herbage quality improves, this does not compensate for the large drop in production.

Direct physical damage to clover plants and soil compaction from wheel tracking are common in farm practice when cutting for silage; reduced growth and plant density occur in the tracks. Damage is worst in wet weather and on sloping fields due to wheel slip and soil smearing. Wheel tracking experiments have shown that damage is minimised by less frequent wheeling during silage making. Damage to young regrowing plants will also be reduced if the operations are not prolonged.

Typical analyses of red clover monocultures freshly cut for silage are: 120 to 150 g/kg for dry matter, 58 to 66 for D value, 9.3 to 10.5 MJ/kg DM for ME and 170 to 190 g/kg DM for CP concentration. A red clover/grass mixture would have a slightly higher DM, D value and ME but a lower CP. Wilting, fine chopping and the addition of an effective additive are all necessary when making silage; formic acid has been used successfully as an additive. Typical analyses of red clover silages are given in Table 22.12. Improved performance from dairy cows, beef cattle and lambs on grazed or ensiled red clover compared with grass has been associated with higher feeding value

**Table 22.12   Analysis of red clover-dominant silage**

| | |
|---|---|
| DM content (g/kg) | 140–270 |
| Crude protein (g/kg DM) | 170–210 |
| D value | 53– 60 |
| ME (MJ/kg DM) | 8.5–9.6 |
| pH | 4.0–4.5 |

*Source:* Frame

and intakes, and more efficient use of the forage nutrients.

Traditionally red clover, with its erect growth habit, was never regarded as a crop suitable for intensive grazing. Repeated severe defoliations are likely to result in loss of plant population, and persistence will be better under lax grazing. Intensive grazing in autumn can depress first cut silage production the following year and cause a reduction in red clover content.

## Management guidelines

Key factors for optimum management of red clover/grass swards are outlined in Table 22.13.

**Table 22.13  Management guidelines for red clover/grass swards**

| | |
|---|---|
| Seed: | Choose compatible companion grass |
| | Sow 15 kg/ha clover plus 5 kg/ha grass |
| | Use clover seed pre-fumigated against stem eelworm |
| Establishment: | Sow shallowly (10 to 15 mm) |
| | Attain soil pH of 6.0 and adequate P and K status |
| | Preferably sow direct in spring |
| Production: | Maintain adequate soil pH, P and K status |
| | Use potash rich slurry in fertiliser programmes |
| | Use clover-safe herbicides for weed control |
| Utilisation: | Take two silage crops and an aftermath grazing |
| | Use an effective additive for silage |
| | Minimise wheel traffic at all times |
| | Keep breeding ewes off clover just before and after mating |
| | Graze off late autumn/early winter growth with sheep |

# LUCERNE

Good establishment is the key to success with lucerne and, to achieve it, direct sowing early in the season at 18 to 20 kg/ha of seed broadcast onto a firm, weed-free seedbed is best. A monoculture has the advantage of being simple to manage since only one species is involved, with weed control simplified, especially if a herbicide against invasive weed grasses is required. However, a companion grass can have beneficial effects on herbage production and quality provided a non-aggressive variety is sown at a low seed rate in the mixture, e.g. meadow fescue at 2 to 4 kg/ha. A variety with resistance to both *Verticillium* wilt and stem eelworm is essential. Seed should be fumigated against eelworm. Inoculation with effective rhizobia is also essential to ensure good root nodulation and subsequent N-fixation.

Lucerne grows best in the drier areas of south-east England where there are well-drained, deep soils. About 120 kg/ha phosphate and 360 kg/ha potash are required annually to replenish the nutrients removed by cropping. A soil pH of 6 to 6.5 is desirable. In English trials, herbage DM production in its early years—15 t/ha—matched that from perennial ryegrass receiving 300 kg/ha N. In Scottish trials at Ayr, production improved from the first to the third harvest year (see Table 22.14) and meadow fescue and tetraploid perennial ryegrass proved suitable companion grasses, allowing 80 to 90 per cent lucerne contribution.

**Table 22.14  Herbage DM production (t/ha) from lucerne and lucerne/grass swards**

| | | Companion grass to lucerne | | |
| --- | --- | --- | --- | --- |
| Harvest year | None | Meadow fescue | Perennial ryegrass (tetraploid) | Timothy |
| 1 | 9.5 | 11.0 | 10.7 | 10.0 |
| 2 | 15.8 | 17.1 | 16.4 | 15.5 |
| 3 | 17.7 | 17.9 | 17.2 | 16.5 |

*Source:* Frame and Harkess

There is some debate over the optimum cutting interval and recommendations have ranged from four to eight weeks. While the shorter interval will improve forage quality, production is reduced; six weeks is a reasonable compromise. Lucerne has the disadvantage of being a special-purpose conservation crop. An autumn rest period of six weeks or so is essential to allow lucerne to build up its root reserves before winter dormancy. Once dormant, in late autumn/early winter, the stand can then be cut or grazed without affecting its persistence.

## Management guidelines

The key points for successful management are summarised in Table 22.15.

**Table 22.15   Management guidelines for lucerne swards**

| | |
|---|---|
| Seed: | Use disease-resistant variety |
| | Use seed pre-fumigated against stem eelworm |
| | Broadcast 18 to 20 kg/ha lucerne alone |
| | Rhizobial inoculation essential |
| | |
| Establishment: | Sow shallowly (10 to 20 cm) |
| | Attain soil pH of 6.0 and adequate P and K status |
| | Preferably sow direct in spring |
| | |
| Production: | Maintain soil pH and P and K status |
| | Use potash rich slurry in fertiliser programmes |
| | Use clover-safe herbicides for weed control |
| | |
| Utilisation | Make silage (or hay) with 6- to 7-week cutting intervals |
| | Use an effective additive for silage |
| | Minimise wheel traffic at all times |
| | Rest swards in late autumn to encourage persistency |

## LEGUME USE IN THE FUTURE

Many of the agronomic limitations to successful production and utilisation of grass/white clover swards have been overcome. However, certain aspects still require refinement, e.g. the proportion of clover desirable in a sward for best agronomic and animal performance. There is a likelihood that grass/clover swards will play an increasing role in grassland farming systems in the UK and EC countries, given the present extensification objectives and the results being achieved in research and practice. Fortunately, a wide range of research and development work has underpinned the principles of white clover management in milk, beef and sheep production systems.

The future of the other forage legumes is less optimistic. Red clover seed usage has declined markedly and its potential in red clover-dominant leys for silage cropping has not been exploited. Lack of longevity is probably the main limiting factor and further plant breeding is necessary. The last British variety to be added to the NIAB Recommended List was Deben in 1977. Lucerne is in almost the same situation as red clover, although the area grown has stabilised. It is unlikely to expand much beyond the south-east because of its lack of flexibility. Owing to the contraction of government-sponsored research and development, the future development of these two species, and of other legumes such as birdsfoot trefoil or sainfoin, lies in the hands of pioneer farming enthusiasts or research sponsored by farmer-dominant organisations.

Longer-term development lies in the hands of the breeder. Genetic manipulation of legume species could lead to the creation of new types in which disadvantageous characteristics such as bloating of stock, oestrogens in red clover and general disease susceptibility are removed and advantageous characteristics such as greater longevity and adaptability to a wider range of environments and managements are brought in (see Chapter 5). However, the challenge of exploiting existing forage legumes should be met by farmers until such time as improved genotypes are introduced.

## Chapter 23

# Grassland Nature Conservation

I'm truly sorry man's dominion
Has broken Nature's social union,
And justifies the ill opinion,
Which makes thee startle
At me, thy poor, earth-born companion,
An' fellow-mortal!

Robert Burns (1759–96). 'To a Mouse'

I have no other wish than to mingle more closely with nature, and I aspire to no other destiny than to work and live in harmony with her laws. . . . Nature is greatness, power and immortality: compared with her, a creature is nothing but a miserable atom.

Claude Monet (1909)

The future is likely to see nature conservation assume a much greater importance than in the past, as its benefits become more widely known, as more land is released from agricultural production and as extensification becomes more widespread. The challenge of incorporating positive nature conservation practices into grassland management should be met head on and the new skills learned. Bodies such as English Nature, Scottish Natural Heritage and the Farming and Wildlife Advisory Group are playing an increasingly significant role in reconciling the sometimes conflicting pressures of productive agriculture and nature conservation. Species rich habitats do not just happen—they require skilled and sympathetic management. Financial incentives are already on offer and are likely to increase.

## LOSS OF SPECIES RICH GRASSLAND

Intensification of land use in recent decades has led to a substantial reduction in the extent of species rich pastures and hay meadows. Many areas of old grass on land hitherto unsuitable and not needed for arable farming have been ploughed up and cropped using modern technology and intensive inputs. This has taken place mainly in the

272

lowlands and extended to chalk downlands, heaths, permanent grass and even marshlands. Semi-natural grassland was ploughed up and reseeded to leys with seed mixtures containing only a few agricultural grass species, especially perennial ryegrass; these respond best to high rates of fertiliser nitrogen application and use for intensive stocking and cutting for silage. In other, traditionally managed species rich pastures and hay meadows, the diversity of the flora was reduced by the introduction of grassland renovation measures such as drainage, liming, fertilisation and pesticide use.

Most indigenous grasses, herbs and wildflowers cannot compete with the aggressive sown grasses under enhanced soil fertility conditions and intensive management. Many are adapted to survival only under conditions of low soil fertility, to marshiness, and to extensive management, by grazing or use as hay meadows. The limited range of plant species found in intensively managed swards results in a correspondingly limited range of insects, birds, wildflowers and other wildlife, earning them the term 'grassland deserts' by nature conservationists.

The extent of loss or significant damage of semi-natural swards in the past is shown in Table 23.1 and, according to other recent local surveys, the loss is continuing. It should be appreciated that farmers had to respond to earlier national policies for intensification and to economic pressures from the marketplace. It is also undeniable that farmers control the management of over 75 per cent of the UK's land area, and thus they have had, and will continue to have, a major formative influence on vegetation and the landscape.

**Table 23.1** **Extent of semi-natural grassland habitats lost or significantly damaged between 1949 and 1984**

| Type of grassland | Per cent loss |
|---|---|
| Limestone grasslands | 80 |
| Lowland wetlands | 50 |
| Lowland heaths | 40 |
| Upland moors | 30 |

*Source:* Nature Conservancy Council

## NATURE CONSERVATION SCHEMES

The increasing public demand for an attractive landscape, conserva-

273

tion of plant and animal wildlife, opportunity for increased recreation, and the desire for food from environmentally friendly farming necessitate new approaches to grassland management in the 1990s. Such concern is not new. The first British conservation society was formed in 1865. Today there is a plethora of societies concerned with land use and wildlife. Extensification of land use is rapidly becoming a feature of British agriculture, and both EC and government policy are geared towards this end. Some of the measures initiated have significant or potential benefit for wildlife habitats, including grassland, and the environment generally. The Farm and Conservation Grant Scheme includes a conservation projects section, for example, with scope for regeneration of heather, bracken control and enclosure of heathland. There are also various extensification schemes designed to encourage farmers to reduce output in return for payments reflecting loss of income incurred.

The creation of Environmentally Sensitive Areas (ESAs), started in 1986, warrants special mention. The scheme is voluntary, but farmers who join are required to agree to continue or adopt specific farming methods which help to protect or enhance the conservation value of their land. In return they receive payments over the duration of the agreement, which is currently five years; further payments can be made for improvements which meet specific nature conservation objectives. The detailed practices agreed are specific for each ESA but essentially maintain the present system of farming, limiting the application of fertiliser, lime and pesticides, restricting new agricultural practices and conserving landscape and historic features. Examples of ESAs are: the low lying, winter flooded wetlands of the Somerset Levels and Moors, rich in flowering herbs and grass species; the grass and heather moors of the Peak District in northern England; the coastal marshes, heaths and meadows of the Suffolk river valleys; and the flower rich heathlands of Breckland, East Anglia. In Scotland, there are ten ESAs with a total area of nearly one million hectares or almost 15 per cent of the land area.

Another long-standing conservation scheme is the Sites of Special Scientific Interest (SSSI). Since 1949 some 4,000 sites, occupying 7 per cent of Britain's land area, have been identified and statutorily notified. The relevant conservation body, e.g. English Nature in the case of England, and the land owners and occupiers then reach agreement on management practices which will maintain the special interest. Negotiations also take place on matters such as the type of grant-aid or compensatory payment which are appropriate, or on any changes in land-use operations being considered.

Apart from wildlife and physical features, there may be grassland or

▲   Grass/white clover sward
showing how well clover foliage
can compete with grass for light
*Chapter 22*

▲   Dairy cows grazing
grass/white clover swards in
continuous stocking system
at SAC, Crichton Royal
*Chapter 22*

Productive grass/white clover being cut for silage at SAC, Aberdeen
▼   *Chapter 22*

Productive red clover sward ▶
for silage at SAC, Auchincruive
*Chapter 22*

▲    Experimental plots of red clover at SAC, Auchincruive: (*left*) showing
reduced vigour and production due to frequent wheel tracking; (*right*)
untracked by wheels    *Chapter 22*

◀    Vigorous white clover
monoculture for seed production,
Estonia    *Chapter 22*

Drift of wild primroses on  ▶
sheep grazed mountain pasture,
Bulgaria    *Chapter 23*

◀    Sown wildflower meadow,
south-west Scotland    *Chapter 23*

▼    Drift of wild crocuses on sheep grazed mountain pasture, Bulgaria
*Chapter 23*

▲  Experimental plots of
different types of wildflower
seed mixtures at SAC,
Auchincruive  *Chapter 23*

▲  Assessing the productivity
of wildflower seed mixtures at
SAC, Auchincruive  *Chapter 23*

Landscape showing many valuable nature conservation features,
▼  central Scotland  *Chapter 23*

other plant communities which depend for their survival on the continuation of extensive farming practices. Species rich habitats require skilled and sympathetic management if they are to survive, preferably the same kind of management and conditions which helped to create them. Some of the consequences of management withdrawal can be seen on the bracken-infested slopes of upland grassland or in the isolated escarpments of scrub or thicket on the chalk downlands in southern England; hitherto covered by flower rich, fescue grass swards, these downland areas, formerly preserved by sheep grazing, were made ungrazable by surrounding arable cultivation. In other areas hay meadows were traditionally cut when mature and the prolonged period of working the hay allowed seed shedding to replenish soil seed reserves. Crops are now often cut at an earlier stage of growth and faster hay making is practised, threatening the persistence of some species.

## GENE SURVIVAL

In addition to landscape enhancement and the encouragement of wildflowers, nature conservation ensures the preservation of genetic diversity. Even on a world scale, agriculture exploits relatively few cultivated plant species out of the total number in existence. Yet dramatic plant breeding advances have been made by drawing upon and incorporating genes from wild, uncultivated plant sources into new improved varieties. Many previously neglected plant species are now making contributions as biotechnology finds innovative uses such as in medicines and drugs. There are also moral and ethical reasons for avoiding the extinction of any wild species.

## WILDFLOWER MIXTURE PRODUCTIVITY

The cost of a wildflower seed mixture is several times higher than a typical agricultural seed mixture, so making it economically unacceptable to sow large areas in the absence of grant-aid. It is possible to establish small, strategically scattered wildflower areas as an adjunct to more intensive grassland by encouraging wildflower and herb associations on difficult slopes, in awkward corners of fields, field margins and hedgerow bottoms, or by enrichment through direct drilling techniques. It could be argued that there is also a 'middle' way for some areas, whereby floral diversity and reasonable but economic production could be combined on land given modest inputs of plant nutrients, whether inorganic or organic, and managed at a moderate intensity of grazing and silage or hay management. This approach is being studied at Ayr in Scotland both on a small plot agronomic basis and in the context of farm systems. Table 23.2 shows some yield and quality results from commercial wildflower mixtures. All but two of the ten mixtures were sown with non-aggressive companion grasses. Within the mixtures, wildflower content, including herbs, ranged from 9 to 21 individual species, legumes from 1 to 6 and grass species from 0 to 9.

**Table 23.2    Yield and composition data from ten commercial wildflower mixtures cut under a hay regime**

| | |
|---|---|
| *Average 3-year data* | |
| Annual yield (t/ha DM) | 7.6–11.0 |
| Annual D value | 56–60 |
| Main hay cut (t/ha DM) | 4.3–6.3 |
| Hay cut D value | 52–56 |
| *Third year composition (g/kg DM)* | |
| N | 25–30 |
| P | 3.4–3.7 |
| K | 18–23 |
| Ca | 11–13 |
| Mg | 2.2–3.3 |

*Source:* Frame and Tiley

Herbage outputs (7 to 11 t/ha DM) were higher than those typical of traditional wildflower meadows (about 2 to 4 t/ha DM). However, the site was of high fertility, some fertiliser nitrogen (50 kg/ha) was applied annually and some of the mixtures contained productive forage legumes. Digestibility values ranged from low to moderate but the material was rich in minerals. The forage legume components, mainly white and red clovers, were abundant in the most productive mixtures. Many of the wildflower constituents of the mixtures failed to establish, while others did not persist for long. Ribwort plantain, yarrow, oxeye daisy, cat's ear and autumn hawkbit were the most successful species. Similar results were achieved in a parallel trial cut under a silage regime.

The commercial mixtures, based upon Nature Conservancy Council guidelines, were devised mainly from work and experience in southern England. This partly explains the poor establishment and persistence under Scottish conditions of some of the wildflowers sown; another potent reason is that wildflower mixtures do not yet carry purity and germination guarantees. The increasing interest in nature conservation and the increase in specialist wildflower seed producers should promote an improvement in the future however. There is a need to devise and test appropriate mixtures adapted to regional areas of the UK.

Other field-scale work in the west of Scotland is making use of simple mixtures of selected grasses and wildflowers in extensive livestock rearing systems, including cutting for silage as well as grazing. In addition to conventional sowing methods, direct drilling into existing swards has been utilised too. The successful establishment of wildflower mixtures requires the same attention to the practices outlined in Chapter 7 for agricultural seed mixtures. It is appropriate to investigate the potential of wildflower mixtures for silage production since this now predominates over hay in the UK. In complementary work, hedgerow margins of intensively managed grass swards and replanted cut-over woodland have been left unfertilised and florally enriched by plugging in wildflower plantlets previously germinated and established in a glasshouse. This method is economical with expensive seed, guarantees initial establishment of individual species and is suitable for strategic small areas on the farm. Another low cost method is the scattering onto the land of seed rich hay made from flora rich meadows.

Table 23.3 shows several criteria which have proved useful in the selection of species for inclusion in wildflower mixtures, while Appendix 4 lists the species, grasses and forbs found suitable by the Institute of Terrestrial Ecology in seed mixtures on differing soil types.

**Table 23.3   General selection criteria for species inclusion in wildflower mixtures**

- Ecologically suitable for particular soil/water conditions
- Common grassland species
- Not rare or locally distributed
- Preferably perennial and long lived
- Those with colourful and attractive flowers
- Attractive to insects as nectar or pollen sources
- Not highly competitive or invasive
- Those with seed which germinate easily over a range of temperature conditions and preferably without dormancy mechanisms

*Source:* Wells *et al.*

## FORAGE HERBS

Before the era of intensification, forage herbs were often traditional components of general purpose seed mixtures, valued for their acceptability to stock, high mineral content and drought resistance. They also contributed to sward yield, though this was derived mainly from grasses and clovers. The use of herbs in seed mixtures reached a peak in the early decades of this century. Robert Elliot at Clifton Park in southern Scotland was a major proponent of herbal leys; Table 23.4 shows one of his typical seed mixtures. Herbal strips have also been sown alongside sown agricultural grass mixtures to avoid competition from the mixture species. There has been a rekindling of interest in the use of herbs for extensively used and 'organic' grassland. Some breeding work is being carried out in New Zealand, and named varieties of herbs are emerging.

**Table 23.4   A Clifton Park herbal ley seed mixture in the late nineteenth century**

|  | lb/ac | (kg/ha) |
|---|---|---|
| Cocksfoot | 10 | (11.2) |
| Tall fescue | 3 | ( 3.4) |
| Crested dogstail | 2 | ( 2.2) |
| Hard fescue | 3 | ( 3.4) |
| Smooth-stalked meadow grass | 2 | ( 2.2) |
| Golden oat grass | 1 | ( 1.1) |
| Burnet | 3 | ( 3.4) |
| Chicory | 1 | ( 1.1) |
| Parsley | 1 | ( 1.1) |
| Ribgrass | 1 | ( 1.1) |
| Yellow suckling clover | 1 | ( 1.1) |
| Kidney vetch | 1 | ( 1.1) |
| Birdsfoot trefoil | 1 | ( 1.1) |
| Lucerne | 2 | ( 2.2) |
| Late flowering red clover | 2 | ( 2.2) |
| White clover | 3 | ( 3.4) |
| Alsike clover | 1 | ( 1.1) |
| Yarrow | 1 | ( 1.1) |
| Total | 39 | 43.5 |

*Source:* Elliot

At Ayr, mixtures of ten herbs were sown with three grass species of differing competitive ability—perennial ryegrass, meadow fescue and crested dogstail—and assessed for output, persistence and quality, using a moderate rate of fertiliser nitrogen, 75 kg/ha/year. Chicory, ribwort plantain, oxeye daisy and burnet performed best. Table 23.5 shows some of the results, including the mineral content of these herbs, in comparison with red clover. The other herbs sown—caraway, wild carrot, self-heal and sheep's parsley—established and persisted poorly. Herb performance was best with the least competitive grasses, crested dogstail and meadow fescue. The relatively high mineral concentrations recorded from some of the herbs would be valuable for raising herbage quality in a sward. Herb D value on an annual basis was generally lower than in the grasses and ranged from 55 for burnet to 72 for chicory. The most promising herbs have since been included in field-scale sowing, along with grasses and forage legumes, for further evaluation.

**Table 23.5**  **Annual herb DM yields and content in grass/forage herb mixtures (3-year means); mean nitrogen and mineral contents in the first harvest year (6 cuts)**

| Species | DM (t/ha) | Content (%) | N | P | K | Ca | Mg |
|---------|-----------|-------------|-----|-----|-----|-----|-----|
|         |           |             |     |     | (g/kg DM) |     |     |
| Chicory | 4.7 | 61 | 23 | 4.2 | 51 | 16 | 2.7 |
| Ribwort plantain | 3.5 | 48 | 20 | 3.5 | 23 | 26 | 1.9 |
| Oxeye daisy | 3.1 | 41 | 22 | 4.3 | 44 | 13 | 2.4 |
| Yarrow | 1.4 | 21 | 27 | 4.3 | 43 | 12 | 2.2 |
| Burnet | 1.2 | 17 | 20 | 3.2 | 18 | 18 | 4.7 |
| Red clover | 9.2 | 77 | 35 | 3.0 | 26 | 15 | 3.8 |

*Source:* Tiley and Frame

## FUTURE NATURE CONSERVATION

Nature conservation is likely to assume a much greater importance in future as agricultural production is progressively reduced. More land will be farmed at lower intensity, such extensification providing a tremendous opportunity for grassland management to promote nature conservation in sustainable, environmentally friendly systems of farming. Key points for farm conservation are indicated in Table 23.6, while the criteria used for judging a typical countryside management competition are shown in Table 23.7. Some forecasts suggest that 2 to 3 million ha of grassland in Britain could be surplus to production

**Table 23.6   Stages in planning and action for countryside conservation**

- Assess the landscape and wildlife interest of the farm, including grassland, hill or rough grazing and peat bogs
- Review past and likely future changes in the landscape and wildlife of the farm
- Determine the aims and objectives for a conservation plan
- Prepare a conservation plan
- Manage existing features of landscape and wildlife interest, including unimproved grassland, heather moorland, scrub, marshes, mosses and peat bogs
- Create new features
- Consider conservation in day to day farm management, including the use of pesticides, fertilisers, slurry and silage effluent

*Source:* Countryside Commission for Scotland

requirements within the next 25 years. Apart from amenity and recreational facilities, and perhaps the creation of wilderness areas in the hills and uplands, this grassland will be available for nature conservation promotion in its widest sense. However, the public purse will have to play a major role.

**Table 23.7   Countryside Management Competition, Northern Ireland**

|  | *Marks* |
|---|---|
| Farming practices | 30 |
| Grassland management | |
| Stock management | |
| Pollution control | |
| Field boundaries | |
| Landscape | 30 |
| Entrance/farm drive | |
| Farmyard/buildings/dwelling | |
| Field boundaries | |
| Trees/woodland | |
| Preservation of features | |
| Habitats | 30 |
| Unimproved grassland | |
| Field boundaries | |
| Trees/woodlands | |
| Streams/open water/wet areas | |
| Farmer knowledge/interest | 10 |
| Total | 100 |

*Source:* Gracey

*Chapter 24*

# Hill Land Improvement

---

The most important single factor which influences animal output from
the hills is nutrition and, although it is more critical prior to mating, in
late pregnancy and during lactation, it needs to be improved
throughout the year if output is to be increased. The key to an
improved nutrition cycle for the grazing animal lies in pasture
management.

P. Newbould (1974)

| Aur dan y rhedyn | Gold under bracken |
| Arian dan yr eithyn | Silver under gorse |
| Newyn dan y grug | Famine under heather |

Welsh Proverb

---

Pasture production on the 6 million ha of hill and upland grazings is
highly seasonal and herbage quality poor. Consequently animal output
from free-range, year-round hill grazings is low. There is a blurred
distinction between hill and upland farm types but it is expressible as
type of enterprise possible and potential output. On hill grazings,
mainly at 350 to 1,000 m above sea level and with farms 800 to
2,000 ha or more in extent, only sheep farming is normally possible.
Farms are extensive and only a small fraction of the land is improvable
by some form of cultivation and reseeding. Upland farms at lower
altitudes (100 to 350 m above sea level) are smaller (200 to 800 ha),
generally have suckler cows in addition to sheep and perhaps some
cropping, and the grassland is generally more productive. Some 80 per
cent of sheep and cattle output originates from upland farms.

The principal factor limiting productivity on hill and upland farms
is the availability of winter feed. This is very low on hill farms but
forage conservation is often possible on upland farms. In recent years,
dramatic increases in sheep output on hill farms have been obtained
following introduction of the two-pasture system. This is based on
improving animal nutrition at critical stages in the livestock breeding
cycle and is achieved on farms mainly by enhancing selected areas and
grazing them strategically to complement the natural pastures.

Techniques to improve pasture are discussed below and include fertilisation, burning, fencing, cultivations and introducing improved herbage species and varieties. Choice of technique depends on individual farm conditions but considerable areas of rough grazing are too difficult to improve because of such factors as altitude, lack of access, terrain or wetness. Pasture improvement is normally only feasible below 500 m above sea level. Other non-technical obstacles include lack of capital, land tenure conditions and defects in farm structure. The ratio of improvable to unimprovable land on individual farms is of fundamental importance in terms of the scale and scope of enterprises, economic viability and suitability for intensification.

## PASTURE RESOURCES

The chief soil and vegetation types of rough grazing land were summarised in Chapter 1. An idealised topographical distribution is shown in Figure 24.1. Climate in the hills and uplands is very variable but characterised by generally low temperatures, severe wind exposure, high rainfall, excessive ground wetness and frequent winter

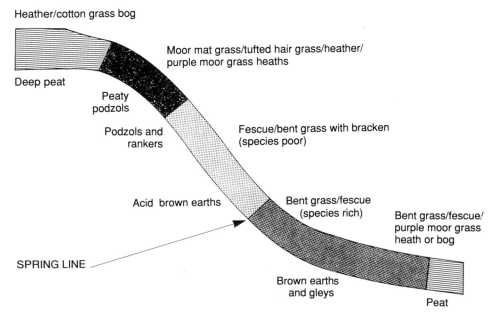

*Figure 24.1   Idealised topographical distribution of soil and vegetation types on hill land*
Source: Hill Farming Research Organisation

282

frosts. Harshness of climate and poverty of soil result in low levels of pasture production on the hills, averaging about 2 t/ha DM but ranging from 1 to 5 t/ha. The low production, and poor inherent quality of most of the indigenous species as illustrated in Figure 24.2, result in a poor utilisation pattern, low ewe stocking rates, poor lambing percentages and inferior lamb growth rates. In upland farms herbage DM ranges from 4 to 9 t/ha but the higher levels are dependent on fertiliser N use.

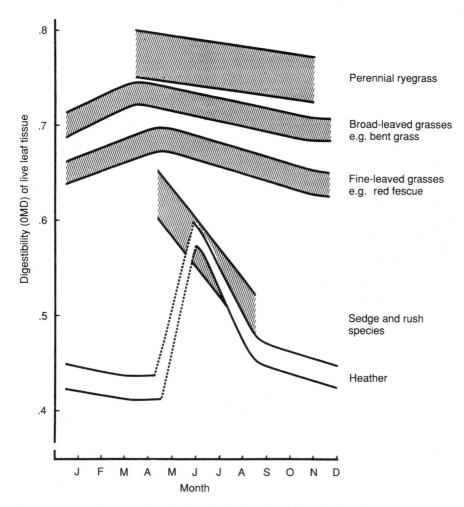

*Figure 24.2    Seasonal variation in the* in vitro *digestibility (OMD) of live leaf tissue from indigenous and sown plants in hill areas*
Source: Hodgson and Grant

The pasture resource supplies 90 to 95 per cent of the annual energy requirements of sheep on the majority of hill and upland farms. However, the traditional free-range grazing systems are geared to winter stock-carrying capacity, resulting in annual levels of utilisation of 20 to 30 per cent. This has an adverse effect on pasture production and quality, in turn affecting animal output. Insufficient utilisation during the late spring/early summer period results in herbage accumulation and senescence. This dead material dilutes the quality of the pasture on offer, and restricts subsequent tiller and leaf initiation by shading. In addition, low stocking and utilisation limit the cycling of ingested nutrients by excretal return.

## THE TWO-PASTURE SYSTEM

Improved animal performance from rough grazing requires upgrading of animal nutrition, notably at critical times in the animal's annual nutritional requirement cycle. The two-pasture system involves the complementary use of fenced, improved areas together with unimproved pastures, with resultant overall benefit to the ewe flock. The beneficial effect on stocking rate and performance is marked, even with a small ratio of improved to unimproved pasture. In turn the higher stocking allows better utilisation of summer herbage production on the hill grazings. Since the stocking rate on rough grazings can vary from approximately one ewe per 0.5 to 4 ha or more, the area of improved pasture needed to make an impact is greater on the more densely stocked upland farms. The two critical times when improved nutrition is required are before and during mating (November/December), so that ewe condition and live weight are improving, and the first three months of lactation (May to July or August). Thus the higher quality and production of the improved pasture are exploited during these periods and the rough grazing at other less critical times.

When the amount of improved pasture is limited, priority use of it is given to the most productive ewes, which have the greatest nutritional demand. During the final six to eight weeks before lambing, also important nutritionally, concentrate supplementation will improve the condition of the pregnant ewes. They may also be supplemented during mid pregnancy. Pregnancy scanning is a valuable management aid for identifying ewes carrying twins or triplets so that they can be selectively fed. Following the introduction and implementation of the two-pasture system on both experimental and commercial farms,

improved animal performance—ewe stocking rates, lambing percentages and weights of weaned lambs—has resulted, together with better
farm profitability. Improved pasture is the lynch pin of the system and
has to be accorded management priority; in recent years the use of
sward surface height has become a valuable grazing management tool
(see Chapter 17).

An example of the benefits to sheep production following the
introduction of the two-pasture system at the SAC Kirkton hill farm
(337 ha), Perthshire, is shown in Table 24.1. In addition, the weight of
weaned calves sold from the farm rose over the period from 3 to
6 tonnes annually.

Table 24.1   **Increase in sheep production at Kirkton Hill Farm, Scottish
Agricultural College, following land improvement and development
of the two-pasture system**

|  | Baseline period | Development phase (After 5 years) | Consolidation (After 10 years plus) |
|---|---|---|---|
| Breeding ewes | 400 | 579 | 622 |
| Lambs weaned/ 100 ewes | 80 | 95 | 122 |
| Average lamb sale weight (kg) | 27 | 28 | 31 |
| Lamb sales (tonnes) | 5.4 | 10.5 | 17.7 |

*Source:* A. Waterhouse (personal communication)

The main pasture improvement techniques are summarised in
Table 24.2. On individual farms, the soil and plant resources must be
critically considered so that the areas most responsive to treatments—
technically and economically—are selected. Preliminary soil analysis
is essential to assess lime and fertiliser needs. Changes in soil
condition and vegetation are slow and initial pre-treatments such
as liming or severe defoliation may be useful before reseeding
operations.

The size, scale and phasing of pasture improvement must be well
planned in relation to the capital, stock and labour resources of the
farm. Siting in relation to farm layout, terrain, water supply and
accessibility also require consideration. It is vital that the resulting
pasture improvement is linked to improved stock management, including animal health considerations, in order to cash in on the invest-

**Table 24.2  Summary of pasture improvement techniques**

| Soil | Sward | Soil/sward |
|------|-------|------------|
| Liming and fertilisation (lime, $P_2O_5$, $K_2O$, N, trace elements) Drainage | Grazing/cutting Fencing Fire Weed control (herbicides, cultural) | Cultivations (ploughing, rotavation discing, harrowing) Surface seeding Direct drilling |

↑_____↑
Selected combinations of
these treatments required
when improving value of
existing sward

↑_____↑_____↑
Selected combinations of these treatments required
when introducing improved grasses and clovers

*Source:* Frame *et al.*

ments made and sustain economic viability. Land improvement is not an end in itself. It is also important to sustain the level of existing improved land by subsequent liming and fertilisation to maintain persistency and productivity of sown species. Table 24.3 illustrates the applicability of techniques for different soil/sward types.

# SOIL IMPROVEMENT

Most hill and upland soils are high in organic matter, acidic, and low in available plant nutrients. Soil N is nearly always deficient, a major factor restricting pasture production. Fertilisation has already been fully discussed in Chapters 11 and 12, so only aspects of specific relevance to land renovation in the hills and uplands are dealt with here.

## Lime

Liming, generally essential in any land improvement package, raises soil pH, provides available calcium and reduces the availability of soluble aluminium. In turn, these changes lead to better mineralisation of N and P from the soil organic matter. Changes in trace element availability can be complex. Liming can increase the availability to

**Table 24.3   Applicability of pasture improvement techniques for differing soil/sward types**

| Soil | Dominant vegetation | Fencing | Fencing + lime + $P_2O_5$ | Fencing + lime + $P_2O_5$ White clover sown | Fencing + lime + $P_2O_5$ White clover and other grasses sown | Fencing + lime + $P_2O_5$ White clover and grasses sown after cultivation |
|---|---|---|---|---|---|---|
| Brown earth | Bent grass fescue | ✓ (slow) | ✓ | ✓ | ✓ | ✓ |
| Peaty podzol (wet) | Moor mat grass/ purple moor grass | ✓ (extremely slow) | (very slow) | (slow) | ✓ | ✓ |
| Peaty podzol (dry) | Heather/ bilberry | ✗ | ✗ | ✗ | ✓ | ✓ |
| Blanket bog | Heather/cotton grass/deer's hair sedge/sphagnum | ✗ | ✗ | ✗ | ✓ | ✗ |

Increasing capital costs per ha →

Increasing pasture production and nutritional quality →

*Source:* Newbould

287

stock of molybdenum and sulphur from the herbage, for example, but this can result in copper being made unavailable. Trace element content in the herbage is also affected by other improvement procedures such as fertiliser application and the introduction of improved plant species. Prophylactic treatment of most trace element disorders is available once the disorders have been identified. Direct treatment of stock, e.g. the use of copper needles, or mineral supplementation, can be effective measures on their own or in addition to pasture treatment. Applications of lime in amounts up to 20 t/ha, which soil analysis can indicate, are impracticable and liming is best phased, with initial limestone applications of 5 to 8 t/ha and subsequently at three- to four-year intervals.

## Phosphorus

Most soils are extremely low in available phosphorus and alleviation is essential for pasture improvement, particularly for the growth of white clover and its N-fixing bacteria. Typical phosphate dressings are 180 kg/ha during cultivations, 60 kg/ha in the seedbed and 160 kg/ha every three to six years, but dependent on soil analysis. A high correlation has been found between soil P status and persistence of sown species in reseeds.

## Potassium

Potash fertiliser is also necessary in a reseeding programme, at a rate determined by soil analysis. Deficiency on hill land is greatest on organic soils which have a poor retention capacity. Typical applications are 80 kg/ha $K_2O$ at establishment, while the amounts thereafter will depend on soil pH and utilisation, higher rates being necessary on acid soils and where the herbage is cut and removed for silage or hay. A balanced input of major nutrients is necessary to prevent adverse interactions between nutrients. The response of white clover to potassium is dependent on adequate phosphorus, for instance.

## Nitrogen

Available nitrogen is limited in hill and upland soils, although substantial amounts are bound up in soil organic matter. The use of applied fertiliser N is mainly tactical, not regular, partly for economic reasons but also because of a lower production response by the grass due to soil, climatic or utilisation restraints compared with lowland conditions. A rate of 40 to 60 kg/ha N is sometimes used on the most responsive bent grass/fescue areas to stimulate growth rate in early season. However, nitrogen is more regularly used (*a*) to promote the establishment of reseeds, using 60 to 90 kg/ha, often in association with lime, phosphate and potash, and also to stimulate mineralisation of nitrogen from soil organic matter; (*b*) on reseeded grass/white clover swards in early spring at 40 to 60 kg/ha to obtain early grazing for lambing ewes; and (*c*) on upland swards used for conservation, at rates of 60 to 90 kg/ha.

## Drainage

Pastures with poor drainage have shorter grazing seasons than well-drained pastures and low sward bearing strengths limit the use of agricultural machinery. The better soil aeration after drainage accelerates the decomposition of organic matter and leads to improved nutrient supply to the sward. The build-up of water snails (*Limnaea truncatula*), which act as secondary hosts for the liver fluke (*Fasciola hepatica*) parasite of sheep and cattle, is inhibited. However, peaty and gley soils, rock outcrops, variable topography and water seepage from springs make hill and upland drainage schemes difficult and costly. Less generous grant-aid is currently available than formerly. The low output from the land rarely justifies the expense of underdrainage, although conventional pipe systems may be warranted on more fertile, lower slopes capable of intensification of use. Open interceptor ditch systems may help on the open hill but should be planned so that the operation of machinery, e.g. lime spreading equipment, is not restricted.

## SWARD IMPROVEMENT

### Grazing/cutting

Stock are used to impose high grazing pressure for a short time on selected areas to improve subsequent leafiness and quality. Heavy grazing also reduces competition from an existing pasture prior to surface seeding or direct drilling improved grass species and clover. The grazing is a useful precursor to encourage young regrowth before herbicides, e.g. glyphosate, are used to desiccate grass swards prior to reseeding. The grazing process also promotes plant nutrient cycling through the return of animal excreta. Grazing stock are sometimes used to trample surface-sown seed to assist establishment.

Cutting is an alternative means of rejuvenating swards by removing accumulations of mature or surplus grass or as a precursor to herbicide treatment and subsequent seeding. The area must be accessible and suitable for machinery, which can be of a simple blade, chain or flail type.

## Fencing

Grazing management implies effective control of stock by the use of adequate fences and boundaries. Forms of cheap fencing, particularly various types of electric fence, have been sought as a substitute for expensive traditional fencing. Grant-aid is available for fencing, including renovation of old stone dykes, hedges and enclosing heathland and heather moorland.

## Fire

Periodic burning—muirburn—is a management tool used to rejuvenate old stands of heather. Burning maintains the vegetation in a more youthful and productive state for grazing, mainly by sheep in winter. Advisory guidelines are available which, if followed, ensure safe, legal and effective burning management and result in the maintenance of a vigorous heather cover. Key factors are to burn heather when 20 to 40 cm tall, usually once every 8 to 18 years, and to burn only a part of the total area in rotation each year, using 30 to 40 m strips. Infrequent burning results in slow heather regeneration with the possibility of colonisation by less desirable plant species.

Muirburn is also practised on purple moor grass pastures which are regenerated by burning once every 3 to 5 years. This species is of value to grazing stock only during a limited spring/early summer period. On heather/flying bent associations, a dual burning management strategy is best: parts are burned on a long-rotation cycle to suit heather and parts on a short-rotation cycle to suit purple moor grass. The mix of species gives a more valuable grazing sward than either species alone.

In recent years, townspeople have developed a perception that the hills and uplands should remain as open 'natural' moorland. The purple-flowered heather dominant on moors is a particularly valued and attractive part of this landscape, as well as being a valuable wildlife habitat. Retention of youthful heather was always an aim on hill farms, since it provides useful winter feed for sheep. Current presumption in favour of conservation rather than production reinforces the desire to retain or even increase the heather area instead of seeking to convert it into improved grassland. Overgrazing weakens the vigour and longevity of heather and overreliance on heather grazing affects sheep performance adversely because of the plant's low feeding value. At least 50 per cent of the sheep's annual diet should be derived from grass to provide adequate nutrition, and optimal stocking rates should be governed by the grassy areas available for complementary grazing in order to avoid overgrazing the heather.

## Weed Control

A number of weed problems exist or can arise in upland areas. Prevention and/or control during pasture establishment was dealt with in Chapter 7, while methods of controlling serious perennial weeds such as rushes and bracken were discussed in Chapter 8. Upland terrain is often difficult for highly mechanised operations but alternative methods are sometimes available, e.g. aerial spraying of asulam by helicopter to control bracken. However, the ability of hill and upland enterprises to bear the operational costs of advanced methods of weed control is often limited.

# SOIL/SWARD IMPROVEMENT

## Cultivations

Ploughing or rotavation are used for total cultivation before reseeding. Rotavation is better than ploughing on peaty soils or land with a thin, fertile topsoil since the upper biologically active soil layer is not buried. Destruction of the surface mat on peaty soils leads to the new sward being easily poached, which in turn is frequently invaded by rushes. Both forms of cultivation are suitable on swards too dense for surface methods, but not on steep rocky or very wet land. Since cultivations usually form part of a high cost combination of techniques, good utilisation and returns are necessary to recoup expenses. Also, because of the sudden change in plant cover, management must be geared to maintain the improvement with adequate liming and fertilisation. On soils with a considerable surface mat, 'pioneer' cropping with fodder turnips or rape one to two years before reseeding with grass and clover has been successful.

Shallow rather than complete, deep rotavation can be used to reduce competition from the existing sward and allow some form of reseeding for partial renovation. Spike rotavation has proved to be successful on a range of soil and vegetation types for this purpose. Discing or harrowing are other surface cultivation methods for reducing the competition from an existing sward before seeding.

## Surface seeding

Surface seeding without prior cultivation is a simple and cheap means of introducing new species or varieties to an existing sward, though an open sward is needed to allow seed establishment. It is a method suited to wet environments and is best preceded by burning, cutting or heavy grazing, and possibly herbicide use to reduce the existing vegetation cover. Sheep or cattle can be used to tread in the sown seeds, but germination occurs at a lower rate than after cultivation.

## Direct drilling

Direct drilling, including strip, slot or sod seeding, offers a means of introducing improved species into existing, possibly unploughable swards by 'partial' cultivation. The different machines produce slits or partially cultivated strips. Chemical pre-treatment to kill or suppress existing herbage may be required if the sward is too dense.

A number of guidelines need to be followed for success (see Chapter 7), and each site must be considered individually as to its suitability for direct drilling.

## Improved forage species

Varietal and species comparisons under hill and upland conditions show that those performing well under lowland conditions are satisfactory, but good winter hardiness is essential and this can vary widely between varieties of perennial ryegrass, the species most frequently sown. Poor winter hardiness and insufficient maintenance fertilisation, especially lime and phosphate, are major reasons for the deterioration of swards. Timothy and cocksfoot, other primary grasses often used in seed mixtures, are both winter hardy species, although cocksfoot is prone to winter burn. Red fescue is suitable for inclusion in seed mixtures for the harsher conditions of low fertility, wet soils at high altitudes, but its lower digestibility and more fibrous leaf make it relatively unattractive to grazing stock, particularly if it is allowed to mature. Recent work at Ayr on red fescue and other secondary grass species such as bent grasses, Yorkshire fog and smooth-stalked meadow grass indicates a useful role in low fertility situations. White clover, with its high intake-high nutritive characteristics which benefit individual animal performance, and its N-fixation ability, is essential in seed mixtures for land reclamation. Small-leaved, highly stoloniferous types have proved the most persistent for sheep grazing in hill conditions (see Chapter 22).

## FORAGE CONSERVATION

Traditionally hay is made in the uplands, on inbye ploughable land close to the farm steading, for feeding to suckler cows and/or sheep as a winter supplement. In the absence of suitable land, hay is bought from lowland farms. However, being sold on a bulk basis rather than by quality, feeding value is often low and transport costs high. Hay making is a difficult and risky operation, dependent on dry weather— at a premium at higher altitudes. Labour-intensive methods of drying the hay overcame these disadvantages in the past but cannot now be used because of cost. Substitute mechanised methods are expensive and more suited to large-scale operations. Cold or warm air blowing systems, which help overcome weather vagaries (also see Chapter 20), are not widely used in marginal land areas.

Being less weather dependent than hay, silage is more adaptable to the wetter conditions and in any case is of higher feeding value since it is cut at an earlier stage of growth. Big bale silage is increasingly being made on upland farms, often by contractor operation, in spite of the difficulties of achieving the desirable high DM content needed (also see Chapter 19).

## FUTURE HILL LAND USE

The main technical constraints to increasing productivity from hill and upland pastures are known, and techniques to overcome them are available. However, in recent times, higher animal output is not necessarily the target on every farm. Because of food overproduction in the EC, resulting economic pressures are such that cost containment through lower inputs is a major objective, together with predictability and quality of animal product to satisfy market demands. Livestock compensatory allowances, a form of production grant, are still paid to farmers to maintain breeding ewes and cows but the amounts are steadily reducing. Grant-aid for environmentally positive measures, such as enclosure of heather moorland and heathland and planting shelter belts, have been introduced. The more gradual and cheaper techniques of pasture improvement are taking precedence over rapid and expensive methods which result in complete and visually dramatic changes of pasture type. More emphasis is likely to be placed on better all-round management of the indigenous sward types.

The hills and uplands will continue to support hardy breeding stock and be a major and traditional source of lambs and calves for fattening on the lowlands, but it is recognised that less land will be needed for food production in the future. A good proportion of the 2 to 3 million ha estimated to be surplus to British agricultural requirements by the year 2000 will be in the hills and uplands. The challenge is to develop viable alternative enterprises. Red deer farming, cashmere production from goats and agroforestry—which involves sheep grazing between spaced trees for timber—are among non-traditional enterprises being tried both experimentally and commercially. Afforestation will also expand. Some of the spare land will play an important role in nature conservation and countryside leisure, as discussed in Chapter 23.

*Appendix 1*

# Common and Latin names of selected plants, pests and diseases

---

GRASSES

| Common name | Latin name |
|---|---|
| Annual meadow grass | *Poa annua* |
| Bent (black) | *Agrostis gigantea* |
| Bent (common) | *Agrostis capillaris* |
| Bent (creeping) | *Agrostis stolonifera* |
| Bent (Highland) | *Agrostis castellana* |
| Black-grass | *Alopecurus myosuroides* |
| Cocksfoot (= orchard grass) | *Dactylis glomerata* |
| Couch grass (= twitch) | *Elymus repens* |
| Couch (onion) | *Arrhenatherum elatius* var. *bulbosus* |
| Creeping soft-grass | *Holcus mollis* |
| Crested dogstail | *Cynosurus cristatus* |
| Chewings fescue | *Festuca rubra* subspecies *commutata* |
| Floating foxtail | *Alopecurus geniculatus* |
| Golden oat grass | *Trisetum flavescens* |
| Italian ryegrass | *Lolium multiflorum* |
| Lyme grass | *Elymus arenarius* |
| Meadow barley | *Hordeum secalinum* |
| Meadow fescue | *Festuca pratensis* |
| Meadow foxtail | *Alopecurus pratensis* |
| Perennial ryegrass | *Lolium perenne* |
| Phalaris | *Phalaris tuberosa* |
| Prairie grass | *Bromus wildenowii* |
| Red fescue | *Festuca rubra* |
| Reed canary grass | *Phalaris arundinacea* |
| Rough-stalked meadow grass | *Poa trivialis* |
| Sea couch | *Elymus pungens* |
| Smaller cat's tail (= timothy) | *Phleum bertolonii* |
| Smooth-stalked meadow grass (= Kentucky bluegrass) | *Poa pratensis* |
| Soft brome (= lop grass) | *Bromus mollis* (= *hordeaceus*) |

| | |
|---|---|
| Sterile (= barren) brome | *Bromus sterilis* |
| Sweet vernal | *Anthoxanthum odoratum* |
| Tall fescue | *Festuca arundinacea* |
| Tall (= false) oat grass | *Arrhenatherum elatius* |
| Timothy (= cat's tail) | *Phleum pratense* |
| Tor grass (= chalk false brome) | *Brachypodium pinnatum* |
| Upright brome | *Bromus erectus* |
| Wall barley | *Hordeum murinum* |
| Westerwold ryegrass | *Lolium multiflorum* subspecies *westerwoldicum* |
| | |
| Wild oat (spring) | *Avena fatua* |
| Wild oat (winter) | *Avena ludoviciana* |
| Yorkshire fog | *Holcus lanatus* |

## FORAGE LEGUMES

| **Common name** | **Latin name** |
|---|---|
| Birdsfoot trefoil | *Lotus corniculatus* |
| Birdsfoot trefoil (marsh) | *Lotus pedunculatus (= uliginosus)* |
| Clover (alsike) | *Trifolium hybridum* |
| Clover (crimson) | *Trifolium incarnatum* |
| Clover (red) | *Trifolium pratense* |
| Clover (subterranean) | *Trifolium subterraneum* |
| Clover (white) | *Trifolium repens* |
| Forage pea | *Pisum arvense* |
| Lucerne (= alfalfa) | *Medicago sativa* |
| Lupin (white) | *Lupinus albus* |
| Lupin (yellow) | *Lupinus luteus* |
| Sainfoin | *Onobrychis viciifolia* |
| Trefoil | *Medicago lupulina* |
| Vetch (common) | *Vicia sativa* |

## HERBS

| **Common name** | **Latin name** |
|---|---|
| Burnet | *Poterium sanguisorba* |
| Caraway | *Carum carvi* |
| Carrot (wild) | *Daucus carota* |
| Cat's ear | *Hypochoeris radicata* |

296

| | |
|---|---|
| Chicory | *Cichorium intybus* |
| Dandelion | *Taraxacum officinale* |
| Pignut | *Conopodium majus* |
| Ribwort plantain | *Plantago lanceolata* |
| Selfheal | *Prunella vulgaris* |
| Sheep's parsley | *Petroselinum crispum* |
| Vetch (kidney) | *Anthyllis vulneraria* |
| Yarrow | *Achillea millefolium* |

## HILL PLANTS

| **Common name** | **Latin name** |
|---|---|
| Bilberry (= blueberry) | *Vaccinium myrtillus* |
| Cloudberry | *Rubus chamaemorus* |
| Cotton drawmoss | *Eriophorum vaginatum* |
| Cotton grass | *Eriophorum augustifolium* |
| Crowberry | *Empetrum nigrum* |
| Deer's hair sedge | *Trichophorum caespitosus* |
| Heath grass | *Danthonia decumbens* |
| Heather (ling) | *Calluna vulgaris* |
| Heather (bell) | *Erica cinerea* |
| Moor mat grass (= white bent) | *Nardus stricta* |
| Moss | *Polytrichum* species |
| Moss (bog) | *Sphagnum* species |
| Purple moor grass (= flying bent) | *Molinia caerulea* |
| Sedges | *Carex* species |
| Sheep's fescue | *Festuca ovina* |
| Tormentil | *Potentilla* species |
| Tufted hair grass (= tussock grass) | *Deschampsia caespitosa* |
| Wavy hair grass | *Deschampsia flexuosa* |

## WEEDS

| **Common name** | **Latin name** |
|---|---|
| Bindweed (black) | *Polygonum convolvulus* |
| Bracken | *Pteridium aquilinum* |
| Broom | *Sarothamnion scoparius* |
| Buttercup (bulbous) | *Ranunculus bulbosus* |
| Buttercup (creeping) | *Ranunculus repens* |

| | |
|---|---|
| Buttercup (meadow) | *Ranunculus acris* |
| Charlock | *Sinapsis arvensis* |
| Chickweed (common) | *Stellaria media* |
| Chickweed (mouse-ear) | *Cerastium arvense* |
| Cleavers | *Galium aparine* |
| Clover (suckling) | *Trifolium dubium* |
| Cowbane | *Cicuta virosa* |
| Cow parsley | *Anthriscus sylvestris* |
| Daisy | *Bellis perennis* |
| Darnel | *Lolium temulentum* |
| Deadly nightshade | *Atropa belladonna* |
| Dead-nettle (red) | *Lamium purpureum* |
| Dock (broad-leaved) | *Rumex obtusifolius* |
| Dock (curled) | *Rumex crispus* |
| Dodder | *Cuscuta* species |
| Fat hen (= goosefoot) | *Chenopodium album* |
| Field horsetail | *Equisetum arvense* |
| Foxglove | *Digitalis purpurea* |
| Fumitory | *Fumaria officinalis* |
| Gorse (= whin) | *Ulex europaeus* |
| Groundsel | *Senecio vulgaris* |
| Hemlock | *Conium maculatum* |
| Hemp-nettle (common) | *Galeopsis tetrahit* |
| Knapweed | *Centaurea nigra* |
| Knotgrass | *Polygonum aviculare* |
| Mayweeds | *Matricaria* species |
| Meadow saffron | *Colchicum autumnale* |
| Meadowsweet | *Filipendula ulmaria* |
| Melilot | *Melilotus* species |
| Nettle (small stinging) | *Urtica urens* |
| Nettle (stinging) | *Urtica dioica* |
| Onion (wild) | *Allium vineale* |
| Pansy (field) | *Viola arvensis* |
| Parsley-piert | *Aphanes arvensis* |
| Penny-cress (field) | *Thlaspi arvense* |
| Plantain (greater) | *Plantago major* |
| Poppy (common) | *Papaver rhoeas* |
| Radish (wild) | *Raphanus raphanistrum* |
| Ragwort (common) | *Senecio jacobea* |
| Ragwort (marsh) | *Senecio aquaticus* |
| Redshank | *Polygonum persicaria* |
| Rush (common or soft) | *Juncus effusus* |
| Rush (hard) | *Juncus inflexus* |

| | |
|---|---|
| Rush (heath) | *Juncus squarrosus* |
| Rush (jointed) | *Juncus articulatus* |
| Shepherd's purse | *Capsella bursa-pastoris* |
| Sorrel (common) | *Rumex acetosa* |
| Sorrel (sheep's) | *Rumex acetosella* |
| Sowthistle (common) | *Sonchus oleraceus* |
| Speedwell (common) | *Veronica persica* |
| Speedwell (ivy-leaved) | *Veronica hederifolia* |
| Spurrey (corn) | *Spergula arvensis* |
| Thistle (creeping) | *Cirsium arvense* |
| Thistle (spear) | *Cirsium vulgare* |
| Water dropwort | *Oenanthe* species |
| Yellow rattle | *Rhinanthus minor* |

## PESTS

| Common name | Latin name |
|---|---|
| Antler moth (= army worm) | *Cerapteryx graminis* |
| Bibionids | *Bibio marci, Dilophus febrilis* |
| Cockchafer | *Melolontha melolontha* |
| Frit fly | *Oscinella frit* |
| Garolen chafer | *Phyllopertha horticola* |
| Grass moths | *Crambus* species |
| Leatherjackets | *Tipula* species |
| Slugs | *Deroceras* species, *Arion* species |
| Springtails | *Onychiurus* species |
| Stem eelworm | *Ditylenchus dipsaci* |
| Swift moth | *Hepialus humuli* |
| Weevils | *Sitona* species |
| Wireworms | *Agriotes* species |

## DISEASES

| Common name | Latin name |
|---|---|
| Blind seed disease of ryegrass | *Gloetinia temulenta* |
| Clover rot | *Sclerotinia trifoliorum* |
| Clover scorch | *Kabatiella caulivora* |
| Crown rust (grasses) | *Puccinia coronata* |
| Downy mildew (clovers) | *Peronospora trifoliorum* |

| | |
|---|---|
| Ergot (grasses) | *Claviceps purpurea* |
| Fusarium root rot (grasses) | *Fusarium culmorum* |
| Leaf blotches (grasses) | *Rhyncosporium* species |
| Leaf fleck (cocksfoot) | *Mastigosporium rubricosum* |
| Leaf spot (clovers) | *Pseudopeziza trifolii* |
| Leaf spots (grasses) | *Drechslera* species |
| Pepper spot (clovers) | *Leptosphaerulina trifolii* |
| Powdery mildew (clovers) | *Erysiphe trifolii* |
| Powdery mildew (grasses) | *Erysiphe graminis* |
| Rusts (clovers) | *Uronyces* species |
| Snow mould (grasses) | *Micronectriella nivalis* syn. *Fusarium nivale* |
| Take-all (grasses) | *Gaeumannomyces graminis* |
| Verticillium wilt (lucerne) | *Verticillium albo-atrum* |

# Appendix 2

## Classification of soil analysis results (mg/litre air dry soil) of samples analysed by standard Scottish and ADAS procedures

| Soil status | Phosphorus | Potassium | Magnesium |
|---|---|---|---|
| *Scottish system* | | | |
| Very low (VL) | Less than 10 | Less than 40 | Less than 20 |
| Low (L) | 10–25 | 40–75 | 20–60 |
| Moderate (M) | 26–75 | 76–200 | 61–200 |
| High (H) | 76–500 | 201–1000 | 201–1000 |
| Excessively high (EH) | More than 500 | More than 1000 | More than 1000 |
| *ADAS index* | | | |
| 0 | 0–9 | 0–60 | 0–25 |
| 1 | 10–15 | 61–120 | 26–50 |
| 2 | 16–25 | 121–240 | 51–100 |
| 3–7 | 26–200 | 241–2400 | 101–1000 |
| 8–9 | 201 to more than 280 | 2401 to more than 3600 | 1001 to more than 1500 |

*Source:* SAC/ADAS

| Approximate equivalents: | Scottish system | ADAS index |
|---|---|---|
| | VL | 0 |
| | L | 1 |
| | M | 2 |
| | H | 3–7 |
| | EH | 8–9 |

## Appendix 3

# Key for identification of vegetative grasses

Note that under practical conditions, the plants are likely to be partially defoliated, or their growth habit modified (e.g. etiolated), by the companion species and weeds. Consequently, not all the key features may be noticeable—in which case check that *most* features are present.

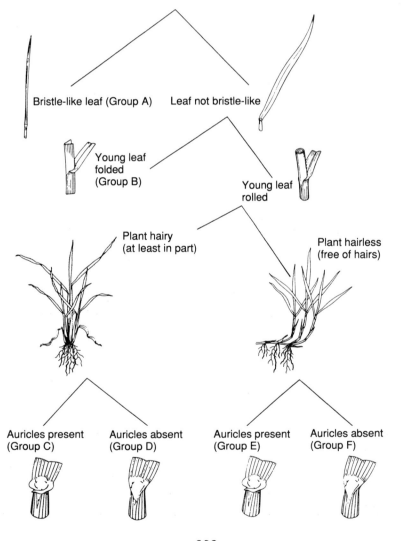

Bristle-like leaf (Group A)   Leaf not bristle-like

Young leaf folded (Group B)   Young leaf rolled

Plant hairy (at least in part)   Plant hairless (free of hairs)

Auricles present (Group C)   Auricles absent (Group D)   Auricles present (Group E)   Auricles absent (Group F)

## GROUP A

### Leaves bristle-like

| Species | Plant characteristics | Leaves | Ligules |
|---|---|---|---|
| Wavy hair-grass (*Deschampsia flexuosa*) | Plant rather stiff | Long, dark green | Short, broad |
| Mat-grass (*Nardus stricta*) | Shoots bunched; tough, shiny sheaths | Outer horizontal | Short, blunt |
| Sheep's fescue (*Festuca ovina*) | Leaf sheath open/split; occasionally pink based | Fine, thin | Minute |
| Red fescue (*Festuca rubra*) | Leaf sheath not split initially; red sheaths; rhizomes | Infolded | Minute |

## GROUP B

### Leaves folded

| Species | Plant characteristics | Leaves | Ligules |
|---|---|---|---|
| Perennial ryegrass (*Lolium perenne*) | Shiny, dark green; red basal sheath | Shiny below | Small, blunt |
| Cocksfoot (*Dactylis glomerata*) | Large succulent base | Broad, puckered | Pronounced |
| Annual meadow grass (*Poa annua*) | Tufted; sheaths green-white | Short, crinkled | Large, white |
| Rough meadow grass (*Poa trivialis*) | Curved tillers; sheaths brownish; stolons; often rough stemmed | Shiny below | Uppers long, pointed |
| Smooth meadow grass (*Poa pratensis*) | Dark green; smooth stemmed; rhizomes | Parallel-sided | Short |
| Heath grass (*Danthonia decumbens*) | Hairs at leaf base | Tramlined | Short hairs |
| Crested dogstail (*Cynosurus cristatus*) | Leaf sheath sometimes yellowish | Trough-like, finely pointed | Short wrap-around |

## GROUP C

### Leaves rolled; plant hairy; auricles present

| Species | Plant characteristics | Leaves | Ligules |
|---|---|---|---|
| Common couch (*Elymus (Agropyron) repens*) | Strong rhizomes; prominent auricles | Long, dull green, downy above | Very short |
| Wall barley (*Hordeum murinum*) | Downy annual | Thin, hairy both sides | Short |
| Meadow barley (*Hordeum secalinum*) | Downy perennial | Hairy but not obvious below | Very short |
| Wheat (*Triticum* spp.) | Auricles fringed | Large, tapering | Large |

## GROUP D

### Leaves rolled; plant hairy; auricles absent

| Species | Plant characteristics | Leaves | Ligules |
|---|---|---|---|
| Barren brome (*Bromus sterilis*) | Annual; weak stems | Soft blades, twisted twice | Long, ragged |
| Soft brome (*Bromus hordeaceus*) (*mollis*) | Tufted, hairy | Thin, flaccid | Short, serrated |
| Yorkshire fog (*Holcus lanatus*) | Tufted, velvety; pink veins on sheaths | Hairy, grey-green | Large, toothed |
| Creeping soft grass (*Holcus mollis*) | Large rhizomes; long hairs on nodes | Softly hairy | Serrated, blunt |
| False oat grass (*Arrhenatherum elatius*) | Tall; often onion-based | Long, pointed | Short, round |
| Sweet vernal grass (*Anthoxanthum odoratum*) | Spreading hairs at sheath blade junction | Aromatic | Long; hairs |

## GROUP E

### Leaves rolled; plant glabrous (hairless); auricles present

| Species | Plant characteristics | Leaves | Ligules |
|---|---|---|---|
| Italian ryegrass (*Lolium multiflorum*) | Bright red base | Dark green; shiny beneath | Short, blunt |
| Meadow fescue (*Festuca pratensis*) | Dark brown basal sheaths | Stiffly erect; spiky | Small, blunt |
| Tall fescue (Festuca *arundinacea*) | Coarse habit; auricles fringed | Broad, dark green | Short |
| Barley (*Hordeum* spp.) | Big clasping auricles | Large, pointed | Small |

## GROUP F

### Leaves rolled; plant glabrous (hairless); auricles absent

| Species | Plant characteristics | Leaves | Ligules |
|---|---|---|---|
| Timothy (*Phleum pratense*) | Stem base sometimes swollen | Large, pale green | Long, white |
| Common bent (*Agrostis capillaris*) (*tenuis*) | Dense; erect; rhizomes cordlike | Triangular, short | Short, blunt |
| Creeping bent (*Agrostis stolonifera*) | Small plant; stolons | Triangular, small | Long, rounded |
| Black bent (*Agrostis gigantea*) | Large plant; rhizomes | Triangular, long | Long, ragged |
| Meadow foxtail (*Alopecurus pratensis*) | Tufted perennial | Smooth, pale green | Small, blunt |

## GROUP F (*continued*)

# Leaves rolled; plant glabrous (hairless); auricles absent

| Species | Plant characteristics | Veins large | Long, white |
| --- | --- | --- | --- |
| Black grass (*Alopecurus myosuroides*) | Tufted annual | Dark green; rough (ribbed) | Long, pointed |
| Tufted hair grass (*Deschampsia caespitosa*) | Tall; tussock; harsh to touch | Trough-like, finely pointed | Short, wrap-around |
| Crested dogstail (*Cynosurus cristatus*) | Leaf sheaths sometimes yellowish | Broad, light, blue-green | Long, blunt |
| Oats (*Avena* spp.) | Annual; basal leaves loosely hairy | | |

*Source:* Williams, 1984.

*Appendix 4*

# Species which have performed well in seed mixtures used in field trials on clay, chalk and limestone, and alluvial soils. Species listed as suitable for dry acid soils have not been tested experimentally

● suitable
○ may be too vigorous on fertile soil

| | | Clay | Chalk & lime-stone | Alluvial | Dry acid |
|---|---|:---:|:---:|:---:|:---:|
| **Grasses** | | | | | |
| *Agrostis capillaris* | Common bent | ● | | ● | ● |
| *Aira caryophyllea* | Silvery hair-grass | | | | ● |
| *Alopecurus pratensis* | Meadow foxtail | ● | | ● | |
| *Anthoxanthum odoratum* | Sweet vernal | ● | | ● | ● |
| *Bromus commutatus* | Meadow brome | ● | | ● | |
| *Bromus erectus* | Upright brome | | ○ | | |
| *Cynosurus cristatus* | Crested dogstail | ● | ● | ● | |
| *Deschampsia flexuosa* | Wavy hair-grass | | | | ● |
| *Festuca ovina* | Sheep's fescue | | ● | | ● |
| *Festuca rubra* ssp. *commutata* | Chewings fescue | ● | ● | ● | ● |
| *Festuca rubra* ssp. *rubra* | Red fescue | ● | ● | ● | ● |
| *Festuca tenuifolia* | Fine-leaved sheep's fescue | | | | ● |
| *Hordeum secalinum* | Meadow barley | ● | | ● | |
| *Phleum nodosum* | Cat's tail (timothy grass) | ● | ● | ● | |
| *Poa pratensis* | Smooth-stalked meadow grass | ● | | ● | |
| *Poa trivialis* | Rough-stalked meadow grass | ● | | ● | |
| *Trisetum flavescens* | Golden oat-grass | ● | ● | ● | |
| **Forbs** | | | | | |
| *Achillea millefolium* | Yarrow | ● | ● | ● | |
| *Anthyllis vulneraria* | Kidney vetch | | ● | | |
| *Centaurea nigra* | Common knapweed | ○ | ○ | | |
| *Centaurea scabiosa* | Greater knapweed | | ○ | | |
| *Clinopodium vulgare* | Wild basil | | ● | | |
| *Conopodium majus* | Pignut | | | ● | |
| *Crepis capillaris* | Smooth hawk's beard | | | | ● |
| *Daucus carota* | Wild carrot | ● | ● | ● | |
| *Erodium cicutarium* | Stork's bill | | | | ● |

▲    Spike rotavation of hill land prior to fertilising and reseeding with improved forage species, west Scotland    (G. E. D. Tiley)

Hunter's Rotary Strip Seeder and slots made by independently mounted cultivators   (J. D. P. Hunter)
▼    *Chapter 24*

Assessing the success of seedling establishment after slot seeding in hill land, Slovakia
▼    *Chapter 24*

▲ Hill grazings showing strips of improved grassland alternating with heather, MLURI, Glensaugh, north-east Scotland   *Chapter 24*

Scottish Blackface ewes grazing on unimproved hill pasture of low nutritive
▼ value, west Scotland   *Chapter 24*

Weaned crossbred hill lambs grazing improved in-bye pasture—
upland pasture in the background—SAC, Kirkton, west Scotland
▼ (A. Waterhouse)   *Chapter 24*

▲ Calved suckler cows grazing unimproved upland pasture of low nutritive value, west Scotland *Chapter 24*

▲ Suckler cows and thriving calves grazing improved upland pasture *Chapter 24*

Fattening beef steer grazing ▶ sown upland sward at controlled sward surface height (I. A. Wright) *Chapter 24*

Evaluating red deer farming
in the uplands at MLURI,
Glensaugh, north-east Scotland
▼    *Chapter 24*

▲    Agro-forestry trial at MLURI,
Glensaugh, north-east Scotland
*Chapter 24*

Assessing cashmere goat
farming in the uplands at
SAC, Kirkton, west of Scotland
▼    (R. F. Gooding)    *Chapter 24*

| | | Clay | Chalk & lime-stone | Alluvial | Dry acid |
|---|---|:---:|:---:|:---:|:---:|
| *Filipendula vulgaris* | Dropwort | ● | ● | | |
| *Galium saxatile* | Heath bedstraw | | | | ● |
| *Galium verum* | Lady's bedstraw | ● | ● | ● | |
| *Geranium molle* | Dove's foot crane's bill | | | | ● |
| *Geranium pratense* | Meadow crane's bill | ○ | | | |
| *Hieracium pilosella* | Mouse-ear hawkweed | | ● | | |
| *Hippocrepis comosa* | Horseshoe vetch | | ● | | |
| *Hypochoeris radicata* | Cat's ear | ● | | ● | ● |
| *Knautia arvensis* | Field scabious | ○ | ○ | | |
| *Leontodon hispidus* | Rough hawkbit | ● | ● | | |
| *Leucanthemum vulgare* (= *Chrysanthemum leucanthemum*) | Oxeye daisy | ● | ● | ● | |
| *Lotus corniculatus* | Birdsfoot trefoil | ○ | ○ | ○ | |
| *Lychnis flos-cuculi* | Ragged robin | ● | | ● | |
| *Malva moschata* | Musk mallow | ● | | | |
| *Medicago lupulina* | Trefoil (black medick) | ● | ● | | |
| *Onobrychis viciifolia* | Sainfoin | | ○ | | |
| *Ononis repens* | Common rest harrow | ● | ● | | |
| *Ononis spinosa* | Spiny rest harrow | ● | ● | | |
| *Pimpinella saxifraga* | Burnet saxifrage | | ● | | |
| *Plantago lanceolata* | Ribwort plantain | ● | ● | ● | |
| *Plantago media* | Hoary plantain | ● | ● | | |
| *Potentilla erecta* | Erect potentilla | | | | ● |
| *Primula veris* | Cowslip | ● | ● | ● | |
| *Prunella vulgaris* | Self-heal | ● | ● | ● | |
| *Ranunculus acris* | Meadow buttercup | ● | | ● | |
| *Ranunculus bulbosus* | Bulbous buttercup | ● | ● | | |
| *Rhinanthus minor* | Yellow rattle | ● | ● | ● | |
| *Rumex acetosa* | Common sorrel | ● | | ● | ● |
| *Rumex acetosella* | Sheep's sorrel | | | | ● |
| *Sanguisorba minor* | Salad burnet | ● | ● | | |
| *Sanguisorba officinalis* | Great burnet | | | ● | |
| *Saxifraga granulata* | Meadow saxifrage | ● | | | |
| *Silaum silaus* | Pepper saxifrage | ● | | ● | |
| *Silene alba* | White campion | ● | ● | ● | |
| *Stachys officinalis* | Wood betony | ● | | | |
| *Stellaria graminea* | Lesser stitchwort | | | ● | |
| *Trifolium arvense* | Hare's foot trefoil | | | | ● |
| *Trifolium campestre* | Hop trefoil | ● | ● | | |
| *Trifolium dubium* | Suckling clover | ● | ● | | |
| *Vicia cracca* | Tufted vetch | ● | | ● | |
| *Vicia sativa* | Winter vetch | ○ | ○ | ○ | |
| *Vicia tetrasperma* | Smooth tare | ● | ● | ● | |

*Source:* Wells *et al.*

*Appendix 5*

# Selected metric conversion factors

---

| *Imperial to Metric* | | | *Metric to Imperial* | | |
|---|---|---|---|---|---|
| *Length* | | | | | |
| inches to mm | × | 25.400 | millimetres to in | × | 0.039 |
| inches to cm | × | 2.540 | centimetres to in | × | 0.394 |
| feet to m | × | 0.305 | metres to ft | × | 3.281 |
| yards to m | × | 0.914 | metres to yd | × | 1.094 |
| miles to km | × | 1.610 | kilometres to miles | × | 0.621 |
| *Area* | | | | | |
| square inches to cm² | × | 6.452 | square centimetres to in² | × | 0.155 |
| square feet to m² | × | 0.093 | square metres to ft² | × | 10.764 |
| square yards to m² | × | 0.836 | square metres to yd² | × | 1.196 |
| acres to ha | × | 0.405 | hectares to ac | × | 2.471 |
| square miles to km² | × | 2.590 | square kilometres to miles | × | 0.386 |
| *Volume* | | | | | |
| fluid ounces to litres | × | 0.028 | litres to fl oz | × | 35.200 |
| pints to litres | × | 0.568 | litres to pints | × | 1.760 |
| gallons to litres | × | 4.546 | litres to gal | × | 0.220 |
| cubic feet to m³ | × | 0.028 | cubic metres to ft³ | × | 35.315 |
| cubic yards to m³ | × | 0.765 | cubic metres to yd³ | × | 1.308 |
| *Weight* | | | | | |
| ounces to g | × | 28.350 | grammes to oz | × | 0.035 |
| pounds to g | × | 453.592 | grammes to lb | × | 0.002 |
| pounds to kg | × | 0.454 | kilogrammes to lb | × | 2.205 |
| hundredweights to kg | × | 50.800 | kilogrammes to cwt | × | 0.020 |
| hundredweights to tonnes (t) | × | 0.051 | tonnes (t) to cwt | × | 19.688 |
| tons to kg | × | 1016.960 | kilogrammes to tons | × | 0.001 |
| tons to tonnes (t) | × | 1.016 | tonnes (t) to tons | × | 0.984 |

*Temperature*

°C to °F $\quad\quad\quad$ $(°C \times 1.8) + 32$ $\quad\quad$ °F to °C $\quad\quad\quad$ $5 (°F - 32) \div 9$

*Double conversions*

| | | | |
|---|---|---|---|
| fluid ounces per acre to litres/ha | × 0.070 | litres per ha to fl oz/ac | × 14.24 |
| pints per acre to litres/ha | × 1.404 | litres per ha to pints/ac | × 0.011 |
| gallons per acre to litres/ha | × 11.233 | litres per ha to gal/ac | × 0.089 |
| | | | |
| ounces per acre to g/ha | × 70.060 | grammes per ha to oz/ac | × 0.014 |
| pounds per acre to kg/ha | × 1.121 | kilogrammes per ha to lb/ac | × 0.892 |
| units per acre to kg/ha | × 1.256 | kilogrammes per ha to units/ac | × 0.796 |
| hundredweights per acre to t/ha | × 0.126 | tonnes (t) per ha to cwt/ac | × 7.966 |
| tons per acre to tonnes (t)/ha | × 2.511 | tonnes (t) per ha to tons/ac | × 0.398 |

*Fertiliser conversions*

| | | | |
|---|---|---|---|
| $P_2O_5$ to P | × 0.4364 | P to $P_2O_5$ | × 2.2915 |
| $K_2O$ to K | × 0.8301 | K to $K_2O$ | × 1.2047 |
| CaO to Ca | × 0.7146 | Ca to CaO | × 1.3994 |
| MgO to Mg | × 0.6031 | Mg to MgO | × 1.6581 |

*Appendix 6*

# Useful addresses

---

Agricultural Development and Advisory Service
  Nobel House, 17 Smith Square, London SW1P 3JR
  Tel: (071) 238 3000

Agricultural and Food Research Council
  Central Office, Polaris House, North Star Avenue, Swindon SN2 1UH
  Tel: (0793) 413200

Agricultural Research Institute of Northern Ireland
  Hillsborough, County Down, Northern Ireland BT26 6DR
  Tel: (0846) 682484

Agricultural Training Board
  Stoneleigh Park Pavilion, National Agricultural Centre, Kenilworth,
  Warwickshire CV8 2UG
  Tel: (0203) 696996

Association of Agriculture
  Victoria Chambers, 16/20 Strutton Ground, London SW1P 2HP
  Tel: (071) 222 6115

British Agrochemicals Association
  4 Lincoln Court, Lincoln Road, Peterborough, Cambs PE1 2RP
  Tel: (0733) 49225

British Crop Protection Council
  Bignor Park Cottage, Bignor, Pulborough, West Sussex RH20 1HQ
  Tel: (0798) 7224

British Deer Society
  Church Farm, Lower Basildon, Reading, Berkshire RG8 9NH
  Tel: (0735) 74094

British Goat Society
  34–36 Fore Street, Bovey Tracey, Newton Abbott, Devon TQ13 9AD
  Tel: (0626) 833168

British Grassland Society
  No. 1 Earley Gate, University of Reading, Reading RG6 2AT
  Tel: (0734) 318189

British Institute of Agricultural Consultants
  Durleigh House, 3 Elm Close, Campton, Shefford, Bedfordshire SG17 5PE
  Tel: (0462) 813380

British Sheep Dairying Association
  Wield Wood Farm, Wield, Alresford, Hampshire SO24 9RU
  Tel: (0420) 63151

British Society of Animal Production
  PO Box 3, Penicuik, Midlothian EH26 OR2
  Tel: (031) 445 4508

British Wool Marketing Board
  Oak Mills, Station Road, Clayton, Bradford, West Yorkshire BD14 6TD
  Tel: (0274) 882091

Country Landowners Association
  16 Belgrave Square, London SW1X 8PQ
  Tel: (071) 235 0511

Countryside Commission
  John Dower House, Crescent Place, Cheltenham GL50 3RA
  Tel: (0242) 521381

Countryside Council for Wales
  Plas Penrhos, Ffordd Penrhos, Bangor LL57 2LQ
  Tel: (0248) 370444

Crofters Commission
  4/6 Castle Wynd, Inverness IV2 3EQ
  Tel: (0463) 237231

Dairy Farmer Magazine
  Farming Press Ltd, Wharfedale Road, Ipswich IP1 4LG
  Tel: (0473) 241122

Department of Agriculture and Fisheries for Scotland
  Pentland House, 47 Robb's Loan, Edinburgh EH14 1TW
  Tel: (031) 556 8400

Department of Agriculture and Food, Ireland
  Agriculture House, Kildare Street, Dublin 2, Ireland
  Tel: (01) 789011

Department of Agriculture, Northern Ireland
  Dundonald House, Upper Newtonwards Road, Belfast BT4 3SB
  Tel: (0232) 650111

EC Information Office
  Abbey Buildings, 8 Storeys Gate, London SW1
  Tel: (071) 222 8122

English Nature
  Northminster House, Peterborough PE1 1WA
  Tel: (0733) 340345

Farmers' Union of Wales
  Llys Amaeth, Queens Square, Aberystwyth SY23 2EA
  Tel: (0970) 612755

Farming and Wildlife Advisory Group
  The Lodge, Sandy, Beds SG19 2DL
  Tel: (0767) 80551

Farming and Wildlife Trust Ltd
  National Agricultural Centre, Stoneleigh, Kenilworth, Warwickshire
  CV8 2RX
  Tel: (0203) 696760

Farming Press Books
  Wharfedale Road, Ipswich IP1 4LG
  Tel: (0473) 241122

Fertiliser Manufacturers' Association Ltd
  Greenhill House, Thorpe Wood, Peterborough PE3 6GF
  Tel: (0733) 331303

Food and Agriculture Organisation of the United Nations
  Via delle Terme di Caracalla, 00100 Rome, Italy
  Tel: 57971

Forestry Commission
  23 Corstorphine Road, Edinburgh EH12 7AT
  Tel: (031) 334 0303

Hannah Research Institute
  Kirkhill, Ayr KA6 5HL
  Tel: (0292) 76013

Highlands and Islands Development Board
  Bridge House, 20 Bridge Street, Inverness IV1 1QR
  Tel: (0463) 234171

Institute of Arable Crop Research
  Rothamsted Experimental Station, Harpenden, Herts AL5 2JQ
  Tel: (0582) 763133

Institute of Grassland and Environmental Research
  Welsh Plant Breeding Station, Plas Gogerddan, Aberystwyth, Dyfed
  SY23 3EB
  Tel: (0970) 828255

Institute of Grassland and Environmental Research
  North Wyke Research Station, Okehampton, Devon EX20 2SB
  Tel: (0837) 82558

Institute of Terrestrial Ecology
  Bush Estate, Penicuik, Midlothian EH26 0QB
  Tel: (031) 445 5719

Liscombe Experimental Husbandry Farm
  Dulverton, Somerset TA22 9PZ
  Tel: (0643) 85291

Livestock Marketing Commission for Northern Ireland
  57 Malone Road, Belfast BT9 6SA
  Tel: (0232) 381022

Macaulay Land Use Research Institute
  Craigiebuckler, Aberdeen AB9 2QJ
  Tel: (0224) 318611

Meat and Livestock Commission
  PO Box 44, Winterhill House, Snowdon Drive, Milton Keynes MK6 1AX
  Tel: (0908) 677577

Meteorological Office
  London Road, Bracknell, Berkshire RG12 2SZ
  Tel: (0344) 420242

Milk Marketing Board for England and Wales
  Thames Ditton, Surrey KY7 0EL
  Tel: (081) 398 4101

Milk Marketing Board for Northern Ireland
  456 Antrim Road, Belfast BT15 5GD
  Tel: (0232) 770123

Ministry of Agriculture, Fisheries and Food
  Whitehall Place, London SW1A 2HH
  Tel: (071) 270 3000

National Agricultural Centre
  Stoneleigh, Kenilworth, Warwickshire CV8 2LZ
  Tel: (0203) 696969

National Cattle Breeders' Association
  24 Courtenay Park, Newton Abbott, Devon TQ12 2HB
  Tel: (0626) 67661

National Farmers' Union
  Agriculture House, 25–31 Knightsbridge, London SW1X 7NJ
  Tel: (071) 235 5077

National Farmers' Union of Scotland
  17 Grosvenor Crescent, Edinburgh EH12 5EN
  Tel: (031) 337 4333

National Institute of Agricultural Botany
  Huntingdon Road, Cambridge CB3 0LE
  Tel: (0223) 276381

National Sheep Association
  106 High Street, Tring, Herts HP23 4AF
  Tel: (0442) 827077

Organic Advisory Service
  Elm Farm Research Centre, Hamstead Marshall, Newbury, Berks
  RG15 0HR
  Tel: (0488) 58298

Red Deer Commission
  Knowsley, 82 Fairfield Road, Inverness IV3 5LH
  Tel: (0463) 231751

Rowett Research Institute
  Greenburn Road, Bucksburn, Aberdeen AB2 9SB
  Tel: (0224) 712751

Royal Agricultural Society of England
  National Agricultural Centre, Stoneleigh, Kenilworth, Warwickshire
  CV8 2LZ
  Tel: (0203) 696969

Royal Association of British Dairy Farmers
  55 Sleaford Street, London SW8 5AB
  Tel: (071) 627 2111

Royal Highland Agricultural Society of Scotland
  Ingliston, Newbridge, Midlothian EH28 8NF
  Tel: (031) 333 2444

Royal Ulster Agricultural Society
  The Kings House, Balmoral, Belfast BT9 6GW
  Tel: (0232) 665225

Royal Welsh Agricultural Society
  Llanelwedd, Builth Wells, Powys LD2 3SY
  Tel: (0982) 553683

Scottish Agricultural College Central Office
West Mains Road, Edinburgh EH9 3JG
Tel: (031) 662 1303

Scottish Crop Research Institute
Invergowrie, Dundee, Tayside DD2 5DA
Tel: (0382) 562731

Scottish Landowners Federation
18 Abercromby Place, Edinburgh EH3 6TY
Tel: (031) 556 4466

Scottish Milk Marketing Board
Underwood Road, Paisley PA3 1JJ
Tel: (041) 887 1234

Scottish Natural Heritage
Battleby, Redgorton, Perth PH1 3EW
Tel: (0738) 27921

Silsoe Research Institute
Wrest Park, Silsoe, Bedford MK45 4HS
Tel: (0525) 60000

Soil Association
86/88 Colston Street, Bristol, Avon BS1 5BB
Tel: (0272) 290661

United Kingdom Agricultural Supply Trade Association Ltd
3 Whitehall Court, London SW1A 2EQ
Tel: (071) 930 3611

Ulster Farmers Union
Dunedin, 475/477 Antrim Road, Belfast BT15 3DA
Tel: (0232) 370222

Welsh Office Agriculture Department
Crown Offices, Cathays Park, Cardiff CF1 3NQ
Tel: (0222) 825111

World Bank
International Bank of Reconstruction and Development
66 Avenue d'Iéna, F-75116, Paris, France
Tel: 723 5421

World Bank
International Bank of Reconstruction and Development
1818 H Street, N.W., Washington D.C. 20433, USA
Tel: 477 1234

*Appendix 7*

# References cited and further reading

---

These have been arranged by chapter for convenience but some books are relevant to several chapters. Cited references are marked with a dagger (†).

## Chapter 1

†Burrell, A., Hill, B. and Medland, J. (1984). *Statistical Handbook of UK Agriculture*. Macmillan, London.

†Forbes, T. J., Dibb, C., Green, J. O., Hopkins, A. and Peel, S. (1980). *Factors Affecting the Productivity of Permanent Grassland: A National Farm Study*. Grassland Research Institute and Agricultural Development and Advisory Service Joint Permanent Pasture Group.

†Hill Farming Research Organisation (1979). *Science and Hill Farming: HFRO 1954–1979*. HFRO, Edinburgh, pp. 9–21.

Hopkins, A., Matkin, E. A., Ellis, J. A. and Peel, S. (1985). South-west England grassland survey 1983. 1. Age structure and sward composition of permanent and arable grassland and their relation to manageability, fertilizer nitrogen and other management features. *Grass and Forage Science*, 40, 349–359.

†Hopkins, A., Wainwright, J., Murray, P. J., Bowling, P. J. and Webb, M. (1988). 1986 survey of upland grassland in England and Wales: changes in age structure and botanical composition since 1970–72 in relation to grassland management and physical features. *Grass and Forage Science*, 43, 185–198.

Lazenby, A. (1981). British grasslands: past, present and future. *Grass and Forage Science*, 35, 243–266.

McAdam, J. H. (1983). Characteristics of grassland on hill farms in N. Ireland—Physical features, botanical composition and productivity. *Report*. Queen's University of Belfast/Department of Agriculture for Northern Ireland.

†Ministry of Agriculture, Fisheries and Food (1992). *Agriculture in the United Kingdom, 1991*. HMSO, London.

Murray, R. B. (1985). *Vegetation Management in Northern Britain*. Monograph, British Crop Protection Council, No. 30, BCPC Publications, Croydon.

†Swift, G., Holmes, J. C., Cleland, A. T. and Fortune, D. (1983). *The*

*Grassland of East Scotland—A Survey 1976–78*. East of Scotland College of Agriculture Bulletin, No. 29.

White, G. (1788). *The Natural History of Selborne*. 4th impression 1924, Arrowsmith, Bristol.

## Chapter 2

Dibb, C. and Haggar, R. J. (1979). Evidence of effect of sward changes on yield. *Occasional Symposium, British Grassland Society*, No. 10, pp. 11–20.

Frame, J. (1989). Herbage productivity of a range of grass species under a silage cutting regime with high fertiliser nitrogen application. *Grass and Forage Science*, 44, 267–276.

Frame, J. (1989). The potential of secondary grasses. *Proceedings XVI International Grassland Congress, Nice, France*, Volume 1, pp. 209–210.

Frame, J. (1990). Herbage productivity of a range of grass species in association with white clover. *Grass and Forage Science*, 45, 57–64.

†Frame, J. (1991). Herbage productivity and quality of a range of secondary grass species at five rates of fertilizer nitrogen application. *Grass and Forage Science*, 46, 139–151.

Frame, J., Hunt, I. V. and Harkess, R. D. (1970). Potentiality studies of tall fescue (*Festuca arundinacea* Schreb). *Proceedings XI International Grassland Congress, Surfer's Paradise, Australia*, pp. 210–214.

Frame, J., Hunt, I. V. and Harkess, R. D. (1971). The role of timothy (*Phleum pratense*) in high yielding conservation leys. *Proceedings 4th General Meeting, European Grassland Federation, Lausanne, Switzerland*, pp. 198–202.

Frame, J. and Morrison, M. W. (1991). Herbage productivity of prairie grass, reed canary grass and phalaris. *Grass and Forage Science*, 46, 417–425.

Harkess, R. D., Hunt, I. V. and Frame, J. (1974). The present-day role of meadow fescue (*Festuca pratensis* Huds.). *Proceedings XIII International Grassland Congress, Moscow, Soviet Union*, Volume 1, Part 1, pp. 375–383.

Harkess, R. D., Morrison, M. W. and Frame, J. (1990). Herbage productivity of brome grass (*Bromus carinatus*). *Grass and Forage Science*, 45, 383–392.

†Hopkins, A., Gilbey, J., Dibb, C., Bowling, P. J. and Murray, P. J. (1990). Responses of permanent and reseeded grassland to fertilizer nitrogen. I. Herbage production and herbage quality. *Grass and Forage Science* 45, 43–55.

Hubbard, C. E. (1984). *Grasses* 3rd edition. Penguin Books, Harmondsworth.

Hunt, W. F. and Easton, H. S. (1989). Fifty years of ryegrass research in New Zealand. *Proceedings New Zealand Grassland Association*, 50, 11–23.

Watt, T. A. (1987). The biology of *Holcus lanatus* and its significance in grassland. *Herbage Abstracts* 48, 195–204.


## Chapter 3

Aldrich, D. T. A. (1987). Developments and procedures in the assessment of grass varieties at NIAB 1950–87. *Journal of the National Institute of Agricultural Botany*, 17, 313–327.

Frame, J. (1988). Recent grassland research and development in the west of Scotland. *Proceedings of Pasture Management Workshop: Current Research and Future Trends*, Nova Scotia Agricultural College, Truro, pp. 36–64.

Munro, J. M. M. (1980). Introduction of improved plant resources. *Occasional Symposium, British Grassland Society*, No. 12, pp. 17–33.

Scottish Agricultural College (1991). Recommended grass and clover varieties 1991–92. *Technical Note*, T269.

Tiley, G. E. D., Swift, G. and Younie, D. (1986). Grass and clover varieties for the hills and uplands. *Scottish Agricultural Colleges Research and Development Note*, No. 30.

National Institute of Agricultural Botany (1989). *Classified List of Herbage Varieties, England and Wales, 1989/90*. NIAB, Cambridge.

†National Institute of Agricultural Botany (1991). Recommended varieties of grasses and herbage legumes, 1991/92. *Farmers Leaflet*, No. 4.

†Scottish Agricultural College (1991). *Classification of Grass and Clover Varieties for Scotland*. SAC, Edinburgh.


## Chapter 4

Harkess, R. D. and Frame, J. (1989). Effect of seed rate on herbage productivity from diploid and tetraploid Italian ryegrasses. *Experimental Record, West of Scotland College*, No. 60.

Harkess, R. D. (1966). Growth characteristics and productivity of tetraploid Italian ryegrasses. *Proceedings X International Grassland Congress, Helsinki, Finland*, pp. 315–318.

Scottish Agricultural College (1989). Seed mixtures for Scotland. *Technical Note*, T168.

Scottish Agricultural College (1989). Grass and clover mixtures for set-aside. *Technical Note*, T170.


## Chapter 5

Department of Scientific and Industrial Research (1991). *Moving into the 1990's: Improving Grassland Through Science*. Grasslands Division, DSIR, Palmerston North, New Zealand.

Institute of Grassland and Environmental Research (1990 *et seq.*) *Reports*.

Moseley, G. and Baker, D. H. (1991). The efficacy of a high magnesium grass cultivar in controlling hypomagnesaemia in grazing animals. *Grass and Forage Science*, 46, 375–380.

†Thomson, A. J. and Wright, A. J. (1972). Principles and problems in grass breeding. *Annual Report, Plant Breeding Institute Cambridge 1971*, pp. 31–67.


## Chapter 6

Kelly, A. Fenwick (1988). *Seed Production of Agricultural Crops*. Longman Scientific and Technical, Harlow.

†Ministry of Agriculture, Fisheries and Food/Welsh Office Agriculture Department/Department of Agriculture and Fisheries for Scotland/ Department of Agriculture for Northern Ireland (1983). *Certification of Seed of Grasses and Herbage Legumes*.

†Ministry of Agriculture, Fisheries and Food (1991). Stocks of seeds in the hands of seedsmen on 31 May 1990 and dispersals during the year ended 31 May 1991. *Summary*.

National Institute of Agricultural Botany (1980). Growing grasses and herbage legumes for seed. *Seed Growers Leaflet*, No. 5.

National Institute of Agricultural Botany (1982). Growing Italian ryegrass and tetraploid hybrid ryegrasses to obtain optimum harvested seed yields and quality. *Seed Growers Leaflet*, No. 8.

†National Institute of Agricultural Botany (1991). *Herbage Seed Production, England and Wales*, No. 13.


## Chapter 7

Allen, H. P. (1981). *Direct Drilling & Reduced Cultivations*. Farming Press Books, Ipswich.

Barker, G. M. (1990). Pasture renovation: interactions of vegetation control with slug and insect infestations. *Journal of Agricultural Science, Cambridge*, 115, 195–202.

Blackshaw, R. P. (1991). Leatherjackets in grassland. In: *Strategies for Weed, Disease and Pest Control in Grassland*. Proceedings British Grassland Society Conference, pp. 6.1–6.12.

Clements, R. O. (1991). Frit fly and slugs. In: *Strategies for Weed, Disease and Pest Control*. Proceedings British Grassland Society Conference, pp. 7.1–7.8.

Clements, R. O. and Bentley, B. R. (1985). Incidence, impact and control of insect pests in newly sown grassland in the UK. *Occasional Symposium, British Grassland Society*, No. 18, pp. 49–56.

Haggar, R. J., Standell, C. J. and Birnie, J. E. (1985). Occurrence, impact and control of weeds in newly sown leys. *Occasional Symposium, British Grassland Society*, No. 18, pp. 11–19.

Macfarlane, M. J. and Bonish, P. M. (1986). Over-sowing white clover into cleared and unimproved North Island hill country—the role of management, fertilizer, inoculation, pelleting and resident Rhizobia. *Proceedings New Zealand Grassland Association*, 47, 43–51.

Naylor, R. E. C., Marshall, A. H. and Matthews, S. (1983). Seed establishment in directly drilled sowings. *Herbage Abstracts*, 53, 73–91.

Pascal, J. A. and Sheppard, B. W. (1985). The development of a strip seeder for sward improvement. *Research and Development in Agriculture*, 2, 125–134.

Schechtner, G. (1991). Austrian experiences on the prevention and solution of problems with grassland swards. In: *Grassland Renovation and Weed Control in Europe. Proceedings European Grassland Federation Conference, Graz, Austria*, pp. 1–9.

Standell, C. J. and Marshall, E. J. P. (1991). Weeds in newly sown grassland. In: *Strategies for Weed, Disease and Pest Control in Grassland. Proceedings British Grassland Society Conference*, pp. 4.1–4.11.

Tiley, G. E. D. and Frame, J. (1980). Methods for the improvement of hill and upland grazing in west Scotland. *Proceedings 8th General Meeting, European Grassland Society, Zagreb, Yugoslavia*, pp. 3.93–3.99.

†Tiley, G. E. D. and Frame, J. (1991). Improvement of upland permanent pasture and lowland swards by surface sowing methods. In: *Grassland Renovation and Weed Control in Europe. Proceedings European Grassland Federation Conference, Graz, Austria*, pp. 89–94.

Younie, D., Frame, J. and Swift, G. (1989). Grass and clover establishment. *Scottish Agricultural Colleges Technical Note*, T31.

Younie, D., Tiley, G. E. D. and Swift, G. (1990). Fertiliser Series No. 3: Recommendations for grass and clover establishment. *Scottish Agricultural Colleges Technical Note*, T209.

## Chapter 8

Cooper, F. B. (1991). Established grassland: developments in weed control. In: *Strategies for Weed, Disease and Pest Control in Grassland. Proceedings British Grassland Society Conference*, pp. 5.1–5.13.

Courtney, A. D. (1985). Impact and control of docks in grassland. *Occasional Symposium, British Grassland Society*, No. 18, pp. 120–127.

Davies, D. H. K. and Frame, J. (1987). Ragwort poisoning in livestock and prevention by control. *Scottish Agricultural Colleges Technical Note*, T96.

Haggar, R. J., Soper, D. and Cormack, W. F. (1990). Weed control in agricultural grassland. In: *Weed Control Handbook: Principles*, eighth edition, Blackwell Scientific Publications, Oxford, pp. 387–405.

Harkess, R. D. and Frame, J. (1981). Weed grasses in sown grassland with special reference to soft brome (*Bromus mollis* L.). *Crop Protection in Northern Britain, SCRI, Dundee*, pp. 231–236.

Hopkins, A. and Peel, S. (1985). Incidence of weeds in permanent grassland. *Occasional Symposium, British Grassland Society*, No. 18, pp. 93–102.

Oswald, A. K. (1985). Impact and control of thistles in grassland. *Occasional Symposium, British Grassland Society*, No. 18, pp. 128–136.

Peel, S. and Hopkins, A. (1980). The incidence of weeds in grassland. *Proceedings 1980 British Crop Protection Conference – Weeds*, pp. 877–890.

Swift, G., Whytock, G. P. and Younie, D. (1988). The control of annual meadow grass in grassland. *Scottish Agricultural Colleges Technical Note*, T140.

Wells, G. J. and Haggar, R. J. (1984). The ingress of *Poa annua* into perennial ryegrass swards. *Grass and Forage Science*, 39, 297–303.

Whytock, G. P., Davies, D. H. K. and Younie, D. (1987). Chickweed control in grassland. *Scottish Agricultural Colleges Technical Note*, T23.

Whytock, G. P., Davies, D. H. K. and Younie, D. (1987). Control of thistles and nettles in grassland. *Scottish Agricultural Colleges Technical Note*, T32.

Whytock, G. P., Davies, D. H. K. and Younie, D. (1991). Control of docks. *Scottish Agricultural Colleges Technical Note*, T33.

Whytock, G. P., Swift, G. and Younie, D. (1988). Control of rushes. *Scottish Agricultural Colleges Technical Note*, T110.

Whytock, G. P., Tiley, G. E. D. and Swift, G. (1988). Control of whins and broom. *Scottish Agricultural Colleges Technical Note*, T142.

Williams, G. H. (1987). Bracken control. *Scottish Agricultural Colleges Technical Note*, T54.

Williams, R. D. (1984). *Crop Protection Handbook – Grass and Clover Swards*. British Crop Protection Council, Croydon.

## Chapter 9

†Avery, B. W. (1980). Soil Classification for England and Wales (Higher Categories). *Soil Survey of England and Wales Technical Monograph*, No. 14.

Baker, A. M., Younger, A. and King, J. A. (1988). The effect of drainage on herbage growth and soil development. *Grass and Forage Science*, 43, 319–336.

Brown, K. R. and Evans, P. S. (1973). Animal treading. A review of the work of the late D. B. Edmond. *New Zealand Journal of Experimental Agriculture*, 1, 217–226.

Charles, A. H. (1979). Treading as a factor in sward deterioration. *Occasional Symposium, British Grassland Society*, No. 10, pp. 137–140.

Curll, M. and Wilkins, R. T. (1983). The comparative effects of defoliation, treading and excreta on a *Lolium perenne—Trifolium repens* pasture grazed by sheep. *Journal of Agricultural Science, Cambridge*, 100, 451–460.

Davies, A. and Armstrong, A. (1986). Field measurements of grassland poaching. *Journal of Agricultural Science, Cambridge*, 106, 67–73.

Douglas, J. T., Campbell, D. J. and Crawford, C. E. (1992). Soil and crop responses to conventional, reduced ground pressure and zero traffic systems for grass silage production. *Soil and Tillage Research*, 23, (in press).

†Douglas, J. T. and Crawford, C. E. (1989). Soil compaction and novel traffic systems in ryegrass grown for silage: effects on herbage yield, quality and nitrogen uptake. *Proceedings XVI International Congress, Nice, France*, Volume 1, pp. 175–176.

†Douglas, J. T. and Crawford, C. E. (1991). Wheel-induced soil compaction. Effects on ryegrass production and nitrogen uptake. *Grass and Forage Science*, 46, 405–416.

Douglas, J. T. and Crawford, C. E. (1992). The response of ryegrass to soil compaction and applied nitrogen. *Grass and Forage Science*, 47, (in press).

Frame, J. (1987). The effect of tractor wheeling on the productivity of red clover and red clover/ryegrass swards. *Research and Development in Agriculture*, 4, 55–60.

Frame, J. and Merrilees, D. (1990). The effect of tractor wheeling on the productivity of perennial ryegrass (*Lolium perenne*) swards. *Proceedings 13th General Meeting, European Grassland Federation, Banská Bystrica, Czechoslovakia*, Volume 1, pp. 170–174.

Frost, J. P. (1988). Effects on crop yield of machinery traffic and soil loosening. Part 1. Effects on grass yield of traffic frequency and date of loosening. *Journal of Agricultural Engineering Research*, 39, 301–312.

Frost, J. P. (1988). Effects on crop yield of machinery traffic and soil loosening. Part 2. Effects on grass yield of soil compaction, low ground pressure tyres and date of loosening. *Journal of Agricultural Engineering Research*, 40, 57–69.

Harkess, R. D. (1970). Fundamentals of grassland management: Grass and the soil. *Scottish Agriculture*, Volume 49, 79–86.

Merrilees, D. (1987). Subsoiling. *Scottish Agricultural Colleges Technical Note*, T83.

Merrilees, D., Speirs, R. B., Morris, R. J. F. and Frost, C. A. (1987). Drainage of soils of low permeability. *Scottish Agricultural Colleges Technical Note*, T93.

Mulholland, B. and Fullen, M. A. (1991). Cattle trampling and soil compaction on loamy sands. *Soil Use and Management*, 7, 189–193.

Patto, P. M., Clements, C. R. and Forbes, T. J. (1978). *Grassland Poaching in England and Wales Permanent Grassland Studies 2*, Permanent Pasture Group, Grassland Research Institute, Hurley.

Simpson, K. (1983). *Soil.* Longman, London.

Smith, C. A., Frame, J., Merrilees, D. and Whytock, G. P. (1990). Alleviation of soil compaction due to animal treading (poaching) in long-term grassland. *Proceedings 13th General Meeting, European Grassland Federation, Banská Bystrica, Czechoslovakia*, Volume 1, pp. 211–214.

Wilkins, R. J. and Garwood, E. A. (1985). Effects of treading, poaching and fouling on grassland production and utilization. *Occasional Symposium, British Grassland Society*, No. 19, pp. 19–31.

## Chapter 10

Bailey, R. (1990). *Irrigated Crops and Their Management*. Farming Press Books, Ipswich.

†Corrall, A. J. (1978). The effect of genotype and water supply on the seasonal pattern of grass production. *Proceedings 7th General Meeting, European Grassland Federation, Ghent, Belgium*, pp. 223–232.

†Down, K. M., Jollans, J. L., Lazenby, A. and Wilkins, R. J. (1981). The distribution of grassland and grassland usage in the UK. *Appendix VIII*. In: *Grassland in the British Economy, CAS Paper 10*, Centre of Agricultural Strategy/University of Reading/Grassland Research Institute.

†Garwood, E. A. (1988). Water deficiency and excess in grassland: the implications for grass production and for the efficiency of use of N. *Proceedings of a colloquium on Nitrogen and Water Use by Grassland*. AFRC Institute for Grassland and Animal Production: Hurley, pp. 24–41.

Jollans, J. L. (1981). *Grassland on the British Economy, CAS Paper 10*, Centre for Agricultural Strategy/University of Reading/Grassland Research Institute.

Ministry of Agriculture, Fisheries and Food (1982). *Irrigation*. Reference Book 138.

†Smith, C. A., Frame, J., Merrilees, D. and Whytock, G. P. (1990). Alleviation of soil compaction due to animal treading (poaching) in long-term grassland. *Proceedings 13th General Meeting, European Grassland Federation, Banská Bystrica, Czechoslovakia*, Volume 1, pp. 211–214.

†Thomas, C., Reeve, A. and Fisher, G. E. J. (1991). *Milk from Grass* (2nd edition). Imperial Chemical Industries/Scottish Agricultural College/Institute of Grassland and Environmental Research.

## Chapter 11

Archer, J. (1985). *Crop Nutrition and Fertilizer Use*. Farming Press Books, Ipswich.

Chestnutt, D. M. B., Murdoch, J. C., Harrington, F. J. and Binnie, R. C. (1977). The effect of cutting frequency and applied nitrogen on production and digestibility of perennial ryegrass. *Journal of the British Grassland Society*, 32, 177–183.

†Dyson, P. W. (1990). Fertiliser Series No. 2: Using the recommendation tables. *Scottish Agricultural Colleges Technical Note*, T208.

†Frame, J., Harkess, R. D. and Talbot, M. (1989). The effect of cutting frequency and fertilizer nitrogen rate on herbage productivity from perennial ryegrass. *Research and Development in Agriculture*, 6, 99–105.

† Harkess, R. D. (1983). Cutting management and sward productivity. *WSAC Agronomy Department Publication*, No. 760.

† Harkess, R. D. and Frame, J. (1986). Efficient use of fertilizer nitrogen on grass swards: effects of timing, cutting management and secondary grasses. *Developments in Plant and Soil Sciences*, No. 25, pp. 29–37.

Lynch, P. B. (1982). *Nitrogen Fertilizers in New Zealand Agriculture*. Ray Richards Publisher, Auckland/New Zealand Institute of Agricultural Science, Wellington.

Ministry of Agriculture, Fisheries and Food/Welsh Office Agriculture Department (1991). *Code of Good Agricultural Practice for the Protection of Water*. MAFF Publications, London.

Morrison, J., Jackson, M. V. and Sparrow, P. E. (1980). The response of perennial ryegrass to fertilizer nitrogen in relation to climate and soil. *Report of the joint GRI/ADAS Grassland Manuring Trial GM20. Technical Report*, No. 27, Grassland Research Institute, Hurley.

Reid, D. (1978). The effect of frequency of defoliation on the yield response of a perennial ryegrass sward to a wide range of nitrogen application rates. *Journal of Agricultural Science, Cambridge*, 90, 447–457.

† Reid, D. (1982). Conservation of herbage as silage. The sward: its composition and management. *Technical Bulletin 2*, National Institute for Research in Dairying/Hannah Research Institute, pp. 39–62.

Ryder, J. C. (1984). The flow of nitrogen in grassland. *Proceedings of the Fertiliser Society*, No. 229.

Scottish Office Agriculture and Fisheries Department (1992). *Prevention of Environmental Pollution from Agricultural Activities*, HMSO, Edinburgh.

Swift, G. (1988). Urea as a grassland fertiliser. *Scottish Agricultural Colleges Technical Note*, T102.

Swift, G., Cleland, A. T. and Franklin, M. F. (1988). A comparison of nitrogen fertilizers for spring and summer grass production. *Grass and Forage Science*, 43, 297–303.

The Royal Society (1983). *The Nitrogen Cycle of the United Kingdom: A Study Group Report*. The Royal Society, London.

† Thomas, C., Reeve, A. and Fisher, G. E. J. (1991). *Milk from Grass* (2nd edition). Imperial Chemical Industries/The Scottish Agricultural College/Institute of Grassland and Environmental Research.

Whitehead, D. C. (1970). *The Role of Nitrogen in Grassland Productivity. A Review of Information from Temperate Regions*. Bulletin 48, CAB, Hurley.

Whitehead, D. C. (1986). Nitrogen in UK grassland agriculture. *Journal of the Royal Agricultural Society of England*, 147, 190–200.

Wilman, D. and Wright, P. T. (1983). Some effects of applied nitrogen on the growth and chemical composition of temperate grasses. *Herbage Abstracts*, 53, 387–393.

† Younie, D., Tiley, G. E. D. and Swift, G. (1990). Fertiliser Series No. 4: Recommendations for grazing and conservation. *Scottish Agricultural Colleges Technical Note*, T210.

## Chapter 12

ADAS (1988). *Fertiliser Recommendations*. MAFF Reference Book 209, HMSO, London.

Chalmers, A., Kershaw, C. and Leech, P. (1990). Fertilizer use on farm crops in Great Britain: Results from the survey of fertilizer practice; 1969–88. *Outlook in Agriculture*, 19, 269–278.

† Hunt, I. V. (1972). Potash and phosphate requirements of a perennial ryegrass sward managed for high yields. *West of Scotland Agricultural College Experimental Record*, No. 28.

Jollans, J. L. (1985). *Fertilizers in UK Farming, CAS Report 9*. Centre for Agricultural Strategy, Reading.

Klessa, D. A., Frame, J., Golightly, R. D. and Harkess, R. D. (1989). The effect of fertilizer sulphur on grass production for silage. *Grass and Forage Science*, 44, 277–281.

Klessa, D. A., Sinclair, A. H. and Speirs, R. B. (1987). Magnesium in soils and crops. *Scottish Agricultural Colleges Technical Note*, T92.

† Mengel, K. and Kirkby, E. A. (1987). *Principles of Plant Nutrition*, 4th edition. International Potash Institute, Bern.

MISR/SAC (1985). Advisory soil analysis and interpretation. *Macaulay Institute for Soil Research/Scottish Agricultural Colleges Bulletin*, 1.

SAC (1987). Phosphate fertilisers. *Scottish Agricultural Colleges Technical Note*, T75.

SAC (1989). Sulphur in soils, fertilisers and crops. *Scottish Agricultural Colleges Technical Note*, T160.

Simpson, K. (1986). *Fertilizers and Manures*. Longmans, London.

Speirs, R. B. (1987). Liming materials. *Scottish Agricultural Colleges Technical Note*, T95.

Syers, J. K., Skinner, R. J. and Curtin, D. (1988). Soil and fertiliser sulphur in UK agriculture. *Proceedings of the Fertiliser Society*, No. 264.

† Younie, D., Tiley, G. E. D. and Swift, G. (1990). Fertiliser Series No. 4: Recommendations for grazing and conservation. *Scottish Agricultural Colleges Technical Note*, T210.

## Chapter 13

Edgar, K. E., Harkess, R. D. and Frame, J. (1983). The manurial value of liquid digested sludge on grassland in the west of Scotland. *Proceedings XIV International Grassland Congress, Lexington, Kentucky, USA*, pp. 539–541.

Frost, J. P., Stevens, R. J. and Laughlin, R. J. (1990). Effect of separation and acidification of cattle slurry on ammonia volatilisation and on the efficiency of slurry nitrogen for herbage production. *Journal of Agricultural Science, Cambridge*, 115, 49–56.

Meer, H. G. van der, Unwin, R. J., Dijk, T. A. van and Ennik, G. C. (1987).

*Animal Manure on Grassland and Fodder Crops: Fertiliser or Waste?* Martinus Nijhoff, Dordrecht.

Pain, B. (1991). Improving the utilisation of slurry and farm effluents. *Occasional Symposium, British Grassland Society*, No. 25, pp. 121–133.

Pain, B. F., Smith, K. A. and Dyer, C. J. (1986). Factors influencing the response of cut grass to the nitrogen content of dairy cow slurry. *Agricultural Wastes*, 19, 189–202.

SAC (1980). Handling and utilisation of animal wastes. *Scottish Agricultural Colleges Publication*, No. 16.

SAC (1986). Disposal of sewage sludge on land—hazards and value. *Scottish Agricultural Colleges Publication*, No. 170.

## Chapter 14

Bartholomew, P. W. and Chestnutt, D. M. B. (1977). The effect of fertilizer nitrogen and defoliation intervals on dry matter production, seasonal response and chemical composition of perennial ryegrass. *Journal of Agricultural Science, Cambridge*, 88, 711–721.

†Frame, J. (1989). The potential of secondary grasses. *Proceedings XVI International Grassland Congress, Nice, France*, Volume 1. pp. 209–210.

†Frame, J. and Dickson, I. A. (1986). The digestibility of rotationally-grazed perennial ryegrass swards. *Occasional Symposium, British Grassland Society*, No. 19, pp. 206–209.

†Frame, J., Bax, J. and Bryden, G. (1992). Herbage quality of perennial ryegrass/white clover and N-fertilized ryegrass swards in intensively managed dairy systems. *Proceedings 14th General Meeting, European Grassland Federation, Lahti, Finland*, pp. 180–184.

†Frame, J., Harkess, R. D. and Talbot, M. (1989). The effect of cutting frequency and fertilizer nitrogen rate on herbage productivity from perennial ryegrass. *Research and Development in Agriculture*, 6, 99–105.

Gill, M., Beever, D. E. and Osbourn, D. F. (1979). The feeding value of grass and grass products. In: *Grass: Its Production and Utilization*. Second edition, Blackwell Scientific Publications, Oxford, pp. 89–129.

†Harkess, R. D. (1969). Grass as a ruminant feed. *Scottish Agriculture*, 48, 327–338.

†National Institute of Agricultural Botany (1988). Grasses and Legumes for Conservation 1988/89. *NIAB Technical Leaflet*, No. 2.

McDonald, P., Edwards, R. A. and Greenhalgh, J. F. D. (1988). *Animal Nutrition*, 4th edition. Longman Scientific and Technical, Harlow.

Moisey, F. R. and Leaver, J. D. (1979). A study of two cutting strategies for the production of grass silage for dairy cows. *Research and Development in Agriculture*, 1, 47–52.

Nicol, A. M. (1987). *Livestock Feeding on Pasture. Occasional Publication, New Zealand Society of Animal Production*, No. 10.

†Terry, R. A. and Tilley, J. M. A. (1964). The digestibility of the leaves and

stems of perennial ryegrass, cocksfoot, timothy, tall fescue, lucerne and sainfoin, as measured by an *in vitro* procedure. *Journal of the British Grassland Society*, 19, 363–372.

†Walters, R. J. K. (1976). The field assessment of digestibility of grass for conservation. *ADAS Quarterly Review*, 23, 323–328.

†Whitehead, D. C. (1966). Nutrient minerals in grassland herbage *Mimeographed Publication*, No. 1/1966. Commonwealth Bureau of Pastures and Field Crops, Hurley.

## Chapter 15

†Bircham, J. S. and Hodgson, J. (1983). The influence of sward conditions on rates of herbage growth and senescense on mixed swards under continuous stocking management. *Grass and Forage Science*, 38, 323–331.

Davies, A. (1977). Structure of the grass sward. *Proceedings International Meeting on Animal Production from Temperate Grassland, Dublin, Ireland*, pp. 36–44.

Grace, N. D. (1983). *The Mineral Requirements of Ruminants. Occasional Publication, New Zealand Society of Animal Production*, No. 9.

Grant, S. A., Barthram, G. T., Torvell, L., King, J. and Smith, H. K. (1983). Sward management, lamina turnover and tiller population density in continuously stocked *Lolium perenne* dominated swards. *Grass and Forage Science*, 38, 333–344.

Hodgson, J. (1985). The significance of sward characteristics in the management of temperate sown pastures. *Proceedings XV International Grassland Congress, Kyoto, Japan*, pp. 31–34.

†Jewiss, O. R. (1979). Better understanding of grass growth can be an aid to management. *Grass Farmer*, 4, 3–7.

Jewiss, O. R. (1981). Shoot development and number. In: *Sward Management Handbook*. British Grassland Society, Hurley, pp. 93–114.

Johnson, I. R. and Parsons, A. J. (1985). Use of a model to analyse the effects of continuous grazing managements on seasonal patterns of grass production. Grass and Forage Science, 40, 449–458.

Jones, M. B. and Lazenby, A. (1988). *The Grass Crop: The Physiological Basis of Production*. Chapman and Hall, London.

Langer, R. H. M. (1990). Pasture plants. In: *Pastures: Their Ecology and Management*. Oxford University Press, Oxford, pp. 39–74.

†Orr, R. J., Parsons, A. J., Treacher, T. J. and Penning, P. D. (1988). Seasonal patterns of grass production under cutting and continuous stocking managements. *Grass and Forage Science*, 43, 199–207.

Parsons, A. J. and Johnson, I. R. (1986). The physiology of grass growth under grazing. *Occasional Symposium, British Grassland Society*, No. 19, pp. 3–13.

†Parsons, A. J., Leafe, E. L., Collett, B. and Lewis, J. (1983). The physiology of grass growth under grazing. 2. Photosynthesis, crop growth and animal

intake of continuously grazed swards. *Journal of Applied Ecology*, 20, 127–139.

Parsons, A. J., Leafe, E. L., Collett, B. and Stiles, W. (1983). The physiology of grass production under grazing. 1. Characteristics of leaf and canopy photosynthesis of continuously grazed swards. *Journal of Applied Ecology*, 20, 117–126.

† Penning, P. D., Parsons, A. J., Orr, R. J. and Treacher, T. T. (1991). Intake and behaviour responses by sheep to changes in sward characteristics under continuous stocking. *Grass and Forage Science*, 46, 15–28.

Thomas, J. O. and Davies, L. J. (1964). *Common British Grasses and Legumes*. Longmans, Green and Co. Ltd, London.

† Williams, R. D. (1984). Key for identification of vegetative grasses. In: *Crop Protection Handbook—Grass and Clover Swards*. R. D. Williams (ed.). BCPC, Croydon, pp. 84–87.

## Chapter 16

Frame, J. (1971). Fundamentals of grassland management: The grazing animal. *Scottish Agriculture*. Volume 50, 28–44.

† Frame, J. (1975). A comparison of herbage production under cutting and grazing (including comments on deleterious factors such as treading). *Occasional Symposium, British Grassland Society*, No. 7, pp. 50–56.

Frame, J. (1982). Plant relationships under grazing. *Proceedings Technical Meeting on Persistency of Pastures, Colonia, Uruguay, Dialogo V*, IICA/Cono Sur BID, pp. 95–105.

Frame, J. and Hunt, I. V. (1971). The effects of cutting and grazing systems on herbage production from grass swards. *Journal of the British Grassland Society*, 26, 163–171.

Hodgson, J. Grazing behaviour and herbage intake. *Occasional Symposium, British Grassland Society*, No. 19, pp. 51–64.

Marsh, R. and Campling, R. C. (1970). Fouling of pastures by dung. *Herbage Abstracts*, 40, 123–130.

† Penning, P. D. (1986). Some effects of sward conditions on grazing behaviour and intake by sheep. In: *Grazing Research at Northern Latitudes. Proceedings of a NATO Advanced Workshop, Hvanneyri, Iceland*, Volume 108, pp. 219–226. Plenum Publishing Corporation, New York.

Penning, P. D., Parsons, A. J., Orr, R. J. and Treacher, T. T. (1991). Intake and behaviour responses by sheep to changes in sward characteristics under continuous stocking. *Grass and Forage Science*, 46, 15–28.

Richards, I. R. (1977). Influence of soil and sward characteristics on the response to nitrogen. *Proceedings of International Meeting, Animal Production from Temperate Grassland, Dublin, June 1977*, pp. 45–49.

Watkin, B. R. and Clements, R. J. (1978). The effects of grazing animals on pastures. In: *Plant Relations in Pastures*, CSIRO, Australia, pp. 273–289.

Wolton, K. M. (1979). Dung and urine as agents of sward change: a review. *Occasional Symposium, British Grassland Society*, No. 19, pp. 131–135.
†Wright, I. A. (1988). Suckler beef production. *Occasional Symposium, British Grassland Society*, No. 22, pp. 51–64.

## Chapter 17

Baker, R. D. (1986). Advances in cow grazing systems. *Occasional Symposium, British Grassland Society*, No. 19, pp. 155–166.
Dickson, I. A. and Frame, J. (1981). Mixed grazing of cattle and sheep versus cattle only in an intensive grassland system. *Animal Production*, 33, 265–272.
Dickson, I. A., Frame, J. and Waterhouse, A. (1986). The potential for mixed grazing systems. *Occasional Symposium, British Grassland Society*, No. 19, pp. 189–198.
Ernst, G., Le Du, Y. L. P. and Carlier, L. (1980). Animal and sward production under rotational and continuous grazing management—a critical review. *Proceedings of International Symposium on the Role of Nitrogen in Intensive Grassland Production, Wageningen, The Netherlands*, pp. 119–126.
Frame, J. (1992). Herbage mass. In: *Sward Management Handbook*, Chapter 3, pp. 39–69. British Grassland Society, Reading.
HFRO (1986). The HFRO sward stick. *Hill Farming Research Organisation Biennial Report, 1984–85*, pp. 9–21.
Hodgson, J. (1990). *Grazing Management: Science into Practice*. Longman Scientific and Technical, Harlow.
†Hodgson, J., Mackie, C. K. and Parker, J. W. G. (1986). Sward surface heights for efficient grazing. *Grass Farmer*, 24, 5–10.
Holmes, W. (1989). Grazing management. In: *Grass: Its Production and Utilization*, Second edition. Blackwell Scientific Publications, Oxford, pp. 130–172.
Holmes, W. (1989). Application on the farm. In: *Grass: Its Production and Utilization*,. Second edition. Blackwell Scientific Publications, Oxford, pp. 258–271.
Leaver, J. D. (1985). Milk production from grazed temperate grassland—review. *Journal of Dairy Research*, 52, 313–344.
Leaver, J. D. (1988). Intensive grazing of dairy cows—what future? *Journal of the Royal Agricultural Society of England*, 149, 176–183.
Mackie, C. K., Lowman, B. and Roberts, D. J. (1987). Grass height and grazing buffers. *Scottish Agricultural Colleges Technical Note*, T36.
Maxwell, T. J. (1986). Systems studies in upland sheep production: Some implications for management and research. *Hill Farming Research Organisation Biennial Report 1984–85*, pp. 155–163.
†Maxwell, T. J. and Treacher, T. T. (1987). Decision rules for grazing manage-

ment. *Occasional Symposium, British Grassland Society*, No. 21, pp. 67–78.

Mayne, C. S., Clements, A. J. and Woodcock, S. C. F. (1990). An evaluation of the effects of grazing management systems involving preferential treatment of high-yielding dairy cows on animal performance and sward utilization. *Grass and Forage Science*, 45, 167–178.

Mayne, C. S., Newberry, R. D. and Woodcock, S. C. F. (1988). The effects of a flexible management strategy and leader/follower grazing on the milk production of grazing dairy cows and on sward characteristics. *Grass and Forage Science*, 43, 137–150.

Mayne, C. S., Newberry, R. D., Woodcock, S. C. F. and Wilkins, R. J. (1987). Effects of grazing severity on grass utilization and milk production of rotationally grazed dairy cows. *Grass and Forage Science*, 42, 59–72.

Mitchell, G. B. B. (1987). Worm control at grass. *Scottish Agricultural Colleges Technical Note*, T70.

Mitchell, G. B. B. (1987). Prevention and Treatment of *Nematodirus* in lambs. *Scottish Agricultural Colleges Technical Note*, T72.

Nolan, T. and Connolly, J. (1977). Mixed grazing by sheep and steers—a review. *Herbage Abstracts*, 47, 367–374.

Smetham, M. C. (1990). Pasture management. In: *Pastures: Their Ecology and Management*. Oxford University Press, Oxford, pp. 197–240.

Syrjälä-Quist, L. and Wilkins, R. J. (1992). Forage utilization. *Proceedings 14th General Meeting, European Grassland Federation, Lahti, Finland*, pp. 60–77.

Wright, I. A. and Whyte, T. K. (1989). Effects of sward surface height on the performance of continuously stocked spring calving beef cows and their calves. *Grass and Forage Science*, 44, 259–266.

## Chapter 18

Baker, A. M. and Leaver, J. D. (1986). Effect of stocking rate in early season on dairy cow performance and sward characteristics. *Grass and Forage Science*, 41, 333–340.

Blackshaw, R. P. and Newbould, P. (1987). Interactions of fertilizer use and leatherjacket control in grassland. *Grass and Forage Science*, 42, 342–346.

Frame, J. (1970). The effect of winter grazing by sheep on spring and early summer pasture production. *Journal of the British Grassland Society*, 25, 167–171.

French, N., Nichols, D. B. R. and Wright, A. J. (1990). Yield response of improved upland pasture to the control of leatherjackets under increasing rates of nitrogen. *Grass and Forage Science*, 45, 99–102.

Hunt, I. V., Frame, J. and Harkess, R. D. (1976). The effect of delayed autumn harvest on the survival of varieties of perennial ryegrass. *Journal of the British Grassland Society*, 31, 181–190.

Jagtenburg, W. D. (1970). Predicting the best times to apply N to grassland. *Journal of the British Grassland Society*, 24, 266–271.

Jones, P. K. and Charles, A. H. (1984). The winter hardiness of *Festuca rubra, Holcus lanatus* and *Agrostis* spp. in comparison with *Lolium perenne. Grass and Forage Science*, 39, 381–389.

McDonald, R. C. (1986). Effect of topping pastures. 1. Pasture accumulation and quality. *New Zealand Journal of Experimental Agriculture*, 14, 279–288.

†Patton, D. L. H. and Frame, J. (1981). The effect of grazing in winter by wild geese on improved grassland in west Scotland. *Journal of Applied Ecology*, 18, 311–325.

Peel, S. and Granfield, R. T. (1991). Creating and maintaining high quality swards. *Occasional Symposium, British Grassland Society*, No. 25, pp. 16–29.

Stakelum, G. and Dillon, P. (1991). Influence of sward structure and digestibility on the intake and performance of lactating and growing cattle. *Occasional Symposium, British Grassland Society*, No. 25, pp. 30–42.

Swift, G., Mackie, C. K., Harkess, R. D. and Franklin, M. F. (1985). Timing of fertilizer nitrogen for spring grass. *Scottish Agricultural Colleges Research and Development Note*, No. 24.

# Chapter 19

†ADAS Liscombe (1989). Silage. *Liscombe Grass Bulletin*, No. 2.

Appleton, M. (1991). Recent developments in the use of additives for sheep and growing cattle. *Occasional Symposium, British Grassland Society*, No. 25, pp. 75–86.

†Bastiman, B. (1977). Factors affecting silage effluent production. *Experimental Husbandry*, 31, 40–46.

†Bastiman, B. and Altman, J. F. B. (1985). Losses at various stages in silage making. *Research and Development in Agriculture*, 2, 19–25.

†British Grassland Society (1991). *Rules and Scoring System.* National Silage Competition (run in association with Kemira Fertilisers, ADAS and SAC).

Castle, M. E. (1982). Production and use of high-quality silage. *Technical Bulletin 2*, National Institute for Research in Dairying/Hannah Research Institute, pp. 105–125.

†Castle, M. E. (1982). Feeding high-quality silage. *Technical Bulletin 2*, National Institute for Research in Dairying/Hannah Research Institute, pp. 127–150.

Gordon, F. J. (1988). The influence of system of harvesting grass for silage on milk output. *Jubilee Report 1926–86, Agricultural Research Institute of Northern Ireland*, pp. 13–22.

Gordon, F. J. (1989). The principles of making and storing high quality, high intake silage. *Occasional Symposium, British Grassland Society*, No. 23, pp. 3–19.

Honig, H. (1976). Schätzung der Verlustre an Trockenmasse und Energie bei verschiedenen Konservierungsverfahren. *KTBL 'Kalkulationsunterlagen' Grunddaten der Futterwirtschaft. Manuskripdruck.*

Johnson, J. and Appleton, M. (1988). Minimising risk. *Occasional Symposium, British Grassland Society,* No. 23, pp. 124–138.

McDonald, P., Henderson, N. and Heron, S. (1991). *The Biochemistry of Silage.* Second edition. Chalcombe Publications, Marlow.

Murdoch, J. C. (1989). The conservation of grass. In: *Grass: Its Production and Utilization.* Second edition. Blackwell Scientific Publications, Oxford, pp. 173–213.

Nash, M. J. (1985). *Crop Conservation and Storage in Cool Temperate Climates.* Pergamon Press, Oxford.

Offer, N. W., Chamberlain, D. G. and Johnston, C. A. (1988). *Occasional Symposium, British Grassland Society,* No. 23, pp. 67–77.

Raymond, F., Redman, P. and Waltham, R. (1986). *Forage Conservation and Feeding.* Farming Press Books, Ipswich.

SAC (1985). Good silage making. *Scottish Agricultural Colleges Publication,* No. 15.

SAC (1989). Silage effluent—its collection and disposal. *Scottish Agricultural Colleges Technical Note,* T181.

Steen, R. W. J. (1991). Recent advances in the use of silage additives for dairy cattle. *Occasional Symposium, British Grassland Society,* No. 25, pp. 87–101.

Weddell, J. R. and Roberts, D. J. (1992). Silage additives 1992. *Scottish Agricultural College Technical Note,* T270.

Wilkins, R. J. (1984). Eurowilt: Efficiency of silage systems—a comparison between unwilted and wilted silages. *Landbauforschung Volkenrode, Sonderheft,* 69, 88 pp.

†Wilkinson, J. M. (1987). Silage: trends and portents. *Journal of the Royal Agricultural Society of England,* 148, 158–167.

†Zimmer, E. and Wilkins, R. J. (1984). Eurowilt: Efficiency of silage systems: a comparison between unwilted and wilted silages. *Landbauforschung Volkenrade, Sonderheft,* 69, 88 pp.

## Chapter 20

†ADAS Liscombe (1974). *Farm Report, 1974.*

†Honig, H. (1976). Schätzung der Verlustre an Trockenmasse und Energie bei verschiedenen Konservierungsverfahren. *KTBL 'Kalkulationsunterlagen' Grunddaten der Futterwirtschaft, Manuskripdruck.*

†Nash, M. J. (1985). *Crop Conservation and Storage in Cool Temperate Climates.* Pergamon Press, Oxford.

SAC (1977). Making and feeding better hay. *Scottish Agricultural Colleges Publication,* No. 21.

Smetham, M. L. (1990). The conservation of herbage as hay or silage.

In: *Pastures: Their Ecology and Management*. Oxford University Press, Oxford, pp. 337–369.


## Chapter 21

†ADAS (1980). *Livestock Units Handbook*. MAFF Booklet 2267, Pinner, Middlesex.

Baker, H. K., Baker, R. D., Deakins, R. M., Gould, J. L., Hodges, J. and Powell, R. A. (1964). Grassland recording V. Recommendations for recording the utilized output of grassland on dairy farms. *Journal of the British Grassland Society*, 19, 160–168.

†Frame, J. (1965). The assessment of utilized-starch-equivalent (USE) output from farm grassland by the farm-recording method. *Journal of the British Grassland Society*, 20, 77–83.

†Frame, J., Tweddle, J. and Bax, J. (1987). UME in dairy costings. *Scottish Agricultural Colleges Technical Note*, T21.

Lucas, R. J. and Thompson, K. F. (1990). Pasture assessment for livestock managers. In: *Pastures: Their Ecology and Management*. Oxford University Press, Oxford, pp. 241–262.

Walsh, A. (1982). The contribution of grass to profitable milk production. *Rex Paterson Memorial Study Publication*.


## Chapter 22

Baker, M. J. and Williams, W. M. (1987). *White clover*. CAB International, Wallingford.

Barney, P. A. (1987). The use of *Trifolium repens, Trifolium subterraneum* and *Medicago lupulina* as overwintering leguminous green manures. *Biological Agriculture and Horticulture*, 4, 225–234.

†Bax, J. and Thomas, C. (1992). Developments in legume use for milk production. *Occasional Symposium, British Grassland Society*, No. 26, pp. 40–53.

Brock, J. L., Caradus, J. R. and Hay, M. J. M. (1989). Fifty years of white clover research in New Zealand. *Proceedings New Zealand Grassland Association*, 50, 25–39.

Charlton, J. F. L. (1973). The potential value of birdsfoot trefoils (*Lotus* spp.) for the improvement of natural pastures in Scotland. 1. Common birdsfoot trefoil (*L. corniculatus* L.). *Journal of the British Grassland Society*, 28, 91–96.

Charlton, J. F. L. (1975). The potential value of birdsfoot trefoils (*Lotus* spp.) for the improvement of natural pastures in Scotland. 2. Marsh birdsfoot trefoil (*L. uliginosus* L.). *Journal of the British Grassland Society*, 30, 251–257.

Douglas, J. A. (1986). The production and utilization of lucerne in New Zealand. *Grass and Forage Science*, 41, 81–128.

Doyle, C. J. and Thomson, D. J. (1985). The future of lucerne in British agriculture: an economic assessment. *Grass and Forage Science*, 40, 57–68.

Doyle, C. J., Thomson, D. J. and Sheehy, J. E. (1984). The future of sainfoin in British agriculture. *Grass and Forage Science*, 39, 43–51.

†Evans, D. R., Hill, J., Williams, T. A. and Rhodes, I. (1985). Effects of co-existence on the performance of white clover/perennial ryegrass mixtures. *Oecologia, Berlin*, 66, 536–539.

Food and Agriculture Organisation of the United Nations (1991). *White Clover Developments in Europe*. REUR Technical Series, 19.

†Frame, J. (1976). The role and potential of tetraploid red clover in the United Kingdom. *Journal of the British Grassland Society*, 31, 139–152.

Frame, J. (1986). The production and quality potential of four forage legumes sown alone and combined in various associations. *Crop Research (Horticultural Research)*, 42, 271–281.

Frame, J. (1987). The role of white clover in United Kingdom pastures. *Outlook on Agriculture*, 16, 28–34.

†Frame, J. (1990). The role of red clover in United Kingdom pastures. *Outlook on Agriculture*, 19, 49–55.

†Frame, J. (1990). Herbage productivity of grass species in association with white clover. *Grass and Forage Science*, 45, 57–64.

†Frame, J., Bax, J. and Bryden, G. (1992). Herbage quality of perennial ryegrass/white clover and N-fertilized ryegrass swards in intensively managed dairy systems. *Proceedings 14th General Meeting, European Grassland Federation. Lahti, Finland*, pp. 180–184.

Frame, J. and Boyd, A. G. (1986). Effect of cultivar and seed rate of perennial ryegrass and strategic fertilizer nitrogen on the productivity of grass/white clover swards. *Grass and Forage Science*, 41, 359–366.

†Frame, J. and Boyd, A. G. (1987). The effect of fertilizer nitrogen rate, white clover variety and closeness of cutting on herbage productivity from perennial ryegrass/white clover swards. *Grass and Forage Science*, 42, 85–96.

†Frame, J. and Boyd, A. G. (1987). The effect of strategic use of fertilizer nitrogen in spring and/or autumn on the productivity of a perennial ryegrass/white clover sward. *Grass and Forage Science*, 42, 429–438.

†Frame, J. and Harkess, R. D. (1987). The productivity of four forage legumes sown alone and with each of five companion grasses. *Grass and Forage Science*, 42, 213–223.

Frame, J. and Newbould, P. (1984). Herbage productivity from grass/white clover swards. *Occasional Symposium, British Grassland Society*, No. 16, pp. 15–35.

†Frame, J. and Newbould, P. (1986). Agronomy of white clover. *Advances in Agronomy*, 40, 1–88.

Frame, J. and Paterson, D. J. (1987). The effect of strategic nitrogen application and defoliation systems on the productivity of a perennial ryegrass/white clover sward. *Grass and Forage Science*, 42, 271–280.

Hay, R. J. M. and Ryan, D. L. (1989). A review of 10 years' research with red clovers under grazing in Southland. *Proceedings New Zealand Grassland Association*, 50, 181–187.

Laidlaw, A. S. and Frame, J. (1988). Maximising the use of the legume in grassland systems. *Proceedings 12th General Meeting, European Grassland Federation, Dublin, Ireland*, pp. 34–46.

Marriott, C. and Rangeley, A. (1984). Nitrogen fixation and transfer by white clover. *Hill Farming Research Organisation Biennial Report*, 1982–83, pp. 109–118.

Novoselova, A. and Frame, J. (1992). The role of legumes in European grassland production. *Proceedings 14th General Meeting, European Grassland Federation, Lahti, Finland*, pp. 87–96.

†Roberts, D. J., Frame, J. and Leaver, J. D. (1989). A comparison of a grass/white clover sward with a grass sward plus fertilizer nitrogen under a three-cut silage regime. *Research and Development in Agriculture*, 6, 147–150.

Stewart, T. A. and Haycock, R. E. (1984). Beef production from low N and high N S24 perennial ryegrass/Blanca white clover swards—a six year farmlet scale comparison. *Research and Development in Agriculture*, 1, 103–113.

Swift, G. and Vipond, J. E. (1991). The beechgrove sheep mixture: Three years' results 1988–90. *Scottish Agricultural College Technical Note*, T274.

Thomson, D. J. (1984). The nutritive value of white clover. *Occasional Symposium, British Grassland Society*, No. 16, pp. 78–92.

†Vipond, J. E. and Swift, G. (1992). Developments in legume use in the hills and uplands. *Occasional Symposium, British Grassland Society*, No. 26, pp. 54–65.

Young, N. E. (1992). Developments in legume use for beef and sheep. *Occasional Symposium, British Grassland Society*, No. 26, pp. 29–39.

Younie, D., Heath, S. B. and Halliday, G. J. (1988). Factors affecting the conversion of a clover based system to organic production from grass/white clover swards. *Occasional Symposium, British Grassland Society*, No. 22, pp. 105–111.

## Chapter 23

Carter, E. S. (1990). Weed control in amenity areas and other non-agricultural land. In: *Weed Control Handbook: Principles*, Eighth edition. Blackwell Scientific Publications, Oxford, pp. 431–455.

†Countryside Commission for Scotland (1986). *Countryside Conservation: A Guide for Farmers*.

Duffey, E., Morris, M. G. , Sheail, J., Ward, L. K., Wells, D. A. and Wells, T. C. E. (1974). *Grassland Ecology and Wildlife Management*. Chapman and Hall, London.

†Elliot, R. H. (1908). *The Clifton Park System of Farming*, 4th edition. Simpkin, Marshall, Hamilton, Kent and Co. Ltd, London.

Farming and Wildlife Advisory Group (1985). Conservation and management of old grassland. *Information*, 19.

Ford, M. A. (1991). Wildflower grasslands. *Scottish Agricultural College Technical Note*, T275.

Foster, L. (1985). Herbs in pastures. Development and research in Britain, 1850–1984. *Biological Agriculture and Horticulture*, 5, 97–133.

Frame, J. (1990). Back to nature with grassland. *Norgrass*, 30, 17–21.

†Frame, J. and Tiley, G. E. D. (1990). Herbage productivity of a range of wildflower mixtures under two management systems. *Proceedings 13th General Meeting, European Grassland Federation, Banská Bystrica, Czechoslovakia*, Volume II, pp. 359–363.

†Gracey, H. I. (1992). *Countryside Management Competition*. Ulster Grassland Society/Fermanagh Grassland Club.

Green, B. H. (1990). Agricultural extensification and the loss of habitat, species and amenity in British grasslands: a review of historical change and assessment of future prospects. *Grass and Forage Science*, 45, 365–372.

Hillier, S. H., Walton, D. W. H. and Wells, D. A. (1990). *Calcareous Grasslands: Ecology and Management*. Bluntisham Books, Huntingdon.

†Nature Conservancy Council (1984). *Nature Conservation in Great Britain*. NCC, Edinburgh.

Nösberger, J. and Charles, J. P. (1990). Herbage production from less intensively managed permanent grassland. *Proceedings 13th General Meeting, European Grassland Federation, Banská Bystrica, Czechoslovakia*, pp. 103–113.

Pawson, H. (1960). *Cockle Park Farm*. Oxford University Press, Oxford.

Stapledon, R. G. (1935). *The Land Today and Tomorrow*. Faber and Faber, London.

Swift, G., Davies, D. H. K., Tiley, G. E. D. and Younie, D. (1990). The nutritive value of broad-leaved weeds and forage herbs in grassland. *Scottish Agricultural College Technical Note*, T223.

Tiley, G. E. D. and Frame, J. (1990). An agronomic evaluation of forage herbs in grassland. *Proceedings 13th General Meeting, European Grassland Federation, Banská Bystrica, Czechoslovakia*, Volume II, pp. 163–166.

†Tiley, G. E. D. and Frame, J. (1992). Evaluation of forage herbs in grass/herb mixtures. *Proceedings 14th General Meeting, European Grassland Federation, Lahti, Finland*, pp. 542–544.

†Wells, T. C. E., Cox, R. and Frost, A. (1989). The establishment and management of wildflower meadows. *Focus on Nature Conservation No. 21*, Nature Conservancy Council.

## Chapter 24

Armstrong, R. H. and McCreath, J. B. (1985). Hill sheep development programme. *Scottish Agricultural Colleges/Hill Farming Research Organisation Report.*

Department of Agriculture and Fisheries for Scotland/Nature Conservancy Council (1977). *A Guide to Good Muirburn Practice.* Handbook, HMSO, Edinburgh.

Eadie, J. (1978). Increasing output in hill farming. *Journal of the Royal Agricultural Society of England,* 139, 103–114.

Floate, M. J. S. (1970). Plant nutrient cycling in hill land. *Fifth Report, Hill Farming Research Organisation, 1967–70,* pp. 15–34.

†Frame, J., Newbould, P. and Munro, J. M. M. (1985). Herbage production in the hills and uplands. *Occasional Publication No. 10, British Society of Animal Production,* pp. 9–37.

Frame, J. and Tiley, G. E. D. (1992). The role of white clover (*Trifolium repens* L.) in hill and upland livestock systems in Scotland. *Proceedings IV International Rangeland Congress, 1991,* Montpellier, France (in press).

Gimingham, C. H. (1985). Muirburn. In: *Vegetation Management in Northern Britain, British Crop Protection Monograph,* No. 30, 71–75.

†Hill Farming Research Organisation (1979). *Science and Hill Farming, HFRO 1954–79.* Silver Jubilee Report, HFRO, Edinburgh, pp. 9–21.

†Hodgson, J. and Grant, S. A. (1981). Grazing animals and forage resources in the hills and uplands. *Occasional Symposium, British Grassland Society,* No. 12, pp. 41–57.

†Macaulay Land Use Research Institute (1989). Heather moorland—a guide to grazing management. *Scottish Agricultural College Technical Note,* T178.

†Newbould, P. (1974). Improvement of hill pastures for agriculture. A review. Part 1. *Journal of the British Grassland Society,* 29, 241–248.

Newbould, P. (1975). Improvement of hill pastures for agriculture. A review. Part 2. *Journal of the British Grassland Society,* 30, 41–44.

Roger, L. (1991). The two pasture system. *Scottish Agricultural College Technical Note,* T291.

Taylor, J. A. (1976). Upland climates. In: *The Climate of the British Isles.* Longmans, London, pp. 264–287.

Tiley, G. E. D. and Frame, J. (1980). Methods for the improvement of hill and upland grazing in west Scotland. *Proceedings 8th General Meeting, European Grassland Federation, Zagreb, Yugoslavia,* pp. 56–60.

# Index

## A

Acceptability, herbage, 177–178
Additive use in silage making, 216–218
  approved scheme in UK, 216
Aerobic deterioration in silage making,
  223
Aerobic treatment of slurry, 142
  biochemical oxygen demand, 142
Agricultural land, UK
  by area, 1
  by grade, 3–4
Agricultural salt, 133
Aluminium calcium phosphate,
    slow-acting fertiliser, 127
Ammonia
  anhydrous, use and application,
    106–107
  aqueous, use and application, 106
Ammonium nitrate
  with calcium, 104–105
  -calcium carbonate mixtures, 122
    limestone needed, 122
  uptake, 104
  use, 104–106
Anaerobic treatment of slurry, 142
  biochemical oxygen demand, 142
Anhydrous ammonia, use and
    application, 106–107
Animal slurry, 136–139
  see also Slurry, animal
Animal unit methods for recording,
    244–245
Annual meadow grass, control, 73
Aqueous ammonia, use and application,
    106
Authenticity of seed, 46
  seals, 46
Available water capacity
  and N requirement, 115–116
  soil, 95

## B

Bales, big, silage, see Big bale silage
Baling in hay making, 238–240
Barn drying systems for hay, 240–241
Basic slag, 127
  phosphate source, 127
Benazolin
  against common chickweed, 72
  against docks, 69
Bent grass
  characteristics, 18, 19, 20, 22
  /fescue swards in rough grazing, 8
Bentazone, against common chickweed, 72
Big bale silage, 223–226
  additives, 226
  bagging and sealing, 225
  golden rules, 227
  target DM, 214
  wrapping, 224–225
Bilberry in rough grazing, 8
Birdsfoot trefoil
  characteristics, 17–18
  future use, 270–271
  for wetland pastures, 45
Bites for grazing, 179
  early, 199–201
  late, 204–205
  see also Grazing
Blended solid fertilisers, 135
Boron, fertilisation, 133
Bracken
  carcinogenic properties, 76
  control, 76–77
    ploughing, 77
    reinvasion, 77
  in rough grazing, 8
  spread, 77
Breeding, forage legumes, 43–45
  matching clover to compatible grass
    companions, 43–44

scope, 43–44
Breeding, plant, 38–45
  present objectives, 41–43
  scientific support, 39
Brome grass, in seed mixtures, 33
Brome, soft, control, 75–76
Broom, control, 77–78
  methods, 78
'Brown patch' infection, 206
Buffer grazing, 187
Buffer silage, 211
Burning heather, 290
Butyric acid fermentation of silage, 230

## C

Calcium ammonium nitrate, 104–105
Calcium carbonate, *see* Lime and liming
Calcium, composition of herbage, *150*
Carbohydrate content of grass, 147, 148
Carbon balance of swards, 169
Cattle
  grazing behaviour, 178–179
  rumination and feeding, 179–180
Chemical composition of grass, 146–150
Chickweed, common, controlling, 72
Chicory, breeding, 45
Chlorine
  composition of herbage, *150*
Clamp silage, 223
  golden rules, *224*
'Clean' grazing, 191
Climate, effects on plant breeding, 42
Clopyralid, against thistles, 70
Clostridia
  and fermentation, 212
  undesirable, in silage making,
    215–216
Clover
  breeding, 43–44
    for compatibility to grass, 43–44
  content, grassland, 5
  seed crops, management, 54–56
  seed production, 44–45
  *see also* Red clover: White clover
Cobalt, fertilisation, 134
Cocksfoot
  for winter grazing, 205
  characteristics, 14
  ear emergence, 25, 154

growth potential in dry summers, 202
  seed crops, harvesting, 55
  in seed mixtures, 33
  winter burn, 205–206
Common vetch, seed, weights harvested,
  49
Compaction, wheel, effect on dry matter
  and crude protein production, 89,
  90
Conservation
  fertiliser N recommendations, *118*
  fertiliser phosphate and potash
    recommendations, 124–125
  *see also* Nature conservation,
    grassland
Continuous stocking, 187–189
  1.2.3 system, 188–189
Conversion factors, class of
  stock/cow-days, 245
Copper, fertilisation, 133, 134
Cotton grass, 7
Cow-days and grassland recording,
  244–245
Creep grazing, 191
Creeping bent grass, characteristics, 19
Crested dogstail, characteristics, 18, 19, 23
Cross-breeding, 38
Crude fibre, content in grass, 147
Crude protein
  content in grass, 147, 148
  production, effect of wheel traffic
    systems, 89
Cutting for hay making, 234
Cutting systems, and N, 116–118

## D

2,4–D against
  docks, 69
  nettles, 73
  ragwort, 71
  rushes, 74
  thistles, 70
D values, *see* Digestibility, values (D
  values)
2,4–DB against docks, 69
Defoliation intensity and N, 111–115
  frequency, 113
    and increased herbage production,
    113–114

Desiccation of old grass, 60
Development and growth, sward, 161–174
Dicamba
  against docks, 69
  /mecoprop combinations, against
    rushes, 74
Digestibility
  digestible nutrients, 151–152
  dry matter, *151*
  and ear emergence, 152–155
  effect of increasing maturity, 151–152
  grass, 150–159
  grazing trials, 157–159
  measurement, 150–151
  and metabolisable energy, 159–160
  organic material, *151*, 152
  values (D values)
    herbal leys, 279
    prediction, 154–157
    of silage, 211
    wildflower mixtures, 276
Direct drilling
  seedbed, 60
  for sward renovation, 65
Disease control
  for red clover, 265–266
  white clover, 258–259
Distinctness, uniformity and stability
    (DUS), 26, 46
DM, *see* Dry matter
Docks, controlling, 68–70
  broad-leaved, 68
  curled, 69
  germination after dormancy, 69
  herbicides, 696–70
  injurious weeds by law, 69
  spraying, 69
Drainage, 97–99
  underdrainage systems, 97–98
Drilling
  direct, for sward renovation, 65
  versus broadcasting, for seed
    establishment, 58–59
Drills for sward renovation, 65
Drinking water and soil nitrogen losses,
    107–108
  measures which reduce losses, *108*
Drought
  resistance, effects on plant breeding,
    43

risk, geographical, 96
Dry matter
  from blanket peats, 10
  contents for baling, 225
  contents in grass, 147, 148
    carbohydrate content, 147–148
  effect of grassland compaction, *90*
  effect of season and N, 110–111
  effect of wheel traffic systems, 89, 90
  grass/clover swards, 259
  grass/forage herb mixtures, *279*
  grass/white clover swards, 252, 253
    effect of fertiliser N, 256
  hay, 233
  hay losses, 238, *239*
  hill pastures, 283
  intensity of defoliation, 111–115
  loss in effluent, 220–221
  lucerne, and lucerne/grass swards,
    269
  magnesium content, 131
  N concentration, 102
    perennial ryegrass, 11–12
  potassium concentration, 128
  productivity, potential, 3
  response to N rates, 108–109, 110
  secondary grasses, 19
  silage density, 211
  sodium concentration, 133
  sulphur content, 130
  target contents for ensiling and field
    wilting time, 214
  without N, 101
    in grass/white clover sward, 101
  yields after winter grazing, 201
Drying processes for hay, 234–238
  mower conditioning, 236
  systems, 240–241
  tedding, 237–238
Dung
  daily excretion, 183–184
  -fouled herbage, acceptability,
    177–178
  nutrients, 182–183

E

Ear emergence
  and digestibility, 152–155
  and silage cutting dates, 24

and spring grazing, 154–155
Early bite, 199–201
Effluent, silage, 220–223
    as fertiliser, 222
    powerful pollutant, 222
    as stock feed, 222
Enterobacteria, and fermentation, 212
Environmental concern, nitrogen losses,
        107–108
    drinking water, 107
    measures which reduce nitrogen
        losses, *108*
Environmentally Sensitive Areas (ESAs),
        practical features, 274
Enzyme additives in silage making, 217
Epsom salts, 131
Ether extract, content in grass, 147
Ethofumesate against soft brome, 76
Evapotranspiration, 95
    Meteorological Office calculation, 96
Excreta
    acceptance of dung-fouled herbage,
        177–178
Excretal return, herbage, 181–186
    amount of dung excreted
        dairy cow, 183
        sheep, 184
    amount of pasture covered daily, 184
    nutrient circulation, 181–182
    nutrients, in dung and urine, 182–183
    pattern, 183–184
    urine
        scorch, 184
        volume daily, 184
Extensification of land use, 274

**F**

Fallow land (set-aside), seed mixtures,
        34–36
Farmyard manure in fertilisation, 143
Feeding
    quality of silage, 226–228
    storage, 193–194
    supplementary, 179
    times of grazing cattle, 179–180
    value of grass, 146–160
Fencing, and hill land improvement, 290
Fermentation
    inhibitors, 216

silage, type, inspection guide, 231
in silage making, 212–213
    results of good fast fermentation
        against badly controlled, 212–213
    stimulants, 217
Fertilisation of seed crops, 53
Fertilisers
    application, seedbed, 60–61
    soil
        blended, solid, 134
        liquid, 135
Fertility and grass production
    lime and mineral nutrients, 119–135
    nitrogen, 101–119
    *see also* Soil fertility
Fescue swards/bent grass in rough
        grazing, 8
Fescues, characteristics, 14–15
Fescues, characteristics, *see also*
        Meadow: Red: Tall
Field records, 244–245
Fine-leaved fescues, characteristics, 18, 19
Fire, and hill land improvement,
        290–291
Fluroxypyr
    against common chickweed, 72
    against docks, 69
Foggage, 205
    and/or autumn grass, deferred, 205
Forage herbs, management, 278–279
Forage legumes
    breeding, 43–45
        matching clover to compatible grass
            companions, 43–44
        scope, 43–44
    characteristics, 16–18
    management, 251–271
    seed production, management, 55–60

**G**

Gene survival and nature conservation,
        275
Genetic manipulation of species, 38–39
Genetic type, 46
Global warming, effects on plant
        breeding, 42–43
Glyphosate
    against docks, 69
    against rushes, 74

GMP fertiliser, 127
Gorse
  control, 77–78
    methods, 78
Grass(es)
  characteristics, 11–15
  chemical composition, 146–150
  digestibility, 150–159
    *see also* Digestibility
  feeding value, 146–160
  growth, 164–174
    most vigorous May to June, 210
    *see also* Leaf development:
      Photosynthesis: Tillering: Tissue
      turnover
  heath, rough grazing, 8
  maturity groups, classification, 24–25
  secondary, characteristics, 18–23
  seeds
    harvesting, 55–56
    management, 53–54
    *see also* Seed
  varietal evaluation, National List,
    25–27
Grass/red clover swards, management
  guidelines, 268
Grass/white clover swards
  cutting for silage, 259–260
  fertiliser N application, daily growth
    rates, 256–258
  fertiliser N usage, 256–258
  grazing, 260–261
  key points for management, 262
  lamb production, 262, *263*
  milk production, 262
  productivity, 3
  soil fertility requirements, 255–256
  systems output, 262–263
  weed, disease and pest control,
    258–259
Grassland
  age and content, 5–6
  area, UK
    over 5 years old, 2–3
    rough grazing, 3
    under 5 years old, 2
  British, types, 1–10
  permanent, upland, botanical
    composition, 6
  recording, 243–250

*see also* Recording, grassland
  stock rearing with arable, survey, 6–7
  surplus by year 2000, 279, 294
  survey information, 4–7
Grazing behaviour, 178–180
  bites, 179
  differences between animal species,
    179
  social factors, 179
  supplementary feeding, 179
  time taken, 179
Grazing/cutting, and hill land
  improvement, 289–290
Grazing methods and systems, 187–198
  choosing, 194
  'clean' grazing, 191–192
  continuous stocking, 187–189
  creep grazing, 191
  mixed grazing, 192–193
  monitoring sward state, 194–198
    *see also* Monitoring sward state
  overgrazing, 190
  rotational grazing, 189–190
  storage feeding, 193
  strip grazing, 190
  undergrazing, 190
  zero grazing, 193
Grazing, midsummer, 202–203
Grazing process, 175–186
  amount of grass needed daily for cow
    maintenance, 176
  herbage acceptability, 177–178
  herbage palatability, 176
  months grazed, UK/New Zealand, 175
  selectivity, 175
Grazing and sward height, 171–172
Ground condition scoring, 100
Growth
  and development, sward, 161–174
  grass, animal grazing behaviour,
    170–171
  and senescence, 169–170
Gut worms, 191
Gypsum for sulphur fertilisation, 131

**H**

Hay
  compositional analyses, *242*
  from hills and uplands, 293

Hay making, 233–242
    bales
        density, 240
        moisture content, 239
    baling, 238–240
    barn drying systems, 240–241
    drying processes, 234–238
    influence of weather, 238
    key points, 241–242
    stage of growth at cutting, 234
    storage losses, 241
    weight of hay per field, 243–244
Hay, undegradable protein content, 149
Hay, vitamin content, 149
Heath rush in rough grazing, 7
Heather
    burning, 290
    management, 7, 290–291
Height management, sward, on
        continuously stocked swards,
        171–172
Herbage
    palatability, 176
    plant breeding, 38–45
        development, 41
        *see also* Breeding, plant
    production, seasonal, 172–174
    variety evaluation, 24–30
        National List, 25–27
Herbal leys, 278–279
    nature conservation, 278
    seed mixture, late nineteenth century,
        *278*
Herbicides
    against
        chickweed, 72
        docks, 69
        nettle, 73
        ragwort, 70
        rushes, 74, 75
        thistles, 70
    spraying before sowing, 60
Herbs, forage, management, 278–279
Higher Voluntary Standard (HVS) for
        certified seed, 47–48
Hill land improvement, 281–294
    drainage, 289
    forage conservation, 293
    future use, 293–294
    pasture resources, 282–284

soil, 286–289
soil/sward, 291–292
    cultivations, 291
    direct drilling, 292
    improved forage species, 292
    surface seeding, 292
sward, 289–291
two-pasture system, 284–286
weed control, 291
Hills, main soil and vegetation types, 9
Horse paddocks, seed mixtures, 36–37
HVS, *see* Higher Voluntary Standard
Hybrid ryegrass
    characteristics, 13
    ear emergence, 25, 154
    for early bite, 199
    varieties for seed mixtures, 32, 33
Hybridising in plant breeding, 38–39


I

Iodine content in grass, 149
Iron, fertilisation, 133
Irrigation, 96–97
    effect on grass production, 96
    scheduling, methods, 96
    to correct soil moisture deficit, 96
Italian ryegrass
    certified seed, 52
    characteristics, 12–13
    digestibility of stem, 42
    ear emergence, 25, 154
    for early bite, 199
    harvesting, 55
    hybrids, 42
    varieties for seed mixtures, 32, 33


K

Kieserite, 131


L

Lactic acid bacteria, 212
*Lactobacilli*, lactic acid bacilli, 212
*Lactobacillus plantarum*, 217
Late bite, 204–205
Leaf area index, 169

Leaf development, 164–166
  effect of temperature, 165–166
Leaf primordia, 164
Leatherjackets, infestation, 206
Legislation, animal slurry application,
  142
Legumes
  -based swards for animal production,
    seed mixtures, 34
  forage
    characteristics, 16–18
    management, 251–271
    symbiotic N-fixation, 103–104
    use in the future, 270–271
  varietal evaluation, National List,
    25–27
  *see also* Birdsfoot trefoil: Lucerne:
    Red clover: Sainfoin: White
    clover
Lime and liming, 119–124
  amounts needed to neutralise soil
    acidity, 122
  application amounts, 121
  application rates, white clover, 256
  benefit, 94–96
  clay soils require more, 121
  dressing of grassland, 60–61
  effect on soil structure, 84
  for hill land improvement, 286–288
  influence of rainfall, 122
  neutralising value, 123
  of seed crops, 53
  soil sampling and analysis, 123
Lipid content of grass, 149
Liquid fertilisers, 135
Liscombe star system for silage
  additives, 218, *219*
*Listeria monocytogenes* and big bale
  silage, 225
Livestock numbers, UK, June 1991, 1
Livestock units and grassland recording
    systems, 245
  grazing days, 245
Lop grass, control, 75–76
Lucerne
  characteristics, 17
  future use, 270
  management, 268–270
    guidelines, 269–270
  in monoculture, 34

**M**

Magnesium, composition of herbage,
  *150*
Magnesium fertilisation, 131–132
  effect of high K levels, 131, 132
  release from clay minerals, 132
Manganese, fertilisation, 133
Manure, farmyard, in fertilisation, 143
Marsh birdsfoot trefoil for wetland
  pastures, 45
Maturity groups, classification, grass
  species, 24–25
MCPA against
  docks, 69
  nettles, 73
  ragwort, 71
  rushes, 74
  thistles, 70
MCPB for docks, 69
Meadow fescue
  for winter grazing, 205
  characteristics, 14
  ear emergence, 154
Meadow grasses, characteristics, 18, 19
Measuring the value of a sward, 243–250
Mecoprop
  against common chickweed, 72
    docks, 69
    nettles, 73
    ragwort, 71
    thistles, 70
  /dicamba combinations, against rushes,
    74
Metabolisable energy, 246–249
  calculations, 246–249
  and digestibility, 159–160
  interpretation, 249–250
Midsummer grazing, 202–203
Milk production systems based on
    grass/white clover swards, 262
Mineral
  composition of herbage, *150*
  content in grass, 149
  nutrients, 124–135
Mixed grazing, 192–193
'Mob stocking', 203
Moling, 100
Molybdenum, fertilisation, 133
Monitoring sward state, 194–198

heights, 195–198
  sward height targets, 196
  rotational grazing systems, 197–198
  standard methods, 195
  sward stick, 195
  walking the field, 196
Moor mat grass in rough grazing, 8
Muirburn, 290

**N**

National Institute of Agricultural Botany
  herbage plant breeding improvement,
    41, 42
  Recommended List, 30
National List and plant breeding, 39
National List varietal evaluation, 25–27
Natrophile, 133
Natrophobe, 133
Nature conservation, grassland, 272–280
  future, 279–280
  gene survival, 275
  loss of species rich grassland, 272–273
  schemes, 273–275
  stages in planning and action, 279
  surplus by year 2000, 279, 294
  wildflower mixture productivity,
    275–277
*Nematodirus*, 191, 192
Nettles, control, 73
Neutralising value of liming materials,
    123
Nitrate, plant metabolism, 102
Nitrogen
  in animal excreta, 104
  application
    for early bite, 199–201
    for late bite, 204
    timing, 200–201
  content in grass, 148
  and cutting systems, 116–118
  cycle, main components, 105
  deficiency, symptoms, 102
  excessive, symptoms, 102
  fertilisation, 101–119
    national amounts applied, 101
  fertiliser
    recommendations for conservation,
      *118*
    recommendations for grazing, 112

types, 104–107
  white clover/grass, repetitive
    application, 256
  fixation by white clover, 254–255
  urinary nitrogen, 255
free extractives, content in grass, 147
high application, and sod pulling,
    87–88
for hill land improvement, 288–289
and intensity of defoliation, 111–115
reducing losses, 107–108
  environmental concern, 107–108
  measures which reduce, *108*
requirement
  and plant breeding, 42
  prediction, 115–116
response of grass swards, 108–109
response of grass/white clover swards,
    109–110
seasonal distribution, 110–111
and silage herbage production,
    116–118
soil status, 102–103
  based on previous crop, 103
  estimation, 102–103
symbiotic N-fixation by legumes, 103
timing of spring application, 117
in undiluted livestock slurry, 137–138,
    139
Non-protein nitrogen content in grass,
    148
Northern Ireland Plant Breeding Station,
    42
Nutrient circulation and excretal return,
    181–182
Nutrients in slurry, 137–139

**O**

Organic manures, 136–145
*Ostertagia*, 191
Overgrazing, 190

**P**

Paraplowing, 100
Peat bog vegetation, 10
Peaty podzols, rough grazing, 8
*Pediococci*, lactic acid bacilli, 212
*Pediococcus acidilactici*, 217

Perennial ryegrass
  average content, 5
  characteristics, 11–12
  digestibility, 158
  ear emergence, 25, 154
  for early bite, 199
  hybrids, 42
  late heading, Recommended List, 28
  N-fertilised, 157
  seed, weights harvested, 49
  in seed mixtures, 33
    diploid, 33
    for long-term swards, 33
    for permanent swards, 33
    tetraploid, 33
  varieties, 11–12
    maturity groups, 24–25
Permanent wilting point, 95
Pest control
  for red clover, 265–266
  white clover, 258–259
Pests, seed establishment, 62–63
pH
  normal for grassland, 60
  soil, sampling, 123–124
Phosphorus (phosphate) fertilisers,
    124–127
  application rates, white clover, 256
  clay soil, high rate of fixation, 124
  composition of herbage, *150*
  and early grass, 125–126
  ground mineral, 127
  for hill land improvement, 288
  and plant metabolism, 124
  recommendations for conservation and
    grazing, 124–125
  release from mineral compounds,
    124–125
  sand/loam/peat soils, rate of fixation,
    124–125
  types, 126–127
  in undiluted livestock slurry, 137–138,
    139
  water soluble/insoluble, 126–127
Photosynthesis, grass, 169–172
Plant breeding, herbage, 38–45
Poaching
  adverse effects, management, 86–87
  effect on soil, 85–87
  key points to minimise, *87*

Podzols, rough grazing, 8
Pollination
  cross-pollination of grass species,
    50–52
  self-, avoidance, 52
  wind, 52
Pollution from slurry application, 141,
    142
Post-sowing management, seedbed, 61
Potassium (potash) fertilisers, 127–129
  and animal excreta, 129
  application rates, white clover, 256
  composition of herbage, *150*
  deficiency, symptoms, 128
  for hill land improvement, 288
  inadequacy in soil, 127–128
  plant metabolism, 128
  recommendations for conservation and
    grazing, 128
  in undiluted livestock slurry, 137–138,
    139
Potentially toxic elements in sewage
    sludge, 144
Prairie grass, characteristics, 15
Preferred species
  average content, 5
  decline with age, 5
Protein
  content in grass, 147, 148
  undegradable, content in grass, 149
Purple moor grass, management, 7, 8

**R**

Ragwort
  controlling, 70
    cutting, 71
    ploughing, 71
    spraying, 71
  injurious weed, 70
  poisonous to livestock, 71
Rainfall
  and grass growth, 92–95
  influence on soil acidity, 122
Recommended List and plant breeding,
    39
Recommended Lists, 27–30
Recording, grassland, 243–250
  animal unit methods, 244–245
  interpretation, 249–250

utilised metabolisable energy, 246–
249
Red clover
  characteristics, 16–17
  classification, maturity, 25
  management, 264–268
    compatibility with grasses, 264
    establishment needs, 264–265
    guidelines, 268
    soil fertility requirements, 265
    utilisation, 267–268
    varieties, 264
    weed, disease and pest control,
      265–266
  seed crops
    harvesting, 56
    management, 54
  in seed mixtures, 34
Red fescue, characteristics, 19–22
Respiration losses in silage making,
  213–214
Rhizobia for N-fixation, white clover
  growth, 167
Rhizobial infection, white clover,
  166–167
Rhizobial inoculant for white clover,
  254–255
  host specific, 254
  pre-pelleted seed, 255
Rhizobially-fixed clover nitrogen, 43,
  44
Ribwort plantain, breeding, 45
Rotational grazing, 189–190
Rough grazing
  characteristics and composition, 7–10
  management, 7–10
  UK area, 3
Rough-stalked meadow grass,
  characteristics, 23
Rumination, cattle, 179–180
Rushes, control, 74–75
  herbicides, 74–75
  spraying, 74
Ryegrass
  characteristics, 11–13
  digestibility, 19
  diploid and tetraploid, enhanced winter
    hardiness, 208
  proneness to winter injury, 208
  seed crops, harvesting, 55

varieties with high magnesium
  content, 29–30

## S

Sainfoin
  characteristics, 17
  future use, 270–271
Salt, see Sodium
Saturation, soil, 97
Sealing silos, 218–220
Season of sowing, seed establishment, 59
Seasonal herbage production, 172–174
Seasonal objectives and management,
  199–208
  early bite, 199–201
  late bite, 204–205
  midsummer grazing, 202–203
  nitrogen application, 200
  topping grassland, 203–204
  winter management, 205–208
Secondary grasses, characteristics, 18–23
Sedges, in rough grazing, 7
Seed
  basic, 46
  certification
    HVS categories, 47
    purity, 47
    standards, 47–48
  crops, management, 53–54
  export trade, UK, 50
  grades and standards, 46–48
  growing
    inspection, 52
    rules and regulations, 50–52
  harvesting, 55–56
  mixtures, 31–37
    advisory services, 31
    choosing, 32–34
    database, 31
    key factors, 32
    medium-term, 35
    rationale, 31–32
    set-aside land, 34–36
    short-term to long-term, 34
    and soil trampling, 86
    specimen mixtures, 34–37
  production and usage, 48–50
  tonnage imported, 50
  weights harvested, 49–50

Seed establishment
  direct drilling, 60
  drilling versus broadcasting, 58–59
  fertiliser application, 60–61
  pests, 62–63
  post-sowing management, 60
  preparation, 58–60
  season of sowing, 59
  weeds, 61–62
Seeds Regulations, 46
Selection of species, 38–39
Selenium content in grass, 149
Senescence and growth, 169–170
Set-aside land, seed mixtures, 34–36
  see also Nature conservation
Sewage sludge, 143–144
  application, 143–144
  grazing after application, 144
  potentially toxic elements, 144
Sheep
  farming, see Hill land improvement
  grazing behaviour, 178–179
  grazing intake, 180
Silage clamp, average density, 211
Silage cutting, white clover, 259–260
Silage effluent
  application, 145
  available nutrients, 145
  biochemical oxygen demand, 145
Silage fermentation, type, inspection
    guide, 231
Silage from hills and uplands, 293
Silage from midseason topping, 203
Silage, grass/white clover, composition,
    259, 260
Silage making, 209–232
  additive use, 216–218
  additives, Liscombe star system, 218,
    219
  aerobic deterioration, 223
  any time during grazing system, 210
  big bale silage, 210, 223–226
  buffer silage, 211
  clamp silage, 223
    golden rules, 224
  competitions, 229–232
  compositional analyses of silage,
    226–228
  early history, 209
  effluent, 220–223

enzyme additives, 217
feeding quality, 226–228
fermentation, 212–213
health and safety, 228–229
inhibitors, 216
inoculants, 217
losses from bad sealing, 220
respiration losses, 213–214
sealing, 218–220
short chopping grass, 214
in systems, 210–212
total amount of silage and hay dry
    matter conserved in UK, 209
undesirable clostridia, 215–216
wilting, 214–215
Silage, red clover, 267
Silage storage feeding, 193–194
Silage systems, herbage production,
    116–117
Sites of Special Scientific Interest,
    practical features, 274
Slurry, animal, 136–139
  effective application, 139–142
    alternative handling, 142
    anaerobic treatment, 142
    atmospheric pollution, 141, 142
    biochemical oxygen demand, 142
    Codes of Practice, 139
    economics of storage, 141
    with fertilisers, 140
    legislation, 142
    water protection, 139
    winter, 140
    winter versus spring application,
      141
  mixture, 136–137
  nutrients, 137–139
    monetary value, 138
  value, late winter/early spring
      application, 207
  volume produced, 137
Smooth-stalked meadow grass,
    characteristics, 22
Snow cover, effect on grassland, 206
Sod pulling, effect on soil, 87–88
Sodium, composition in herbage, 150
Sodium fertilisation, 132–133
  agricultural salt, 132
  loving/hating grasses/legumes, 133
Soft brome, control, 75–76

Soil
  acidity, lime and liming, 119–124
    *see also* Lime and liming
  components, 81–82
  drainage, 97–99
    underdrainage systems, 97–98
  effect of sod pulling, 87–88
  effect of stock trampling and sward
      poaching, 85–87
    grazing curtailed or zero, 86–87
    removal of stock, 86–87
    seed mixtures, 86
  factors, effect on grass, 79–90
  fertility requirements
    for red clover, 265
    and white clover management,
      255–256
  formation
    climatic factors, 81
    and farming operations, detrimental
      and beneficial, 80–81
    geological factors, 80
    organic matter, 81
  ground condition scoring, 100
  maps, 84–85
    classification of types, 84–85
    for forecasting problems, 85
    land capability, 84
  nitrogen, *see* Nitrogen
  sampling, 123
  texture and structure, 82–84
    effect of gravel and stones, 82–83
    effect of liming, 84
    mineral particles, 83
    porosity and drainage, 83–84
  texture triangle, 183
  water capacity, 95
    available, 95
    moisture deficit, 95, 96
  wheel-induced effects on, 88–90
Sowing seed, 58–60
Spraying herbicide before sowing, 60
Stocking, continuous, 187–189
  1.2.3 system, 188–189
Stolons, and white clover growth,
    167–168
Stomach worm, 191
Storage feeding, 193
Storage losses of hay, 241
*Streptococci*, lactic acid bacilli, 212

Strip grazing, 190
Subsoiling, 99–100
Sulphur, composition of herbage, *150*
Sulphur fertilisation, 129–131
  deposition from industrial areas,
      129–130
  N:S ratios, 130
  soil reserves, 130
Sward, age and species content, 5–6
Sward establishment, 57–62
  fertiliser application, 60–61
  key points, *63*
  pests, 62–63
  post-sowing management, 61
  seedbed, 58–60
  weeds, 61–62
Sward growth and development, 161–174
Sward poaching, effect on soil, 85–87
Sward productivity, 3–4
Sward renovation, 63–67
  direct drilling, 65
  guidelines, *67*
  moisture, 66
  nutrient supply, 66
  partial reseeding techniques, 65–67
  post-sowing management, 66
  seed germination, 66
  seedling emergence, 66
  strategies, *64*
Sward state, monitoring, 194–198
  *see also* Monitoring sward state
Sward surface height, 196
Sward stick, 195
Sweet brome, characteristics, 15
Sweet vernal, characteristics, 18–20, 23

## T

Tall fescue
  characteristics, 15
  ear emergence, 154
  growth potential in dry summers, 202
  in seed mixtures, 33
Temperature
  and early grazing, 200
  effect on leaf development, 165–166
  and grass growth, 91–92
  winter, effects, 206–207
Thistles
  controlling, 70

Thistles (*contd.*)
  injurious weed, 70
Tillering process, 161–164
  creeping grass species, 163
  and defoliation, 161–162
  following silage cut, 163
  numbers of tillers, 162–163
Tillers
  characteristics, 161
  density under continuous stocking, 187
  density under rotational grazing,
      189–190
Timothy
  for winter grazing, 205
  characteristics, 13–14
  ear emergence, 25, 154
  harvesting, 55
  /meadow fescue/white clover sward,
      for midsummer grazing, 202
  in seed mixtures, 33
Tissue turnover
  grass, 169–172
  and pattern of growth, 166
Topping grassland, 203–204
Topsoil, importance, 79
Toxic plants, effects, 178
Trace elements
  adequacy in soil, 133–134
  concentrations in grasses and legumes,
      133
  fertilisation, 133–134
    deficiencies, treatment, 134
Trampling (treading) by stock, 184–186
  damage to sward, 185
  effect on soil, 85–87
    management, 86–87
  estimates of extent, 185
  tolerance level of grasses, 185, 186
Transpiration, 95
*Trichostrongylus*, 191
Triclopyr
  against nettles, 73
  against ragwort, 71
  against docks, 69
Two-pasture system, 284–286

U

Underdrainage systems, 97–98
Undergrazing, 190, 210

United Kingdom
  agricultural land area, 1
  livestock numbers, 1
  Seed Certification Scheme, 46
Unsown grasses, 18–23
Uplands, main soil and vegetation types,
      9
Urea, fertiliser, use and application,
      105–106
Urine
  daily volume, 184
  nutrients, 182–183
Utilised metabolisable energy, 246–249
  calculations, 246–249
  interpretation, 249–250

V

Value for cultivation and use (VCU)
      trials, 26, 46
Varietal evaluation, national list, 25–27
Varietal purity, 46
Vitamin content of hay, 149

W

Water capacity, available, and N
      requirement, 115–116
Water-soluble carbohydrates, content in
      grass, 147, 148
Water supply for grass growth, 92–95
  available capacity, 95
  permanent wilting point, 95
  soil moisture deficit, 95
  *see also* Irrigation
Weather and water control, 91–100
Weed control, 68–78
  hill land, 291
  for red clover, 265–266
  white clover, 258–259
Weeds, seedbed, 61–62
Weighted-disc meters, 195
Welsh Plant Breeding Station, herbage
      plant breeding, 41, 42
Westerwold ryegrass
  characteristics, 13
  in seed mixtures, 32–33
Wheel-induced effects on soil, 88–90
White clover
  breeding for pest resistance, 44

characteristics, 16
classification, maturity, 25
development, 166–168
   N-fixing rhizobia, 166–167
dynamics, 258
ear emergence, 154
growth, 168
longevity, 168
management, 252–263
   compatibility with grasses, 252
   cutting for silage, 259–260
   effect of fertiliser N, 256
   establishment needs, 253–254
   grazing swards, 260–262
   nitrogen fixation, 254–255
   rhizobial inoculant, 254–255
     seed pre-pelleted, 255
   soil fertility requirements, 255–256
   system output, 262–263
   varieties, 252
   weed, disease and pest control, 258–259
   wintering, 254
midsummer growth capability, 202
in rough grazing, 8
in seed mixtures, 34

Wildflower mixtures
   general selection criteria for species inclusion, 277
   productivity, 275–277
     cut under hay regime, 276
Wilting in silage making, 214–215
'Winter burn', 205
Winter grassland management, key points, 208
Winter hardiness ratings, 208
Winter kill of sward shoots and roots, 206
Winter management of grassland, 205–208
Worm infection and grazing, 191–192

**Y**

Yarrow, breeding, 45
Yorkshire fog, characteristics, 18, 19, 20, 23

**Z**

Zero grazing, 193
Zinc, fertilisation, 133

# Farming Press

Below is a sample of the wide range of agricultural and veterinary books we publish. For more information or for a free illustrated catalogue of all our publications please contact:

Farming Press
Miller Freeman UK Ltd
Miller Freeman House
Sovereign Way
Tonbridge, Kent TN9 1RW, United Kingdom
Telephone 01732 364422    Fax 01732 361534
E-mail: farmingpress@unmf.com
www.farmingpress.co.uk

## A Veterinary Book for Dairy Farmers (Third Edition)    ●    ROGER W. BLOWEY

The standard text for a wide range of college courses around the world. Now in full colour and completely updated, the third edition is an invaluable text for dealing with the sick animal and an essential tool in the daily fight to keep intensely managed stock in first-class condition and to optimise productivity. *Also available on CD-ROM.*

## Genetic Improvement of Cattle and Sheep    ●    GEOFF SIMM

A fascinating and practical introduction to genetics for the farmer and student. Geoff Simm describes the origins of today's livestock breeds and the scientific principles of livestock improvement, the basic principles of genetics and the application of these principles to livestock farming today. *Also available on CD-ROM.*

## Organic Farming    ●    NICOLAS LAMPKIN

The leading English language book on organic farming. Divided into two sections, it covers the principles of organics as well as offering practical advice for crops, livestock and marketing.

## Marketing for Farm and Rural Enterprise    ●    MICHAEL HAINES

A lively step-by-step guide to the complete marketing process, full of practical insights on how to deal with the challenges facing the modern farmer. Essential reading for anyone wanting to market farm produce effectively, the book is also invaluable to anyone wanting to embark on a new farm enterprise with confidence.

Farming Press is a division of Miller Freeman UK Ltd which provides a wide range of media services in agriculture and allied businesses. Among the magazines published by the group are *Arable Farming*, *Dairy Farmer* and *Farming News*. For a specimen copy please contact the address above.